William Chauvenet

A treatise on elementary geometry

William Chauvenet

A treatise on elementary geometry

ISBN/EAN: 9783742891570

Manufactured in Europe, USA, Canada, Australia, Japa

Cover: Foto ©berggeist007 / pixelio.de

Manufactured and distributed by brebook publishing software (www.brebook.com)

William Chauvenet

A treatise on elementary geometry

A TREATISE

ON

ELEMENTARY GEOMETRY

WITH APPENDICES CONTAINING

A COLLECTION OF EXERCISES FOR STUDENTS

AND

AN INTRODUCTION TO MODERN GEOMETRY.

BY

WILLIAM CHAUVENET, LL.D.,

PROFESSOR OF MATHEMATICS AND ASTRONOMY IN WASHINGTON UNIVERSITY.

PHILADELPHIA
J. B. LIPPINCOTT & CO.
1870.

PREFACE.

The invention of Analytic Geometry by DESCARTES in the early part of the seventeenth century, quickly followed by that of the Infinitesimal Calculus by NEWTON and LEIBNITZ, produced a complete revolution in the mathematical sciences themselves and accelerated in an astonishing degree the progress of all the sciences in which mathematics are applied, but arrested for a time the progress of pure geometry. The new methods, characterized by great generality and facility in their application to problems of the most varied kinds, offered to the succeeding generations of investigators more inviting fields of research and promises of surer and richer reward than the special and apparently more restricted methods of the ancients. During the eighteenth century hardly any important addition to geometry was made that was not the direct product, either of the Cartesian method alone, or of that method in alliance with the Infinitesimal Calculus.

With the present century, however, a new era commenced in pure geometry. The first impulse was given by the Descriptive Geometry of MONGE; then followed CARNOT's Theory of Transversals, PONCELET's Projective Properties of Figures and Method of Reciprocal Polars, the researches of STEINER, POINSOT, GERGONNE, CAYLEY, MACCULLAGH, and many others, crowned by the brilliant discoveries of CHASLES.

All this progress, it is true, has been chiefly in the higher departments of pure geometry, and has not yet essentially changed the substance or form of what is known as Elementary Geometry, which is little more than the Geometry of EUCLID in a modern dress, with certain necessary additions in solid geometry; for, although some of the recent discoveries are of a remarkably simple character and (if simplicity were the only requisite) might be introduced into the elements, it is generally conceded that in elementary instruction it is most expedient to commence with the Euclidian geometry, and to reserve the new developments for subsequent study under the name of the *Modern Geometry*.

Nevertheless, this advance in the general science has not failed to produce its legitimate effect upon the primary branch; and the modern treatises on the elements, especially in France, from that of LEGENDRE in 1794 to that of ROUCHÉ and COMBEROUSSE in 1868, exhibit a gradual and marked improvement both in matter and method.

In the following treatise, designed especially for use in colleges and schools, I have endeavored to set forth the elements with all the rigor and completeness demanded by the present state of the general science, without seriously departing from the established order of the propositions, or sacrificing the simplicity of demonstration required in a purely elementary work. Some subjects, not usually included in elementary works, are so placed that they may be omitted without breaking the chain of demonstration, and the remainder may be used as an abridged course in those schools where the time allotted to the study does not suffice for the perusal of the whole. Such, for example, are the articles on Maxima and Minima at the end of Book V. and those on Similar Polyedrons and the Regular Polyedrons at the end of Book VII.

As the student can make no solid acquisitions in geometry without frequent practice in the application of the principles he has acquired, a copious collection of exercises is given in the Appendix. The discouraging difficulties which the young student commonly experiences in his first attempts at demonstrating new theorems, or solving new problems, are here obviated in a great degree by giving him such suggestions for the solution of many of the exercises as may fairly be presumed to be necessary for him at the successive stages of his progress. These suggestions are given with less and less frequency as he advances, and he is finally left to rely entirely upon his own resources when he may be supposed to have acquired by practice considerable familiarity with principles, and dexterity in their application.

The Appendix on the *Modern Geometry*, although restricted to the properties of the straight line and circle, will serve a good purpose, it is hoped, either as an introduction to such works as those of PONCELET and CHASLES in which the methods of pure geometry are employed, or as a companion to the works of SALMON and others in which the new geometry is treated by the analytic method.

In the preparation of this work, I have derived valuable aid from a number of the more recent French treatises on Elementary Geometry, and especially from those of BOBILLIER, BRIOT, COMPAGNON, LEGENDRE (edited by Blanchet), and the very complete *Traité de Géométrie Élémentaire* of ROUCHÉ and COMBEROUSSE. The last named work has furnished many of the exercises of Appendix I. and much of the matter of Appendix II.

WASHINGTON UNIVERSITY,
    ST. LOUIS, June 1, 1869.

CONTENTS.

INTRODUCTION.. 9

PLANE GEOMETRY.

BOOK I.
RECTILINEAR FIGURES 12

BOOK II.
THE CIRCLE 52

BOOK III.
PROPORTIONAL LINES. SIMILAR FIGURES.. 91

BOOK IV.
COMPARISON AND MEASUREMENT OF THE SURFACES OF RECTILINEAR FIGURES 126

BOOK V.
REGULAR POLYGONS. MEASUREMENT OF THE CIRCLE. MAXIMA AND MINIMA OF PLANE FIGURES 142

GEOMETRY OF SPACE.

BOOK VI.
THE PLANE. POLYEDRAL ANGLES 171

CONTENTS.

BOOK VII.

POLYEDRONS.. 196

BOOK VIII.

THE THREE ROUND BODIES. THE CYLINDER. THE CONE. THE SPHERE... 238

BOOK IX.

MEASUREMENT OF THE THREE ROUND BODIES................... 271

APPENDIX I.

EXERCISES IN ELEMENTARY GEOMETRY............................ 291

APPENDIX II.

INTRODUCTION TO MODERN GEOMETRY............................. 333

ELEMENTS OF GEOMETRY.

INTRODUCTION.

1. EVERY person possesses a conception of *space* indefinitely extended in all directions. Material bodies occupy finite, or limited, portions of space. The portion of space which a body occupies can be conceived as abstracted from the matter of which the body is composed, and is called a *geometrical solid*. The material body filling the space is called a *physical solid*. A geometrical solid is, therefore, merely the *form*, or *figure*, of a physical solid. In this work, since only geometrical solids will be considered, we shall, for brevity, call them simply solids.

2. *Definitions.* In geometry, then, a *solid* is a limited, or bounded, portion of space.

The limits, or boundaries, of a solid are *surfaces*.

The limits, or boundaries, of a surface are *lines*.

The limits of a line are *points*.

3. A solid has extension in all directions; but for the purpose of measuring its magnitude, it is considered as having three specific *dimensions*, called *length*, *breadth* and *thickness*.

A surface has only two dimensions, length and breadth.

A line has only one dimension, namely, length. The intersection of two surfaces is a line.

A point has no extension, and therefore neither length, breadth nor thickness. The intersection of two lines is a point.

4. Although our first notion of a surface, as expressed in the definition above given, is that of the boundary of a solid, we can suppose

such boundary to be abstracted and considered separately from the solid. Moreover, we may suppose a surface of indefinite extent as to length and breadth; such a surface has no limits.

Similarly, a line may be considered, not only as the limit of a surface, but as abstracted from the surface and existing separately in space. Moreover, we may suppose a line of indefinite length, or without limits.

Finally, a point may be considered, not merely as a limit of a line, but abstractly as having only position in space.

5. *Definitions.* A *straight line* is the shortest line between two points; as AB.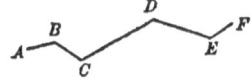

Since our first conception of a straight line may be regarded as derived from a comparison of all the lines that can be imagined to exist between two points, *i.e.*, of lines of *limited* length, this definition (which is the most common one) may be admitted as expressing such a first conception; but since we can suppose straight lines of indefinite extent, a more general definition is the following:

A straight line is a line of which every portion is the shortest line between the points limiting that portion.

A *broken line* is a line composed of different successive straight lines; as $ABCDEF$.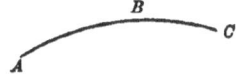

A *curved line*, or simply a *curve*, is a line no portion of which is straight; as ABC.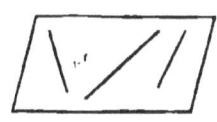

If a point moves along a line, it is said to *describe* the line.

6. *Definitions.* A *plane surface*, or simply a *plane*, is a surface in which, if any two points are taken, the straight line joining these points lies wholly in the surface.

A *curved surface* is a surface no portion of which is plane.

7. Solids are classified according to the nature of the surfaces which limit them. The most simple are bounded by planes.

8. *Definitions.* A *geometrical figure* is any combination of points, lines, surfaces, or solids, formed under given conditions. Figures formed by points and lines in a plane are called *plane figures*. Those formed by straight lines alone are called *rectilinear*, or *right-lined*, figures; a straight line being often called a *right* line.

9. *Definitions.* Geometry may be defined as the science of extension and position. More specifically, it is the science which treats of the *construction* of figures under given conditions, of their *measurement*, and of their *properties*.

Plane geometry treats of plane figures.

The consideration of all other figures belongs to the *geometry of space*, also called the *geometry of three dimensions*.

10. Some terms of frequent use in geometry are here defined.

A *theorem* is a truth requiring demonstration. A *lemma* is an auxiliary theorem employed in the demonstration of another theorem. A *problem* is a question proposed for solution. An *axiom* is a truth assumed as self-evident. A *postulate* (in geometry) assumes the possibility of the solution of some problem.

Theorems, problems, axioms and postulates are all called *propositions*.

A *corollary* is an immediate consequence deduced from one or more propositions. A *scholium* is a remark upon one or more propositions, pointing out their use, their connection, their limitation, or their extension. An *hypothesis* is a supposition, made either in the enunciation of a proposition, or in the course of a demonstration.

PLANE GEOMETRY.

BOOK I.

RECTILINEAR FIGURES.

THE STRAIGHT LINE.

1. *Axiom.* There can be but one straight line between the same two points.

2. *Postulate.* A straight line can be drawn between any two points; and any straight line can be *produced* (*i. e.*, prolonged) indefinitely.

3. *Axiom.* If two indefinite straight lines coincide in two points, they coincide throughout their whole extent, and form but one line.

Hence two points *determine* a straight line; and a straight line may be designated by any two of its points.

4. Different straight lines drawn from the same point are said to have different *directions;* as OA, OD, etc. The point *from* which they are drawn, or *at* which they commence, is often called the *origin*.

If any one of the lines, as OA, be produced through O, the portions OA, OB, on opposite sides of O, may be regarded as two different lines having *opposite directions* reckoned from the common origin O.

Hence, also, every straight line AB has two opposite directions, namely, from A toward B (A being regarded as its origin) expressed by AB, and from B toward A (B being regarded as its origin) expressed by BA. If a line AB is to be produced through B, that is, toward C, we should express this by saying that AB is to be produced; but if it is to be produced through *A*, that is toward *D*, we should express this by saying that *BA* is to be produced.

ANGLES.

5. Definition. An *angle* is a figure formed by two straight lines drawn from the same point; thus *OA*, *OB* form an angle at *O*. The lines *OA*, *OB* are called the *sides* of the angle; the common point *O*, its *vertex*.

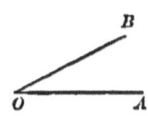

An isolated angle may be designated by the letter at its vertex, as "the angle *O*;" but when several angles are formed at the same point by different lines, as *OA*, *OB*, *OC*, we designate the angle intended by three letters; namely, by one letter on each of its sides, together with the one at its vertex, which must be written between the other two. Thus, with these lines there are formed three different angles, which are distinguished as *AOB*, *BOC* and *AOC*.

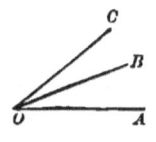

Two angles, such as *AOB*, *BOC*, which have the same vertex *O* and a common side *OB* between them, are called *adjacent*.

6. Definition. Two angles are *equal* when one can be placed upon the other so that they shall coincide. Thus, the angles *AOB* and *A'O'B'* are equal, if *A'O'B'* can be superposed upon *AOB* so that while *O'A'* coincides with *OA*, *O'B'* shall also coincide with *OB*. The equality of the two angles is not affected by producing the sides; for the coincident sides continue to coincide when produced indefinitely (3).* Thus the *magnitude* of an angle is independent of the length of its sides.

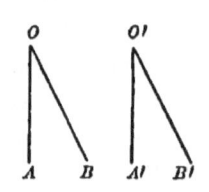

7. A clear notion of the magnitude of an angle will be obtained by supposing that one of its sides, as *OB*, was at first coincident with the other side *OA*, and that it has *revolved* about the point *O* (turning upon *O* as the leg of a pair of dividers turns upon its hinge) until it has arrived at the position *OB*. During this revolution the movable side makes with the fixed side a *varying* angle, which increases by insensible degrees, that is, *continuously*; and the revolving line is

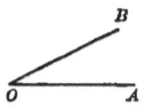

---

* An Arabic numeral alone refers to an article in the same Book; but in referring to articles in another Book, the number of the Book is also given.

said to *describe*, or to *generate*, the angle AOB. By continuing the revolution, an angle of any magnitude may be generated.

It is evident from this mode of generation, as well as from the definition (6), that the magnitude of an angle is independent of the length of its sides.

PERPENDICULARS AND OBLIQUE LINES.

8. *Definition.* When one straight line meets another, so as to make two adjacent angles equal, each of these angles is called a *right angle;* and the first line is said to be *perpendicular* to the second.

Thus, if AOC and BOC are equal angles, each is a right angle, and the line CO is perpendicular to AB.

Intersecting lines not perpendicular are said to be *oblique* to each other.

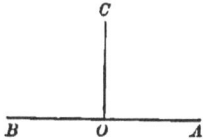

PROPOSITION I.—THEOREM.

9. *At a given point in a straight line one perpendicular to the line can be drawn, and but one.*

Let O be the given point in the line AB. Suppose a line OD, constantly passing through O, to revolve about O, starting from the position OA. In any one of its successive positions, it makes two different angles with the line AB; one, AOD, with the portion OA; and another, BOD, with the portion OB. As it revolves from the position OA around to the position OB, the angle AOD will continuously increase, and the angle BOD will continuously decrease. There will therefore be one position, as OC, where the two angles become equal; and there can evidently be but one.

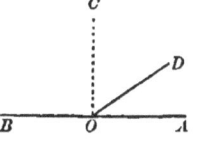

10. *Corollary.* All right angles are equal. That is, the right angles AOC, BOC made by a line CO meeting AB, are each equal to each of the right angles $A'O'C'$, $B'O'C'$, made by a line $C'O'$ meeting any other line $A'B'$. For, the line $A'O'B'$ can be applied to the line AOB, so that O' shall

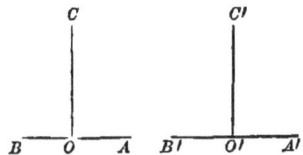

fall upon O, and then $O'C'$ will fall upon OC, unless there can be two perpendiculars to AB at O, which by the preceding proposition is impossible. The lines will therefore coincide and the angles will be equal (6).

PROPOSITION II.—THEOREM.

11. *The two adjacent angles which one straight line makes with another are together equal to two right angles.*

If the two angles are equal, they are right angles by the definition (8), and no proof is necessary.

If they are not equal, as AOD and BOD, still the sum of AOD and BOD is equal to two right angles. For, let OC be drawn at O perpendicular to AB. The angle AOD is the sum of the two angles AOC and COD. Adding the angle BOD, the sum of the two angles AOD and BOD is the sum of the three angles AOC, COD and BOD. The first of these three is a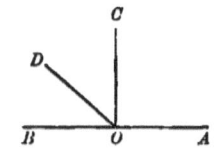
right angle, and the other two are together equal to the right angle BOC; hence the sum of the angles AOD and BOD is equal to two right angles.

12. *Corollary* I. If one of the two adjacent angles which one line makes with another is a right angle, the other is also a right angle.

13. *Corollary* II. If a line CD is perpendicular to another line AB, then, reciprocally, the line AB is perpendicular to CD. For, CO being perpendicular to AB at O, AOC is a right angle, hence (Cor. I.) AOD is a right angle, and AO or AB is perpendicular to CD.
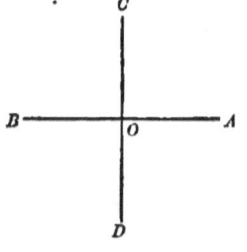

14. *Corollary* III. The sum of all the consecutive angles, AOB, BOC, COD, DOE, formed on the same side of a straight line AE, at a common point O, is equal to two right angles. For, their sum is equal to the sum of the two adjacent angles AOB, BOE, which by the proposition is equal to two right angles.
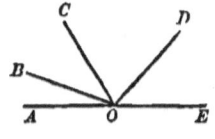

16. GEOMETRY.

15. *Corollary* IV. The sum of all the consecutive angles *AOB*, *BOC*, *COD*, *DOE*, *EOA*, formed about a point *O*, is equal to four right angles. For, if two straight lines are drawn through *O*, perpendicular to each other, the sum of all the consecutive angles formed about *O* will be equal to the four right angles formed by the perpendiculars.

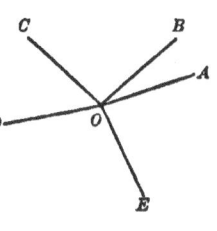

16. *Scholium.* A straight line revolving from the position *OA* around to the position *OB* describes the two right angles *AOC* and *COB*; hence *OA* and *OB*, regarded as two different lines having opposite directions (4), are frequently said to make an angle with each other equal to two right angles.

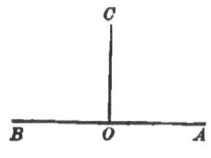

A line revolving from the position *OA from right to left*, that is, successively into the positions *OC*, *OB*, *OD*, when it has arrived at the position *OD* will have described an angle greater than two right angles. On the other hand, if the position *OD* is reached by revolving *from left to right*, that is, successively into the positions *OE*, *OD*, then the angle *AOD* is less than two right angles. Thus, any two straight lines drawn from a common point make two different angles with each other, one less and the other greater than two right angles. Hereafter the angle which is less than two right angles will be understood, unless otherwise expressly stated.

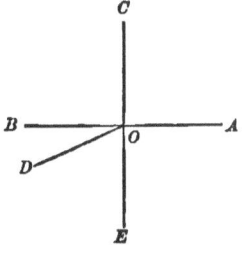

17. *Definitions.* An *acute* angle is an angle less than a right angle; as *AOD*. An *obtuse* angle is an angle greater than a right angle; as *BOD*.

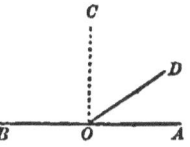

18. When the sum of two angles is equal to a right angle, each is called the *complement* of the other. Thus *DOC* is the complement of *AOD*, and *AOD* is the complement of *DOC*.

19. When the sum of two angles is equal to two right angles, each is called the *supplement* of the other. Thus *BOD* is the supplement of *AOD*, and *AOD* is the supplement of *BOD*.

20. It is evident that the complements of equal angles are equal to each other; and also that the supplements of equal angles are equal to each other.

PROPOSITION III.—THEOREM.

21. Conversely, *if the sum of two adjacent angles is equal to two right angles, their exterior sides are in the same straight line.*

Let the sum of the adjacent angles AOD, BOD, be equal to two right angles; then, OA and OB are in the same straight line.

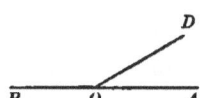

For BOD is the supplement of AOD (19), and is therefore identical with the angle which OD makes with the prolongation of AO (11). Therefore OB and the prolongation of AO are the same line.

22. Every proposition consists of an *hypothesis* and a *conclusion*. The *converse* of a proposition is a second proposition of which the hypothesis and conclusion are respectively the conclusion and hypothesis of the first. For example, Proposition II. may be enunciated thus:

Hypothesis—if two adjacent angles have their exterior sides in the same straight line, then—*Conclusion*—the sum of these adjacent angles is equal to two right angles.

And Proposition III. may be enunciated thus:

Hypothesis—if the sum of two adjacent angles is equal to two right angles, then—*Conclusion*—these adjacent angles have their exterior sides in the same straight line.

Each of these propositions is therefore the converse of the other.

A proposition and its converse are however not always both true.

PROPOSITION IV.—THEOREM.

23. *If two straight lines intersect each other, the opposite (or vertical) angles are equal.*

Let AB and CD intersect in O; then will the opposite, or vertical, angles AOC and BOD be equal. For, each of these angles is the supplement of the same angle BOC, or AOD, and hence they are equal (20).

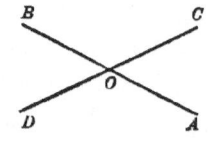

In like manner it is proved that the opposite angles AOD and BOC are equal.

24. *Corollary* I. The straight line EOF which bisects the angle AOC also bisects its vertical angle BOD. For, the angle FOD is equal to its vertical angle EOC, and FOB is equal to its vertical angle EOA; therefore if EOC and EOA are equal, FOD and FOB are equal.

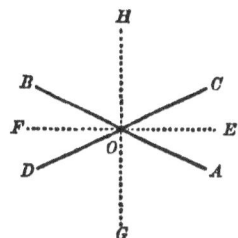

25. *Corollary* II. The two straight lines EOF, HOG, which bisect the two pairs of vertical angles, are perpendicular to each other. For, $HOC = HOB$ and $COE = BOF$; hence, by addition, $HOC + COE = HOB + BOF$; that is, $HOE = HOF$; therefore, by the definition (8), HO is perpendicular to FE.

PROPOSITION V.—THEOREM.

26. *From a given point without a straight line, one perpendicular can be drawn to that line, and but one.*

Let AB be the given straight line and P the given point.

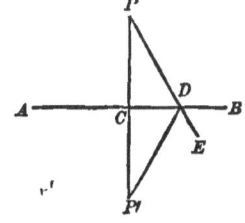

The line AB divides the plane in which it is situated into two portions. Let the portion containing P, which we suppose to be the upper portion, be revolved about the line AB (*i.e., folded over*) until the point P comes into the lower portion; and let P' be that point in the plane with which P coincides after this revolution. Restoring P to its original position, join PP', cutting AB in C, and again revolve the upper portion of the plane about AB until P again coincides with P'. Since the line AB is fixed during the revolution, the point C is fixed; therefore PC will coincide with $P'C$, and the angle PCD with the angle $P'CD$. These angles are therefore equal (6), and BC is perpendicular to PP' (8), or PC perpendicular to AB (13). There can therefore be one perpendicular from the point P to the line AB.

Moreover, PC is the only perpendicular. Let PD be any other

BOOK I. 19

line drawn from P to AB, and join $P'D$. Then, when the upper portion of the plane is revolved until P coincides with P', D being fixed, PD coincides with $P'D$, and consequently the angle PDC with the angle $P'DC$. Hence the angles PDC and $P'DC$ are equal. Now PP' being the only straight line that can be drawn from P to P' (1), PDP' is not a straight line; and if PD is produced to E, PDE and DP' are different straight lines. Hence the angle PDP' is less than two right angles, and its half, PDC, is less than one right angle; that is, PD is an oblique line. Therefore PC is the only perpendicular.

27. *Corollary.* Of the two angles which any oblique line drawn from P makes with AB, that one is acute within which the perpendicular from P upon AB falls; thus, PDC is acute.

PROPOSITION VI.—THEOREM.

28. *The perpendicular is the shortest line that can be drawn from a point to a straight line.*

Let PC be the perpendicular, and PD any oblique line, from the point P to the line AB. Then $PC < PD$.

For, produce PC to P', making $CP' = CP$, and join $P'D$. When the portion of the plane which contains P is revolved about AB, as in the preceding proposition, until P coincides with P', PD also coincides with $P'D$; and hence $PD = P'D$. But the straight line PP', being the shortest distance 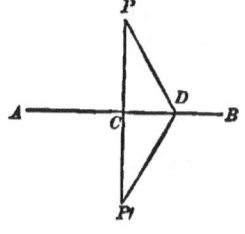 between the points P and P', is less than the broken line PDP'. Therefore PC, the half of the straight line, is less than PD, the half of the broken line.

29. *Definition.* By the *distance of a point from a line* is always understood the *shortest distance*. By the preceding proposition, therefore, the perpendicular measures the distance of a point from a straight line.

Also, by the *distance of one point from another* is understood the shortest distance, that is, the straight line between the points.

PROPOSITION VII.—THEOREM.

30. *Two oblique lines drawn from the same point to a straight line, cutting off equal distances from the foot of the perpendicular, are equal.*

Let the oblique lines PD, PE, meet the line AB in the points D and E, cutting off the equal distances CD and CE from the foot of the perpendicular. Then $PD = PE$.

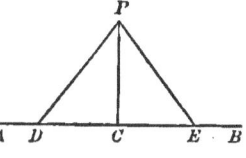

For, DCE being perpendicular to PC, and $CD = CE$, the figure PCD may be revolved about PC into coincidence with PCE; and since the point D will fall on E, PD will coincide with PE. Therefore $PD = PE$.

31. *Corollary.* The angles PDC and PEC are equal; that is, *two equal straight lines from a point to a straight line make equal acute angles with that line.*

32. *Definition.* A broken line, as $ABCDE$, is called *convex*, when no one of its component straight lines, if produced, can enter the space enclosed by the broken line and the straight line joining its extremities.

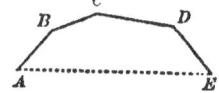

PROPOSITION VIII.—THEOREM.

33. *A convex broken line is less than any other line which envelops it and has the same extremities.*

Let the convex broken line $AFGE$ have the same extremities A, E, as the line $ABCDE$, and be *enveloped* by it; that is, wholly included within the space bounded by $ABCDE$ and the straight line AE. Then $AFGE < ABCDE$.

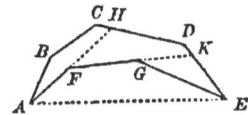

For, produce AF and FG to meet the enveloping line in H and K. Imagine $ABCDE$ to be the path of a point moving from A to E. If the straight line AH be substituted for ABC, the path $AHDE$ will be shorter than the path $ABCDE$, the portion HDE being common to both. If, further, the straight line FK be substituted for $FHDK$, the path $AFKE$ will be a still shorter path from A to E. And if, finally, GE be substituted for

BOOK I.

GKE, *AFGE* will be a still shorter path. Therefore, *AFGE* is less than any enveloping line.

34. *Scholium.* The preceding demonstration applies when the enveloping line is a *curve*, or any species of line whatever.

PROPOSITION IX.—THEOREM.

35. *Of two oblique lines drawn from the same point to the same straight line, that is the greater which cuts off upon the line the greater distance from the perpendicular.*

Let *PC* be the perpendicular from *P* to *AB*, and suppose $CE > CD$; then $PE > PD$.

For, produce *PC* to *P'*, making $CP' = CP$, and join *DP'*, *EP'*. Then, as in Proposition VI., we have $PD = P'D$, and $PE = P'E$. But (33), the broken line *PDP'* is less than the enveloping line *PEP'*; therefore *PD*, the half of *PDP'*, is less than *PE*, the half of *PEP'*.

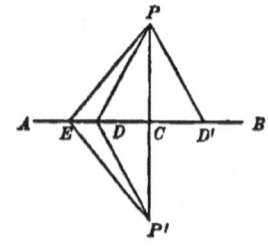

If the two oblique lines are on opposite sides of the perpendicular, as *PE* and *PD'*, and if $CE > CD'$, take $CD = CD'$, and join *PD*. Then, as above $PE > PD$; and, by Proposition VII., $PD = PD'$; hence $PE > PD'$.

36. *Corollary* I. (Converse of Proposition VII.). Two equal oblique lines cut off equal distances from the perpendicular.

37. *Corollary* II. (Converse of Proposition IX.). Of two unequal oblique lines, the greater cuts off the greater distance from the perpendicular.

PROPOSITION X.—THEOREM.

38. *If a perpendicular is erected at the middle of a straight line, then,*

1st. *Every point in the perpendicular is equally distant from the extremities of the line;*

2d. *Every point without the perpendicular is unequally distant from the extremities of the line.*

Let AB be a finite straight line, and C its middle point; then,

1st. Every point P in the perpendicular erected at C is equally distant from A and B. For, since $CA = CB$, we have (30) $PA = PB$.

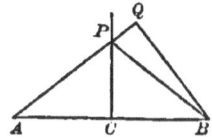

2d. Any point Q without the perpendicular is unequally distant from A and B. For, Q being on one side or the other of the perpendicular, one of the lines QA, QB must cut the perpendicular; let it be QA and let it cut in P; join PB. The straight line QB is less than the broken line QPB, that is, $QB < QP + PB$. But $PB = PA$; therefore $QB < QP + PA$, or $QB < QA$.

39. *Corollary.* Every point equally distant from the extremities of a straight line lies in the perpendicular erected at the middle of the line.

40. *Definition.* A geometric *locus* is the assemblage of all the points which possess a common property.

In this definition, points are understood to have a common property when they satisfy the same geometrical *conditions*.

Thus, since all the points in the perpendicular erected at the middle of a line possess the common property of being equally distant from the extremities of the line (that is, satisfy the *condition* that they shall be equally distant from those extremities), and *no other points possess this property*, the perpendicular is the locus of these points; so that the preceding proposition and its corollary are fully covered by the following brief statement:

The perpendicular erected at the middle of a straight line is the locus of all the points which are equally distant from the extremities of that line.

41. *Scholium.* Two points are sufficient to determine a straight line (3); hence any two points each of which is equally distant from the extremities of a straight line determine the perpendicular at the middle of the line. Thus if P and P' are known to be each equally distant from A and B, the line PP' joining these points is known to be perpendicular to AB at its middle point.

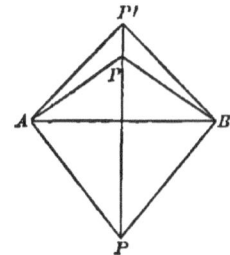

PARALLEL LINES.

42. *Definition.* *Parallel lines* are straight lines which lying in the same plane cannot meet, though indefinitely produced: as AB, CD.

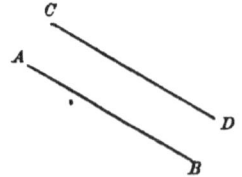

43. *Axiom.* Through the same point there cannot be two parallels to the same straight line.

Thus, if through a point P, one line CD is drawn parallel to AB, the axiom assumes that any other line drawn through P, as EPF, will not be parallel to AB, but will meet it, if both EF and AB be sufficiently produced.

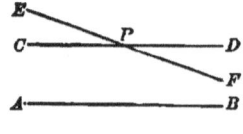

PROPOSITION XI.—THEOREM.

44. *Two straight lines perpendicular to the same straight line are parallel.*

Let AB and CD be perpendicular to AC; then, they are parallel.

For, if they could meet when produced, we should have from one point (their point of meeting) two perpendiculars to the same straight line AC, which (26) is impossible. Therefore they cannot meet, and by the definition (42) are parallel.

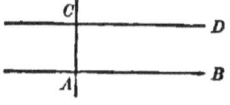

45. *Corollary* I. Through a given point a parallel to a given straight line can always be drawn. For, let C be the given point, and AB the given line. From C a perpendicular CA can be drawn to AB (26); and at C a perpendicular CD to CA can be drawn (9); and by the preceding proposition CD will be parallel to AB.

46. *Corollary* II. A straight line perpendicular to one of two parallels is perpendicular to the other.

Let AC be a perpendicular to AB; it will also be perpendicular to the parallel CD. In the first place it is to be observed that AC being a different line from AB cannot also be parallel to CD (43), and must therefore meet CD in some point, as C. Moreover the perpendicular to AC at C is parallel to AB (44) and must coincide with CD (9) and (43). Hence AC is perpendicular to CD.

PROPOSITION XII.—THEOREM.

47. *Two straight lines parallel to a third are parallel to each other.*

Let CD and EF be parallel to AB; then, they are parallel to each other. For, if they could meet, there would be drawn through their point of meeting two straight lines parallel to the same straight line, which (43) is impossible. Hence they cannot meet, and are parallel to each other.

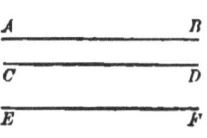

48. *Definitions.* When two straight lines AB, CD, are cut by a third EF, the eight angles formed at their points of intersection are named as follows:

The four angles, 1, 2, 3, 4, without the two lines, are called *exterior* angles.

The four angles, 5, 6, 7, 8, within the two lines, are called *interior* angles.

Two exterior angles on opposite sides of the secant line and not adjacent—as 1, 3—or 2, 4—are called *alternate-exterior* angles.

Two interior angles on opposite sides of the secant line and not adjacent—as 5, 7—or 6, 8—are called *alternate-interior* angles.

Two angles similarly situated with respect both to the secant and to the line intersected by it, are called *corresponding* angles; as 1, 5—2, 6—3, 7—4, 8.

PROPOSITION XIII.—THEOREM.

49. *If two parallel lines are cut by a third straight line, the alternate-interior angles are equal.*

Let the parallels AB, CD, be cut by the straight line EF in the points G and H; then, the *alternate-interior* angles, HGB and GHC, are equal.

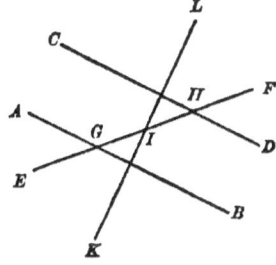

For, through I, the middle point of GH, suppose the indefinite line KIL to be drawn perpendicular to AB; it will also (46) be perpendicular to CD. Conceive the portion IGB of the figure, including the perpendicular *IK*, to be revolved *in its own plane* about *I* (as upon a pivot), until *IG* comes into coincidence with its equal *IH*. The angle *GIK* being equal to its vertical angle *HIL*, the indefinite line *IK* will fall upon *IL* and form with it but one line. Moreover, the point *G* being then at *H*, the line *GB* which is perpendicular to *IK* will then coincide with *HC* which is perpendicular to *IL*, and consequently the angles *IGB* and *IHC* will coincide. Therefore the angles *HGB* and *GHC* are equal.

Hence, also, their supplements, *HGA* and *GHD*, are equal.

50. *Corollary* I. The alternate-exterior angles, *AGE* and *DHF*, being equal to their vertical angles, *HGB* and *GHC*, are also equal to each other.

51. *Corollary* II. Any one of the eight angles is equal to its corresponding angle. Thus, since $HGB = GHC$ and $GHC = FHD$, there follows $HGB = FHD$; etc.

52. *Corollary* III. The sum of the two interior angles on the same side of the secant line is equal to two right angles. For, $GHD + HGB = GHD + GHC =$ two right angles (11).

53. *Scholium.* When the secant line is oblique to the parallels, there are formed four equal acute angles and four equal obtuse angles, and each acute angle is the supplement of each obtuse angle. But if any one of the eight angles is a right angle, they are all right angles.

PROPOSITION XIV.—THEOREM.

54. Conversely, *when two straight lines are cut by a third, if the alternate-interior angles are equal, these two straight lines are parallel.*

Let *EF* cut *AB* and *CD* in the points *G* and *H*, and let *HGB* and *GHC* be equal; then, *AB* and *CD* are parallel.

For, a parallel to *AB* drawn through *H* makes with *GH* an interior angle, alternate to *HGB*, which is equal to *HGB* (49); this angle must therefore coincide with the angle *GHC*, and the parallel drawn through *H* must coincide with *CD*. That is, *CD* is parallel to *AB*.

55. *Corollary* I. If the alternate-exterior angles are equal, or if the corresponding angles are equal, the two lines are parallel.

56. *Corollary* II. If the sum of the two interior angles on the same side of the secant line is equal to two right angles, the two lines are parallel.

57. *Corollary* III. From (52) and (56) it follows that, when two straight lines are cut by a third, if the sum of two interior angles on the same side of the secant line is *not* two right angles, the two straight lines are *not* parallel; and it is evident that they will meet, if produced, on that side of the secant line on which the two interior angles are together less than two right angles.

PROPOSITION XV.—THEOREM.

58. *Two parallels are everywhere equally distant.*

Let AB and CD be two indefinitely extended parallels; G and H any two points in CD; GE and HF the perpendiculars from G and H upon AB. Then, GE and HF are also perpendicular to CD (46), and measure the distance between the parallels at G and H, or at E and F. 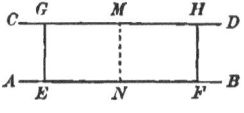 We are to prove that $GE = HF$.

Let M be the middle of GH, and suppose MN drawn perpendicular to GH and consequently also to EF. The portion of the figure on the right of MN may be revolved upon the line MN (*i.e.*, folded over); the angles at M and N being right angles the indefinite lines MD and NB will fall upon MC and NA; and since $MH = MG$, the point H will fall upon G, so that HF and ~~HG~~ (being then perpendiculars from the same point G upon the same straight line NA), will coincide (26). Therefore $GE = HF$.

59. *Corollary.* The *locus* (40) of all the points at a given distance, MN, from a given straight line AB, consists of two parallel lines, CD and $C'D'$, drawn on opposite sides of AB, at the given distance from it.

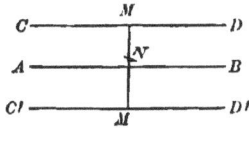

PROPOSITION XVI.—THEOREM.

60. *If two angles have their sides respectively parallel and lying in the same direction, they are equal.*

Let the angles ABC, DEF, have their sides BA and ED parallel and in the same direction, and also their sides BC and EF parallel and in the same direction. Then $ABC = DEF$.

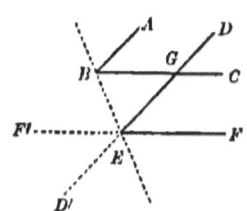

For, let DE, produced if necessary, intersect BC in G. The angle DGC is equal to its corresponding angle ABC and also to its corresponding angle DEF (51); therefore $ABC = DEF$.

Note. Two parallels, as BA and ED, are said to be in the same direction when they lie on the same side of the indefinite straight line joining the *origins*, B and E, of these parallels.

61. *Corollary* I. Two angles, as ABC and $D'EF'$, having their sides parallel and lying in opposite directions (that is ED' opposite to BA and EF' opposite to BC), are equal. For we have $D'EF' = DEF = ABC$.

62. *Corollary* II. Two angles, as ABC and DEF', having two of their sides, BA and ED, parallel and in the same direction, while their other two sides, BC and EF', are parallel and in opposite directions, are supplements of each other.

63. *Corollary* III. *If two angles, ABC, DEF, have their sides perpendicular each to each, that is, AB to ED and BC to EF, they are either equal or supplementary.* For, suppose the angle DEF to be revolved into the position HEK, by revolving ED and EF each through a right angle; that is, ED through the right angle DEH and EF through the right angle FEK. Then EH

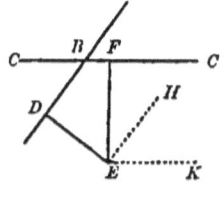

being perpendicular to ED is parallel to AB, and EK being perpendicular to EF is parallel to BC (44); therefore HEK, or DEF, is either equal to ABC by (60) or (61), or it is the supplement of ABC by (62).

TRIANGLES.

64. *Definitions.* A plane *triangle* is a portion of a plane bounded by three intersecting lines; as ABC. The *sides* of the triangle are the portions of the bounding lines included between the points of intersection; viz., AB, BC, CA. The angles of the triangle are the angles formed by the sides with each other; viz., CAB, ABC, BCA. The three angular points, A, B, C, which are the vertices of the angles, are also called the vertices of the triangle.

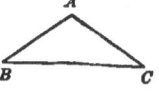

If a side of a triangle is produced, the angle which the prolongation makes with the adjacent side is called an *exterior angle;* as ACD.

65. A triangle is called *scalene* (ABC) when no two of its sides are equal; *isosceles* (DEF) when two of its sides are equal; *equilateral* (GHI) when its three sides are equal.

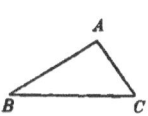 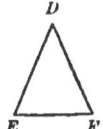 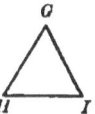 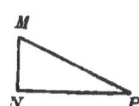

A *right triangle* is one which has a right angle; as MNP, which is right-angled at N. The side MP, opposite to the right angle, is called the *hypotenuse.*

The *base* of a triangle is the side upon which it is supposed to stand. In general any side may be assumed as the base; but in an isosceles triangle DEF, whose sides DE and DF are equal, the third side EF is always called the base.

When any side BC of a triangle has been adopted as the base, the angle BAC opposite to it is called the *vertical angle*, and its angular point A the vertex of the triangle. The perpendicular AD let fall from the vertex upon the base is then called the *altitude* of the triangle.

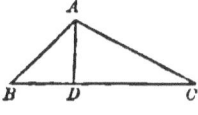

PROPOSITION XVII.—THEOREM.

66. *Any side of a triangle is less than the sum of the other two.*

Let BC be any side of a triangle whose other two sides are AB and AC; then $BC < AB + AC$. For, the straight line BC is the shortest distance between the points B and C.

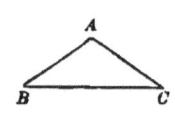

67. *Corollary.* Any side of a triangle is greater than the difference of the other two. For, if from each member of the inequality

$$BC < AB + AC$$

we subtract AB, we shall have

$$BC - AB < AC, \text{ or } AC > BC - AB.$$

PROPOSITION XVIII.—THEOREM.

68. *The sum of the three angles of any triangle is equal to two right angles.*

Let ABC be any triangle; then, the sum of its three angles, A, B and C, is equal to two right angles.

For, produce BC to D, and draw CE parallel to BA. The angle ACE is equal to its alternate angle BAC (49), and the angle ECD is equal to its corresponding angle

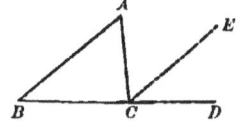

ABC (51). Therefore the sum of the three angles of the triangle is equal to $ECD + ACE + BCA$, which is two right angles (14).

69. *Corollary* I. Any exterior angle, as ACD, is equal to the sum of the two opposite interior angles, A and B; and consequently greater than either of them.

70. *Corollary* II. If one angle of a triangle is a right angle, or an obtuse angle, each of the other two angles must be acute; that is, a triangle cannot have two right angles, or two obtuse angles.

71. *Corollary* III. In a right triangle, the sum of the two acute angles is equal to one right angle; that is, each acute angle is the complement of the other (18).

72. *Corollary* IV. If two angles of a triangle are given, or only their sum, the third angle will be found by subtracting their sum from two right angles.

73. *Corollary* V. If two angles of one triangle are respectively equal to two angles of another triangle, the third angle of the one is also equal to the third angle of the other.

PROPOSITION XIX.—THEOREM.

74. *The angle contained by two straight lines drawn from any point within a triangle to the extremities of one of the sides is greater than the angle contained by the other two sides of the triangle.*

From any point D, within the triangle ABC, let DB, DC be drawn; then, the angle BDC is greater than the angle BAC.

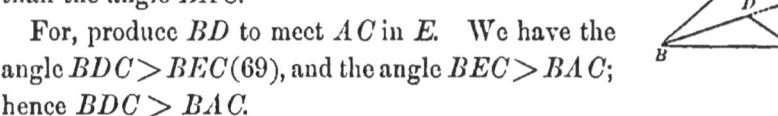

For, produce BD to meet AC in E. We have the angle $BDC > BEC$ (69), and the angle $BEC > BAC$; hence $BDC > BAC$.

75. *Definition.* Equal triangles, and in general equal figures, are those which can exactly fill the same space, or which can be applied to each other so as to coincide in all their parts.

PROPOSITION XX.—THEOREM.

76. *Two triangles are equal when two sides and the included angle of the one are respectively equal to two sides and the included angle of the other.*

In the triangles ABC, DEF, let AB be equal to DE, BC to EF, and the included angle B equal to the included angle E; then, the triangles are equal.

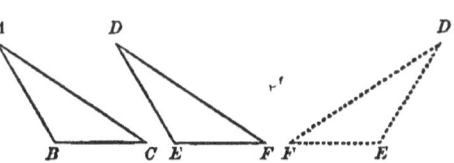

For, the triangle ABC may be superposed upon the triangle DEF, by applying the angle B to the equal angle E, the side BA upon its equal ED, and the side BC upon its equal EF. The points A and C then coinciding with the points D and F, the side AC will coincide with the side DF, and the triangles will coincide in all their parts; therefore they are equal (75).

77. *Corollary.* If in two triangles ABC, DEF, there are given $B = E$, $AB = DE$ and $BC = EF$, there will follow $A = D$, $C = F$, and $AC = DF$.

PROPOSITION XXI.—THEOREM.

78. *Two triangles are equal when a side and the two adjacent angles of the one are respectively equal to a side and the two adjacent angles of the other.*

In the triangles ABC, DEF, let BC be equal to EF, and let the angles B and C adjacent to BC be respectively equal to the angles E and F adjacent to EF; then, the triangles are equal.

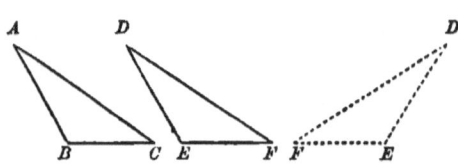

For, the triangle ABC may be superposed upon the triangle DEF, by applying BC to its equal EF, the point B upon E, and the point C upon F. The angle B being equal to the angle E, the side BA will take the direction of ED, and the point A will fall somewhere in the line ED. The angle C being equal to the angle F, the side CA will take the direction of FD, and the point A will fall somewhere in the line FD. Hence the point A, falling at once in both the lines ED and FD, must fall at their intersection D. Therefore the triangles will coincide throughout, and are equal.

79. *Corollary.* If in two triangles ABC, DEF, there are given $B = E$, $C = F$, and $BC = EF$, there will follow $A = D$, $AB = DE$, and $AC = DF$.

PROPOSITION XXII.—THEOREM.

80. *Two triangles are equal when the three sides of the one are respectively equal to the three sides of the other.*

In the triangles ABC, DEF, let AB be equal to DE, AC to DF, and BC to EF; then, the triangles are equal.

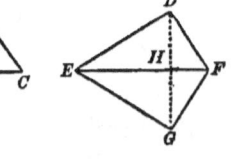

For, suppose the triangle ABC to be placed so that its base BC coincides with its equal EF, but so that the vertex A falls on the opposite side of EF from D, as at G; and join DG which intersects EF in H.

Then, by hypothesis, $EG = ED$ and $FG = FD$; therefore, E and F being two points equally distant from D and G, the line EF is perpendicular to DG at its middle point H. Hence, if the figure DEF be revolved upon the line EF, H being a fixed point, HD will fall upon its equal HG, and the triangle DEF will coincide entirely with the triangle GEF. Therefore, the triangle DEF is equal to the triangle GEF, or to the triangle ABC.

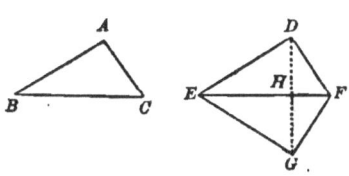

81. *Corollary.* If in two triangles ABC, DEF, there are given $AB = DE, AC = DF, BC = EF$, there will follow $A = D, B = E, C = F$.

82. *Scholium.* In two equal triangles, the equal angles lie opposite to the equal sides.

PROPOSITION XXIII.—THEOREM.

83. *Two right triangles are equal, 1st, when the hypotenuse and a side of the one are respectively equal to the hypotenuse and a side of the other; or, 2d, when the hypotenuse and an acute angle of the one are respectively equal to the hypotenuse and an acute angle of the other.*

1st. In the right triangles ABC, DEF, let the hypotenuse AB be equal to DE, and the side AC to DF; then, the triangles are equal.

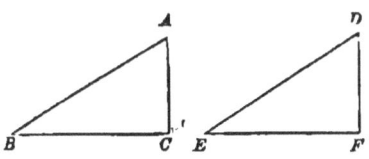

For, applying AC to its equal DF, the angles C and F being equal, the side CB will take the direction FE, and B will fall somewhere in the line FE. But AB being equal to DE, will cut off on FE the same distance from the perpendicular (36), and hence B will fall at E. The triangles will therefore coincide, and are equal.

2d. Let $AB = DE$, and the angle $ABC =$ the angle DEF; then, the triangles are equal.

For, the third angles BAC and EDF are equal (73), and hence the triangles are equal by (78).

PROPOSITION XXIV.—THEOREM.

84. *If two sides of a triangle are respectively equal to two sides of another, but the included angle in the first triangle is greater than the included angle in the second, the third side of the first triangle is greater than the third side of the second.*

Let ABC and ABD be the two triangles in which the sides AB, AC are respectively equal to the sides AB, AD, but the included angle BAC is greater than the included angle BAD; then, BC is greater than BD.

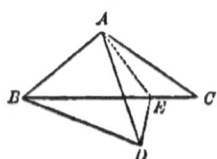

For, suppose the line AE to be drawn, bisecting the angle CAD and meeting BC in E; join DE. The triangles AED, AEC are equal (76), and therefore $ED = EC$. But in the triangle BDE we have

$$BE + ED > BD,$$

and substituting EC for its equal ED,

$$BE + EC > BD, \text{ or } BC > BD.$$

85. *Corollary.* Conversely, if in two triangles ABC, DEF, we have $AB = DE$, $AC = DF$, but $BC > EF$; then, $A > D$.

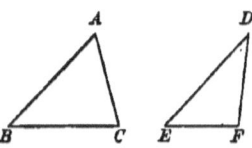

For, if A were equal to D, we should have $BC = EF$ (76); and if A were less than D, we should have $BC < EF$ (by the above proposition); but as both these conclusions are absurd, being contrary to the hypothesis, we can only have $A > D$.

PROPOSITION XXV.—THEOREM.

86. *In an isosceles triangle, the angles opposite to the equal sides are equal.*

Let AB and AC be the equal sides of the isosceles triangle ABC; then, the angles B and C are equal.

For, let D be the middle point of BC, and draw AD. The triangles ABD and ADC are equal (80); therefore the angle $ABD =$ the angle ACD (82).

87. *Corollary* I. From the equality of the triangles ABD and ACD, we also have the angles $ADB = ADC$,

and $BAD = CAD$; that is, *the straight line joining the vertex and the middle of the base of an isosceles triangle is perpendicular to the base and bisects the vertical angle.*

Hence, also, *the straight line which bisects the vertical angle of an isosceles triangle bisects the base at right angles.*

88. *Corollary* II. Every equilateral triangle is also equiangular; and by (68), each of its angles is equal to one-third of two right angles, or to two-thirds of one right angle.

PROPOSITION XXVI.—THEOREM.

89. *If two sides of a triangle are unequal, the angles opposite to them are unequal, and the greater angle is opposite to the greater side.*

In the triangle ABC, let AB be greater than AC; then, the angle ACB is greater than the angle B.

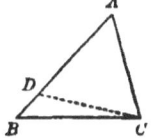

For, from the greater side AB cut off a part $AD = AC$, and join CD. The triangle ADC is isosceles, and therefore the angles ADC and ACD are equal (86). But the whole angle ACB is greater than its part ACD, and therefore greater than ADC; and ADC, an exterior angle of the triangle BDC, is greater than the angle B (69); still more, then, is ACB greater than B.

PROPOSITION XXVII.—THEOREM.

90. *If two angles of a triangle are equal, the sides opposite to them are equal.*

In the triangle ABC, let the angles B and C be equal; then, the sides AB and AC are equal.

For, if the sides AB and AC were unequal, the angles B and C could not be equal (89).

91. *Corollary.* Every equiangular triangle is also equilateral.

PROPOSITION XXVIII.—THEOREM.

92. *If two angles of a triangle are unequal, the sides opposite to them are unequal, and the greater side is opposite to the greater angle.*

In the triangle ABC let the angle C be greater than the angle B; then, AB is greater than AC.

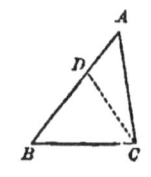

For, suppose the line CD to be drawn, cutting off from the greater angle a part $BCD = B$. Then BDC is an isosceles triangle, and $DC = DB$. But in the triangle ADC, we have $AD + DC > AC$; or, putting DB for its equal DC, $AD + DB > AC$; or $AB > AC$.

POLYGONS.

93. *Definitions.* A *polygon* is a portion of a plane bounded by straight lines; as $ABCDE$. The bounding lines are the *sides;* their sum is the *perimeter* of the polygon. The angles which the adjacent sides make with each other are the angles of the polygon; and the vertices of these angles are called the vertices of the polygon.

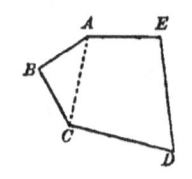

Any line joining two vertices not consecutive is called a *diagonal;* as AC.

94. *Definitions.* Polygons are classed according to the number of their sides:

A *triangle* is a polygon of three sides.

A *quadrilateral* is a polygon of four sides.

A *pentagon* has five sides; a *hexagon*, six; a *heptagon*, seven; an *octagon*, eight; an *enneagon*, nine; a *decagon*, ten; a *dodecagon*, twelve; etc.

An *equilateral* polygon is one all of whose sides are equal; an *equiangular* polygon, one all of whose angles are equal.

95. *Definition.* A *convex* polygon is one no side of which when produced can enter within the space enclosed by the perimeter, as $ABCDE$ in (93). Each of the angles of such a polygon is less than two right angles.

It is also evident from the definition that the perimeter of a convex polygon cannot be intersected by a straight line in more than two points.

A *concave* polygon is one of which two or more sides, when produced, will enter the space enclosed by the perimeter; as $MNOPQ$, of which OP and QP when produced will enter within the polygon. The angle OPQ, formed by two adjacent re-entrant sides, is called a *re-entrant angle;* and hence a concave polygon is sometimes called a *re-entrant polygon.*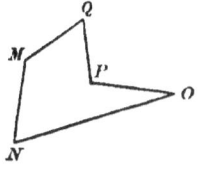

All the polygons hereafter considered will be understood to be convex.

96. A polygon may be divided into triangles by diagonals drawn from one of its vertices. Thus the pentagon $ABCDE$ is divided into three triangles by the diagonals drawn from A. The number of triangles into which any polygon can thus be divided is evidently equal to the number of its sides, less two. The number of diagonals so drawn is equal to the number of sides, less three.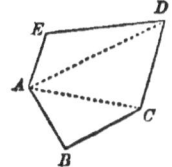

97. Two polygons $ABCDE$, $A'B'C'D'E'$, are equal when they can be divided by diagonals into the same number of triangles, equal each to each, and similarly arranged; for the polygons can evidently be superposed, one upon the other, so as to coincide.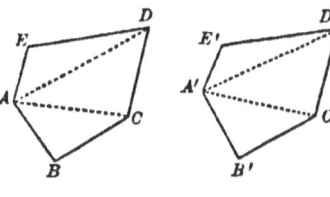

98. *Definitions.* Two polygons are *mutually equiangular* when the angles of the one are respectively equal to the angles of the other, taken in the same order; as $ABCD$, $A'B'C'D'$, in which $A = A'$, $B = B'$, etc.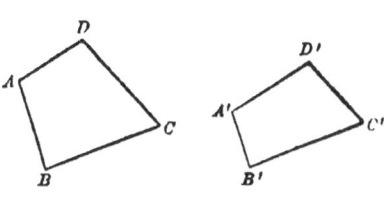

The equal angles are called *homologous angles;* the sides containing equal angles, and similarly placed, are *homologous sides;* thus A and A' are homologous angles, AB and $A'B'$ are homologous sides, etc.

Two polygons are *mutually equilateral* when the sides of the one are respectively equal to the sides of the other, taken in the same order; as $MNPQ$, $M'N'P'Q'$, in which $MN = M'N'$, $NP = N'P'$, etc. The equal sides are homologous; and angles contained by equal sides similarly placed, are homologous; thus MN and $M'N'$ are homologous sides; M and M' are homologous angles.

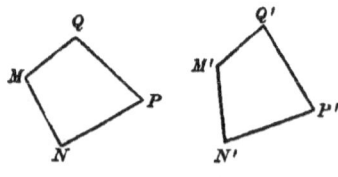

Two mutually equiangular polygons are not necessarily also mutually equilateral. Nor are two mutually equilateral polygons necessarily also mutually equiangular, except in the case of triangles (80).

If two polygons are mutually equilateral and also mutually equiangular, they are equal; for they can evidently be superposed, one upon the other, so as to coincide.

PROPOSITION XXIX.—THEOREM.

99. *The sum of all the angles of any polygon is equal to two right angles taken as many times less two as the polygon has sides.*

For, by drawing diagonals from any one vertex, the polygon can be divided into as many triangles as it has sides, less two (96). The sum of the angles of all the triangles is the same as the sum of the angles of the polygon, and the sum of the angles of each triangle is two right angles (68). Therefore, the sum of the angles of the polygon is two right angles taken as many times less two as the polygon has sides.

100. *Corollary* I. If N denotes the number of the sides of the polygon, and R a right angle, the sum of the angles is $2R \times (N-2) = (2N-4)R = 2NR - 4R$; that is, twice as many right angles as the polygon has sides, less four right angles.

For example, the sum of the angles of a quadrilateral is four right angles; of a pentagon, six right angles; of a hexagon, eight right angles, etc.

101. *Corollary* II. If all the sides of any polygon $ABCDE$, be produced so as to form one exterior angle at each vertex, the sum of these exterior angles, a, b, c, d, e, is four right angles. For, the sum of each interior and its adjacent exterior angle, as $A + a$, is two right angles (11); therefore, the sum of all the angles, both interior and exterior, is twice as many right angles as the polygon has sides. 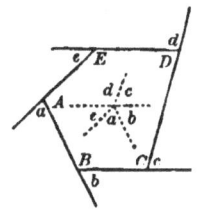 But the sum of the interior angles alone is twice as many right angles as the polygon has sides, less four right angles (100); therefore the sum of the exterior angles is equal to four right angles.

This is also proved in a very simple manner, by drawing, from any point in the plane of the polygon, a series of lines respectively parallel to the sides of the polygon and in the same directions as their prolongations. The consecutive angles formed by these lines will be equal to the exterior angles of the polygon (60), and their sum is four right angles (15).

QUADRILATERALS.

102. *Definitions.* Quadrilaterals are divided into classes as follows:

1st. The *trapezium* (A) which has no two of its sides parallel.

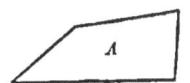

2d. The *trapezoid* (B) which has two sides parallel. The parallel sides are called the *bases*, and the perpendicular distance between them the *altitude* of the trapezoid.

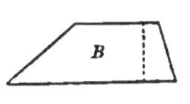

3d. The *parallelogram* (C) which is bounded by two pairs of parallel sides.

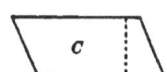

The side upon which a parallelogram is supposed to stand and the opposite side are called its lower and upper *bases*. The perpendicular distance between the bases is the *altitude*.

103. *Definitions.* Parallelograms are divided into species, as follows:

1st. The *rhomboid* (*a*), whose adjacent sides are not equal and whose angles are not right angles.

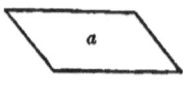

2d. The *rhombus*, or *lozenge* (*b*), whose sides are all equal.

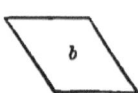

3d. The *rectangle* (*c*), whose angles are all equal and therefore right angles.

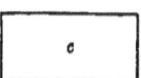

4th. The *square* (*d*), whose sides are all equal and whose angles are all equal.

The square is at once a rhombus and a rectangle.

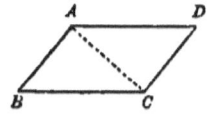

PROPOSITION XXX.—THEOREM.

104. *In every parallelogram, the opposite angles are equal, and the opposite sides are equal.*

Let *ABCD* be a parallelogram.

1st. The opposite angles *B* and *D*, contained by parallel lines lying in opposite directions, are equal (61); and for the same reason the opposite angles *A* and *C* are equal.

2d. Draw the diagonal *AC*. Since *AD* and *BC* are parallel, the alternate angles *CAD* and *ACB* are equal (49), and since *DC* and *AB* are parallel, the alternate angles *ACD* and *CAB* are equal. Therefore, the triangles *ADC* and *CBA* are equal (78), and the sides opposite to the equal angles are equal, namely, $AD = BC$, and $DC = AB$.

105. *Corollary* I. A diagonal of a parallelogram divides it into two equal triangles.

106. *Corollary* II. If one angle of a parallelogram is a right angle, all its angles are right angles, and the figure is a rectangle.

PROPOSITION XXXI.—THEOREM.

107. *If the opposite angles of a quadrilateral are equal, or if its opposite sides are equal, the figure is a parallelogram.*

1st. Let the opposite angles of the quadrilateral $ABCD$ be equal, or $A = C$ and $B = D$. Then, by adding equals, we have
$$A + B = C + D;$$
therefore, each of the sums $A + B$ and $C + D$ is equal to one-half the sum of the four angles. But the sum of the four angles is equal to four right angles (100); therefore, $A + B$ is equal to two right angles, and the lines AD and BC are parallel (56). In like manner it may be proved that AB and CD are parallel. Therefore the figure is a parallelogram.

2d. Let the opposite sides of the quadrilateral $ABCD$ be equal, or $BC = AD$ and $AB = DC$. Then, drawing the diagonal AC, the triangles ABC, ACD are equal (80); therefore, the angles CAD and ACB are equal, and the lines AD and BC are parallel (54). Also since the angles CAB and ACD are equal, the lines AB and DC are parallel. Therefore $ABCD$ is a parallelogram.

PROPOSITION XXXII.—THEOREM.

108. *If two opposite sides of a quadrilateral are equal and parallel, the figure is a parallelogram.*

Let the opposite sides BC and AD of the quadrilateral $ABCD$ be equal and parallel. Draw the diagonal AC. The alternate angles CAD and ACB are equal (49), and hence the triangles ADC and CBA are equal (76). Therefore, the sides AB and CD are equal and the figure is a parallelogram (107).

PROPOSITION XXXIII.—THEOREM.

109. *The diagonals of a parallelogram bisect each other.*

Let the diagonals AC, BD of the parallelogram $ABCD$ intersect in E; then, $AE = EC$ and $ED = EB$.

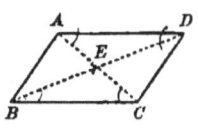

For, the side AD and the angles EAD, ADE, of the triangle EAD, are respectively equal to the side CB and the angles ECB, EBC of the triangle ECB; hence these triangles are equal (78), and the sides respectively opposite the equal angles are equal, namely, $AE = EC$ and $ED = EB$.

110. *Corollary* I. The diagonals of a rhombus $ABCD$ bisect each other at right angles in E. For, since $AD = CD$ and $AE = EC$, ED is perpendicular to AC (41).

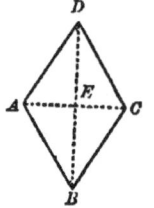

111. *Corollary* II. The diagonals of a rhombus bisect its opposite angles. For, in each of the isosceles triangles ADC, ABC, BCD, DAB, the line drawn from the vertex to the middle of the base bisects the vertical angle (87).

PROPOSITION XXXIV.—THEOREM.

112. *If the diagonals of a quadrilateral bisect each other, the figure is a parallelogram.*

Let the diagonals of the quadrilateral $ABCD$ bisect each other in E. Then, the triangles AED and CEB are equal (76), and the angles EAD, ECB, respectively opposite the equal sides, are equal. Therefore AD and BC are parallel (54). In like manner AB and DC are shown to be parallel, and the figure is a parallelogram.

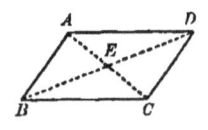

113. *Corollary.* If the diagonals of a quadrilateral bisect each other at right angles, the figure is a rhombus.

PROPOSITION XXXV.—THEOREM.

114. *The diagonals of a rectangle are equal.*

Let $ABCD$ be a rectangle; then its diagonals, AC and BD, are equal.

For, the right triangles ABC and DCB are equal (76); therefore, $AC = BD$.

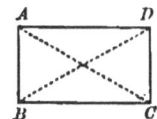

115. *Corollary* I. The diagonals of a square are equal, and, since the square is also a rhombus, they bisect each other at right angles (110), and also bisect the angles of the square (111).

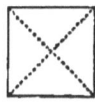

116. *Corollary* II. A parallelogram is a rectangle if its diagonals are equal.

117. *Corollary* III. A quadrilateral is a square, if its diagonals are equal and bisect each other at right angles.

118. *Scholium.* The rectangle, being a species of parallelogram, has all the properties of a parallelogram.

The square, being at once a parallelogram, a rectangle and a rhombus, has the properties of all these figures.

PROPOSITION XXXVI.—THEOREM.

119. *Two parallelograms are equal when two adjacent sides and the included angle of the one are equal to two adjacent sides and the included angle of the other.*

Let AC, $A'C'$, have $AB = A'B'$, $AD = A'D'$, and the angle $BAD = B'A'D'$; then, these parallelograms are equal.

For they may evidently be applied the one to the other so as to coincide throughout.

120. *Corollary.* Two rectangles are equal when they have equal bases and equal altitudes.

APPLICATIONS.

PROPOSITION XXXVII.—THEOREM.

121. *If a straight line drawn parallel to the base of a triangle bisects one of the sides, it also bisects the other side; and the portion of it intercepted between the two sides is equal to one-half the base.*

Let DE be parallel to the base BC of the triangle ABC, and bisect the side AB in D; then, it bisects the side AC in E, and $DE = \frac{1}{2}BC$.

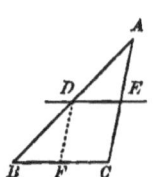

1st. Through D suppose DF to be drawn parallel to AC. In the triangles ADE, DBF, we have $AD = DB$, and the angles adjacent to these sides equal, namely $DAE = BDF$, and $ADE = DBF$ (51); therefore these triangles are equal (78), and $AE = DF$. Also, since $DECF$ is a parallelogram, $DF = EC$ (104); and hence $AE = EC$.

2d. The triangles ADE and BDF being equal, we have $DE = BF$, and in the parallelogram $DECF$ we have $DE = FC$; therefore $BF = FC$. Hence F is the middle point of BC, and $DE = \frac{1}{2}BC$.

122. *Corollary* I. The straight line DE, joining the middle points of the sides AB, AC, of the triangle ABC, is parallel to third side BC, and is equal to one-half of BC. For, the straight line drawn through D parallel to BC, passes through E (121), and is therefore identical with DE. Consequently, also, $DE = \frac{1}{2}BC$.

123. *Corollary* II. The straight line drawn parallel to the bases of a trapezoid, bisecting one of the non-parallel sides, also bisects the opposite side.

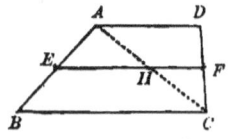

Let $ABCD$ be a trapezoid, BC and AD its parallel bases, E the middle point of AB, and let EF be drawn parallel to BC or AD; then, F is the middle of DC. For, draw the diagonal AC, intersecting EF in H. Then in the triangle ABC, EH is drawn through the middle of AB parallel to BC; therefore H is the middle of AC. In the triangle ACD, HF is drawn through the middle of AC parallel to AD; therefore F is the middle of DC.

124. *Corollary* III. In a trapezoid, the straight line joining the middle points of the non-parallel sides is parallel to the bases, and is equal to one-half their sum.

Let EF join the middle points, E and F, of AB and DC. Then, 1st, EF is parallel to BC. For, by Cor. II. the straight line drawn through E parallel to BC passes through F and is therefore identical with EF.

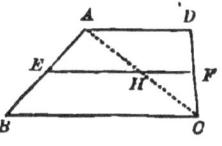

2d. Drawing the diagonal AC, intersecting EF in H, we have, in the triangle ABC,

$$EH = \tfrac{1}{2} BC,$$

and in the triangle ACD,

$$HF = \tfrac{1}{2} AD,$$

the sum of which gives

$$EF = \tfrac{1}{2}(BC + AD).$$

PROPOSITION XXXVIII.—THEOREM.

125. *If a series of parallels cutting any two straight lines intercept equal distances on one of these lines, they also intercept equal distances on the other line.*

Let MN, $M'N'$, be two straight lines cut by a series of parallels AA', BB', CC', DD'; then, if AB, BC, CD are equal, $A'B'$, $B'C'$, $C'D'$ are also equal.

For, through the points A, B, C, draw Ab, Bc, Cd, parallel to $M'N'$. In the triangles ABb, BCc, CDd, we have $AB = BC = CD$; and the corresponding angles adjacent to these sides are equal (51), namely, $BAb = CBc = DCd$, and $ABb = BCc = CDd$; therefore, these triangles are equal to each other (78), and $Ab = Bc = Cd$. But, the figures $A'b$, $B'c$, $C'd$, being parallelograms, we have $Ab = A'B'$, $Bc = B'C'$, $Cd = C'D'$; therefore, $A'B' = B'C' = C'D'$.

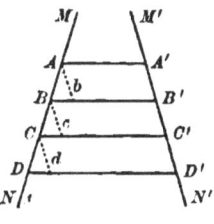

PROPOSITION XXXIX.—THEOREM.

126. *Every point in the bisector of an angle is equally distant from the sides of the angle; and every point not in the bisector, but within the angle, is unequally distant from the sides of the angle.*

1st. Let AD be the bisector of the angle BAC, P any point in it, and PE, PF, the perpendicular distances of P from AB and AC; then, $PE = PF$.

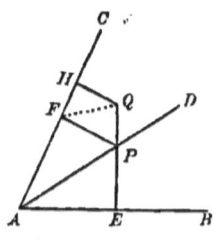

For, the right triangles APE, APF, having the angles PAE and PAF equal, and AP common, are equal (83); therefore, $PE = PF$.

2d. Let Q be any point not in the bisector, but within the angle; then, the perpendicular distances QE and QH are unequal.

For, suppose that one of these distances, as QE, cuts the bisector in some point P: from P let PF be drawn perpendicular to AC, and join QF. We have $QH < QF$; also $QF < QP + PF$, or $QF < QP + PE$, or $QF < QE$; therefore, $QH < QE$.

When the angle BAC is obtuse, the point Q, not in the bisector, may be so situated that the perpendicular on one of the sides, as AB, will fall at the vertex A; the perpendicular QH is then less than the oblique line QA. Or, a point Q' may

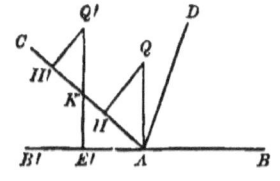

be so situated that the perpendicular $Q'E'$, let fall on one of the sides, as AB, will meet that side produced through the vertex A; this perpendicular must cut the side AC in some point, K, and we then have $Q'H' < Q'K < Q'E'$.

127. *Corollary.* The bisector of an angle is the *locus* (40) of all the points within the angle which are equally distant from its sides.

PROPOSITION XL.—THEOREM.

128. *The three bisectors of the three angles of a triangle meet in the same point.*

Let AD, BE, CF, be the bisectors of the angles A, B, C, respectively, of the triangle ABC.

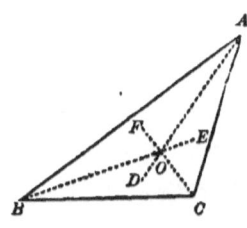

Let the two bisectors AD, BE, meet in O. The point O, being in AD, is equally distant from AB and AC (126); and being in BE, it is equally distant from AB and BC;

therefore, the point *O* is equally distant from
AC and *BC*, and must lie in the bisector of
the angle *C* (127). That is, the bisector *CF*
also passes through *O*, and the three bisectors meet in the same point.

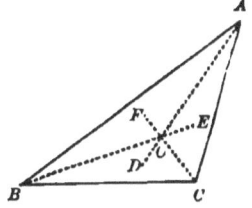

129. *Corollary.* The point in which the three bisectors of the
angles of a triangle meet is equally distant from the three sides of
the triangle.

PROPOSITION XLI.—THEOREM.

130. *The three perpendiculars erected at the middle points of the
sides of a triangle meet in the same point.*

Let *DG*, *EH*, *FK*, be the perpendiculars
erected to *BC*, *CA*, *AB*, respectively, at their
middle points, *D*, *E*, *F*.

It is first necessary to prove that any two of
these perpendiculars, as *DG*, *EH*, meet in some
point. If they did not meet, they would be
parallel, and then *CB* and *CA* being perpen-

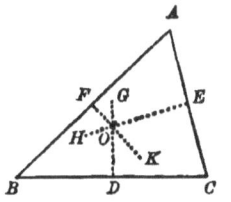

diculars to these parallels from the same point *C*, would be in one
straight line, which is impossible, since they are two sides of a triangle. Therefore, *DG* and *EH* are not parallel, and must meet in
some point, as *O*.

Now the point *O* being in the perpendicular *DG* is equally distant
from *B* and *C* (38), and being also in the perpendicular *EH*, it is
equally distant from *A* and *C*; therefore it is equally distant from *A*
and *B*, and must lie in the perpendicular *FK* (39). That is, the
perpendicular *FK* passes through *O*, and the three perpendiculars
meet in the same point.

131. *Corollary.* The point in which the three perpendiculars meet
is equally distant from the three vertices of the triangle.

BOOK I. 47

PROPOSITION XLII.—THEOREM.

132. *The three perpendiculars from the vertices of a triangle to the opposite sides meet in the same point.*

Let AD, BE, CF, be the perpendiculars from the vertices of the triangle ABC to the opposite sides, respectively.

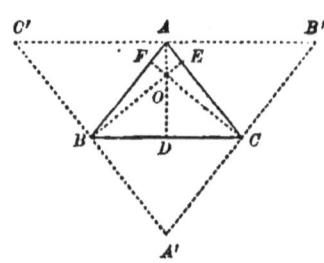

Through the three vertices, A, B, C, draw the lines $B'C'$, $A'B'$, $A'C'$, respectively parallel to BC, AB, AC. Then the two quadrilaterals $ABCB'$ and $ACBC'$ are parallelograms, and we have $AB' = BC$ and $AC' = BC$; therefore $AB' = AC'$, or A is the middle of $B'C'$. But AD being perpendicular to BC is perpendicular to the parallel $B'C'$; therefore AD is the perpendicular to $B'C'$ erected at its middle point A. In like manner, it is shown that BE and CF are the perpendiculars to $A'C'$ and $A'B'$ at their middle points; therefore, by (130), the three perpendiculars meet in the same point.

133. *Definition.* A straight line drawn from any vertex of a triangle to the middle point of the opposite side is called a *medial line* of the triangle. Thus, D being the middle point of BC, AD is the medial line to BC.

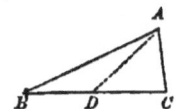

PROPOSITION XLIII.—THEOREM.

134. *The three medial lines of a triangle meet in the same point.*

Let D, E, F, be the three middle points of the sides of the triangle ABC; AD, BE, CF, the three medial lines.

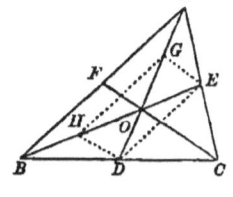

Let the two medial lines, AD and BE, meet in O. Let G be the middle point of OA, and H the middle point of OB; join GH, HD, DE, EG. In the triangle AOB, GH is parallel to AB, and $GH = \frac{1}{2}AB$: and in the triangle ABC, ED is parallel to AB, and $ED = \frac{1}{2}AB$ (122). Therefore, HG and ED, being parallel to AB, are parallel to each other; and each being equal to $\frac{1}{2}AB$, they are equal to each other; consequently, $EGHD$ is a parallelogram (108), and its diagonals bisect each other (109). Therefore $OD = OG = GA$, or $OD = \frac{1}{3}AD$; that is, the medial line BE cuts the medial line AD at a point O whose distance from D is one-third of AD. In the same way it is proved that the medial line CF cuts AD at a point whose distance from D is one-third of AD, that is, at the same point O; and therefore the three medial lines meet in the same point.

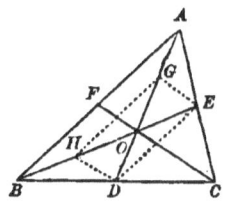

SYMMETRICAL FIGURES.

a. *Symmetry with respect to an axis.*

135. Definition. Two points are *symmetrical* with respect to a fixed straight line, called the *axis of symmetry*, when this axis bisects at right angles the straight line joining the two points.

Thus, A and A' are symmetrical with respect to the axis MN, if MN bisects AA' at right angles at a.

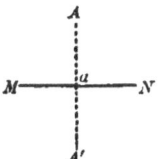

If the portion of the plane containing the point A on one side of the axis MN, is revolved about this axis (or *folded* over) until it coincides with the portion on the other side of the axis, the point A' at which A falls is the symmetrical point of A.

136. Definition. Any two figures are symmetrical with respect to an axis when every point of one figure has its symmetrical point on the other.

Thus, $A'B'$ is the symmetrical figure of the straight line AB, with respect to the axis MN, every point, as C, of the one, having its symmetrical point C' in the other.

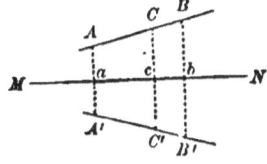

The symmetrical figure of an indefinite straight line, AB, is an indefinite straight line, $A'B'$, which intersects the first in the axis and makes the same angle with the axis as the first line.

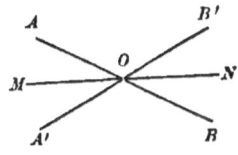

BOOK I.   49

137. *Definition.* In two symmetrical figures the corresponding symmetrical lines are called *homologous*.

Thus, in the symmetrical figures $ABCDE$, $A'B'C'D'E'$, the homologous lines are AB and $A'B'$, BC and $B'C'$, etc.

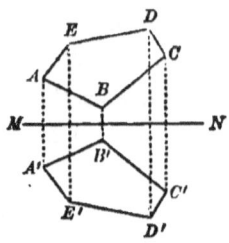

In all cases, two figures, symmetrical with respect to an axis, can be brought into coincidence by the revolution of either about the axis.

b. *Symmetry with respect to a centre.*

138. *Definition.* Two points are symmetrical with respect to a fixed point, called the *centre of symmetry*, when this centre bisects the straight line joining the two points.

Thus, A and A' are symmetrical with respect to the centre O, if the line AA' passes through O and is bisected at O.

The distance of a point from the centre is called its *radius of symmetry*. A point A is brought into coincidence with its symmetrical point A', by revolving its radius OA through two right angles in its own plane (16).

139. *Definition.* Any two figures are symmetrical with respect to a centre, when every point of one figure has its symmetrical point on the other.

Thus, $A'B'$ is the symmetrical figure of the straight line AB with respect to the centre O.

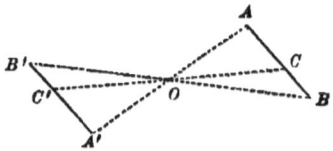

Since the triangles AOB, $A'OB'$, are equal (76), the angle B is equal to the angle B'; therefore, AB and $A'B'$ are parallel. In general, the homologous lines of two figures, symmetrical with respect to a centre, are parallel. Thus, in the symmetrical figures $ABCD$, $A'B'C'D'$, the homologous lines AB and $A'B'$ are parallel, BC and $B'C'$ are parallel, etc.

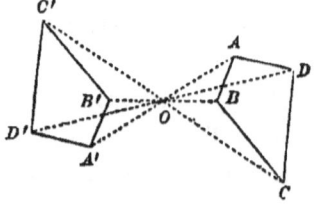

Two figures symmetrical with respect to a centre can be brought into coincidence by revolving one of them, in its own plane, about the centre; every radius of symmetry revolving through two right angles at the same time.

140. *Definition.* Any single figure is called *a symmetrical figure*, either when it can be divided by an axis into two figures symmetrical with respect to that axis, or when it has a centre such that every straight line drawn through it cuts the figure in two points which are symmetrical with respect to this centre.

Thus, $ABCDC'B'$ is a symmetrical figure with respect to the axis MN, being divided by MN into two figures, $ABCD$ and $AB'C'D$, which are symmetrical with respect to MN.

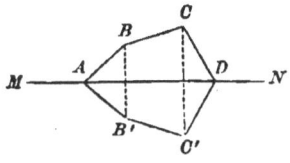

Also, the figure $ABCDEF$ is symmetrical with respect to the centre O, its vertices, taken two and two, being symmetrical with respect to O. In this case, any straight line KL drawn through the centre and terminated by the perimeter, is called a *diameter*.

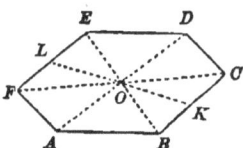

PROPOSITION XLIV.—THEOREM.

141. *If a figure is symmetrical with respect to two* AXES *perpendicular to each other, it is also symmetrical with respect to the intersection of these axes as a* CENTRE *of symmetry.*

Let the figure $ABCDEFGH$ be symmetrical with respect to the two perpendicular axes MN, PQ, which intersect in O; then, the point O is also the centre of symmetry of the figure.

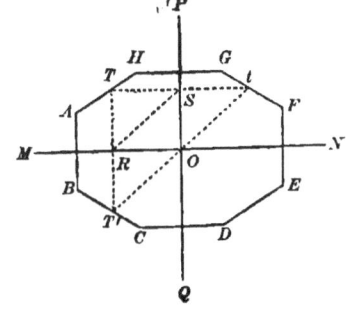

For, let T be any point in the perimeter of the figure; draw TRT' perpendicular to MN, and TSt perpendicular to PQ; join $T'O$, Ot and RS.

Since the figure is symmetrical with respect to MN, we have $RT' = RT$; and since $RT = OS$, it follows that $RT' = OS$; therefore,

$RT'OS$ is a parallelogram (108), and RS is equal and parallel to OT'.

Again, since the figure is symmetrical with respect to PQ, we have $St = ST = OR$; therefore, $SROt$ is a parallelogram, and RS is equal and parallel to Ot. Hence, T, O and t, are in the same straight line, since there can be but one parallel to RS drawn through the same point O.

Now we have $OT' = RS$ and $Ot = RS$, and consequently $OT' = Ot$; therefore, any straight line $T'Ot$, drawn through O, is bisected at O; that is, O is the centre of symmetry of the figure.

BOOK II.

THE CIRCLE.

1. DEFINITIONS. A *circle* is a portion of a plane bounded by a curve, all the points of which are equally distant from a point within it called the *centre*.

The curve which bounds the circle is called its *circumference*.

Any straight line drawn from the centre to the circumference is called a *radius*.

Any straight line drawn through the centre and terminated each way by the circumference is called a *diameter*.

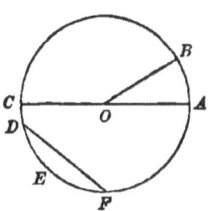

In the figure, O is the centre, and the curve $ABCEA$ is the circumference of the circle; the circle is the space included within the circumference; OA, OB, OC, are radii; AOC is a diameter.

By the definition of a circle, all its radii are equal; also all its diameters are equal, each being double the radius.

If one extremity, O, of a line OA is fixed, while the line revolves in a plane, the other extremity, A, will *describe* a circumference, whose radii are all equal to OA.

2. *Definitions.* An *arc* of a circle is any portion of its circumference; as DEF.

A *chord* is any straight line joining two points of the circumference; as DF. The arc DEF is said to be *subtended* by its chord DF.

Every chord subtends two arcs, which together make up the whole circumference. Thus DF subtends both the arc DEF and the arc $DCBAF$. When an arc and its chord are spoken of, the arc less than a semi-circumference, as *DEF*, is always understood, unless otherwise stated.

A *segment* is a portion of the circle included between an arc and its chord; thus, by the segment *DEF* is meant the space included between the arc *DF* and its chord.

A *sector* is the space included between an arc and the two radii drawn to its extremities; as *AOB*.

3. From the definition of a circle it follows that every point *within* the circle is at a distance from the centre which is less than the radius; and every point *without* the circle is at a distance from the centre which is greater than the radius. Hence (I. 40), *the locus of all the points in a plane which are at a given distance from a given point is the circumference of a circle described with the given point as a centre and with the given distance as a radius.*

4. It is also a consequence of the definition of a circle, *that two circles are equal when the radius of one is equal to the radius of the other,* or when (as we usually say) they have the *same* radius. For if one circle be superposed upon the other so that their centres coincide, their circumferences will coincide, since all the points of both are at the same distance from the centre.

If when superposed the second circle is made to turn upon its centre as upon a pivot, it must continue to coincide with the first.

5. *Postulate.* A circumference may be described with any point as a centre and any distance as a radius.

ARCS AND CHORDS.

PROPOSITION I.—THEOREM.

6. *A straight line cannot intersect a circumference in more than two points.*

For, if it could intersect it in three points, the three radii drawn to these three points would be three equal straight lines drawn from the same point to the same straight line, which is impossible (I. 36).

PROPOSITION II.—THEOREM.

7. *Every diameter bisects the circle and its circumference.*

Let $AMBN$ be a circle whose centre is O; then, any diameter AOB bisects the circle and its circumference.

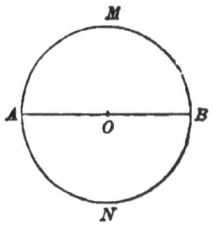

For, if the figure ANB be turned about AB as an axis and superposed upon the figure AMB, the curve ANB will coincide with the curve AMB, since all the points of both are equally distant from the centre. The two figures then coincide throughout, and are therefore equal in all respects. Therefore, AB divides both the circle and its circumference into equal parts.

8. *Definitions.* A *segment* equal to one half the circle, as the segment AMB, is called a *semi-circle*. An *arc* equal to half a circumference, as the arc AMB, is called a *semi-circumference*.

PROPOSITION III.—THEOREM.

9. *A diameter is greater than any other chord.*

Let AC be any chord which is not a diameter, and AOB a diameter drawn through A: then $AB > AC$.

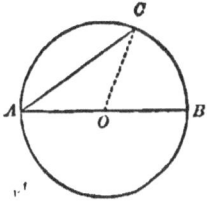

For, join OC. Then, $AO + OC > AC$ (I. 66); that is, since all the radii are equal, $AO + OB > AC$, or $AB > AC$.

PROPOSITION IV.—THEOREM.

10. *In equal circles, or in the same circle, equal angles at the centre intercept equal arcs on the circumference, and conversely.*

Let O, O', be the centre of equal circles, and AOB, $A'O'B'$, equal angles at these centres; then, the intercepted arcs, AB, $A'B'$, are equal. For, one of the angles, together with its arc, may be superposed upon the other; and when

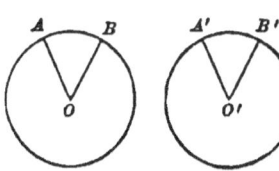

the equal angles coincide, their intercepted arcs will evidently coincide also.

Conversely, if the arcs AB, $A'B'$ are equal, the angles AOB, $A'O'B'$ are equal. For, when one of the arcs is superposed upon its equal, the corresponding angles at the centre will evidently coincide.

If the angles are in the same circle, the demonstration is similar.

11. *Definition.* A fourth part of a circumference is called a *quadrant.* It is evident from the preceding theorem that a right angle at the centre intercepts a quadrant on the circumference.

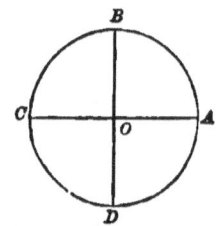

Thus, two perpendicular diameters, AOC, BOD, divide the circumference into four quadrants, AB, BC, CD, DA.

PROPOSITION V.—THEOREM.

12. *In equal circles, or in the same circle, equal arcs are subtended by equal chords, and conversely.*

Let O, O', be the centres of equal circles, and AB, $A'B'$, equal arcs; then, the chords AB, $A'B'$, are equal.

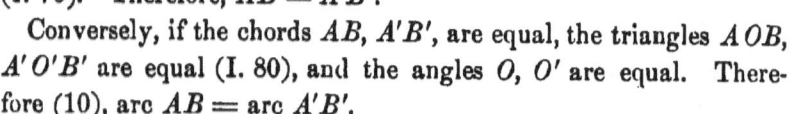

For, drawing the radii to the extremities of the arcs, the angles O and O' are equal (10), and consequently the triangles AOB, $A'O'B'$, are equal (I. 76). Therefore, $AB = A'B'$.

Conversely, if the chords AB, $A'B'$, are equal, the triangles AOB, $A'O'B'$ are equal (I. 80), and the angles O, O' are equal. Therefore (10), arc $AB =$ arc $A'B'$.

If the arcs are in the same circle, the demonstration is similar.

PROPOSITION VI.—THEOREM.

13. *In equal circles, or in the same circle, the greater arc is subtended by the greater chord, and conversely; the arcs being both less than a semi-circumference.*

Let the arc AC be greater than the arc AB; then, the chord AC is greater than the chord AB.

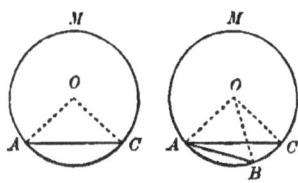

For, draw the radii OA, OB, OC. In the triangles AOC, AOB, the angle AOC is obviously greater than the angle AOB; therefore, (I. 84), chord $AC >$ chord AB.

Conversely, if chord $AC >$ chord AB, then, arc $AC >$ arc AB. For, in the triangles AOC, AOB, the side $AC >$ the side AB; therefore (I. 85), angle $AOC >$ angle AOB; and consequently, arc $AC >$ arc AB.

14. *Scholium*. If the arcs are greater than a semi-circumference, the contrary is true; that is, the arc AMB, which is greater than the arc AMC, is subtended by the less chord; and conversely.

PROPOSITION VII.—THEOREM.

15. *The diameter perpendicular to a chord bisects the chord and the arcs subtended by it.*

Let the diameter DOD' be perpendicular to the chord AB at C; then, 1st, it bisects the chord. For, the radii OA, OB being equal oblique lines from the point O to the line AB, cut off equal distances from the foot of the perpendicular (I. 36); therefore, $AC = BC$.

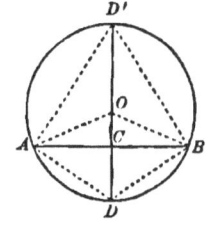

2d. The subtended arcs ADB, $AD'B$, are bisected at D and D', respectively. For, every point in the perpendicular DOD' drawn at the middle of AB being equally distant from its extremities A and B (I. 38), the chords AD and BD are equal; therefore, (12), the arcs AD and BD are equal. For the same reason, the arcs AD' and BD' are equal.

16. *Corollary* I. The perpendicular erected upon the middle of a chord passes through the centre of the circle, and through the middle of the arc subtended by the chord.

Also, the straight line drawn through any two of the three points O, C, D, passes through the third and is perpendicular to the chord AB.

17. *Corollary* II. The middle points of any number of parallel chords all lie in the same diameter perpendicular to the chords.

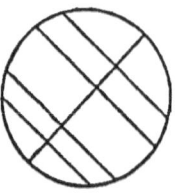

In other words, the *locus* of the middle points of a system of parallel chords is the diameter perpendicular to these chords.

PROPOSITION VIII.—THEOREM.

18. *In the same circle, or in equal circles, equal chords are equally distant from the centre; and of two unequal chords, the less is at the greater distance from the centre.*

1st. Let AB, CD, be equal chords; OE, OF, the perpendiculars which measure their distances from the centre O; then, $OE = OF$.

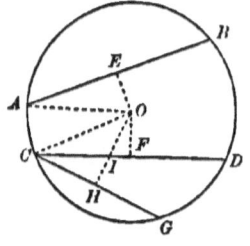

For, since the perpendiculars bisect the chords (15), $AE = CF$; hence (I. 83), the right triangles AOE and COF are equal, and $OE = OF$.

2d. Let CG, AB, be unequal chords; OE, OH, their distances from the centre; and let CG be less than AB; then, $OH > OE$.

For, since chord $AB >$ chord CG, we have arc $AB >$ arc CG; so that if from C we draw the chord $CD = AB$, its subtended arc CD, being equal to the arc AB, will be greater than the arc CG. Therefore the perpendicular OH will intersect the chord CD in some point I. Drawing the perpendicular OF to CD, we have, by the first part of the demonstration, $OF = OE$. But $OH > OI$, and $OI > OF$ (I. 28); still more, then, is $OH > OF$, or $OH > OE$.

If the chords be taken in two equal circles, the demonstration is the same.

19. *Corollary* I. The converse of the proposition is also evidently true, namely: *in the same circle, or in equal circles, chords equally distant from the centre are equal; and of two chords unequally distant from the centre, that is the greater whose distance from the centre is the less.*

20. *Corollary* II. The *least chord* that can be drawn in a circle through a given point P is the chord, AB, perpendicular to the line OP joining the given point and the centre. For, if CD is any other chord drawn through P, the perpendicular OQ to this chord is less than OP; therefore, by the preceding corollary, CD is greater than AB.

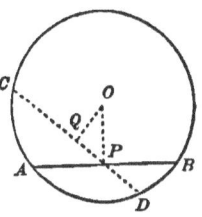

PROPOSITION IX.—THEOREM.

21. *Through any three points, not in the same straight line, a circumference can be made to pass, and but one.*

Let A, B, C, be any three points not in the same straight line.

1st. A circumference can be made to pass through these points. For, since they are not in the same straight line, the lines AB, BC, AC, joining them two and two, form a triangle, and the three perpendiculars DE,

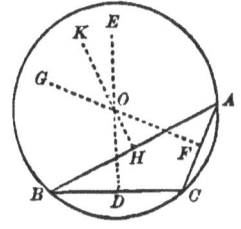

FG, HK, erected at the middle points of the sides, meet in a point O which is equally distant from the three points A, B, C, (I. 131). Therefore a circumference described from O as a centre and a radius equal to any one of the three equal distances OA, OB, OC, will pass through the three given points.

2d. Only one circumference can be made to pass through these points. For the centre of *any* circumference passing through the three points must be at once in two perpendiculars, as DE, FG, and therefore at their intersection; but two straight lines intersect in only one point, and hence O is the centre of the only circumference that can pass through the three points.

22. *Corollary*. Two circumferences can intersect in but two points; for, they could not have a third point in common without having the same centre and becoming in fact but one circumference.

TANGENTS AND SECANTS.

23. *Definitions.* A *tangent* is an indefinite straight line which has but one point in common with the circumference; as *ACB*. The common point, *C*, is called the *point of contact*, or the *point of tangency.* The circumference is also said to be tangent to the line *AB* at the point *C*.

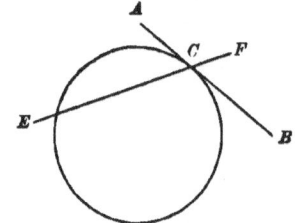

A *secant* is a straight line which meets the circumference in two points; as *EF*.

24. *Definition.* A rectilinear figure is said to be *circumscribed* about a circle when all its sides are tangents to the circumference.

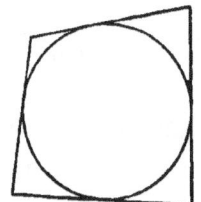

In the same case, the circle is said to be *inscribed* in the figure.

PROPOSITION X.—THEOREM.

25. *A straight line oblique to a radius at its extremity cuts the circumference.*

Let *AB* be oblique to the radius *OC* at its extremity *C*; then, *AB* cuts the circumference at *C*, and also in a second point *D*.

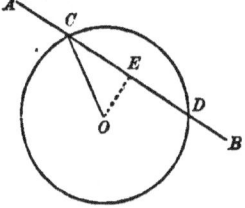

For, let *OE* be the perpendicular from *O* upon *AB*; then $OE < OC$, and the point *E* is within the circumference. Therefore *AB* cuts the circumference in *C*, and must evidently cut it in a second point *D*.

PROPOSITION XI.—THEOREM.

26. *A straight line perpendicular to a radius at its extremity is a tangent to the circle.*

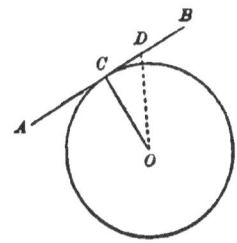

Let AB be perpendicular to the radius OC at its extremity C; then, AB is a tangent to the circle at the point C.

For, from the centre O draw the oblique line OD to any point of AB except C. Then, $OD > OC$, and D is a point without the circumference. Therefore AB having all its points except C without the circumference, has but the point C in common with it, and is a tangent at that point (23).

27. *Corollary.* Conversely, *a tangent AB at any point C is perpendicular to the radius OC drawn to that point.* For, if it were not perpendicular to the radius it would cut the circumference (25), and would not be a tangent.

28. *Scholium.* If a secant EF, passing through a point C of the circumference, be supposed to revolve upon this point, as upon a pivot, its second point of intersection, D, will move along the circumference and approach nearer and nearer to C. When the second point comes into coincidence with C, the revolving line ceases to be strictly a secant, and becomes the tangent AB; but, continuing the revolution,

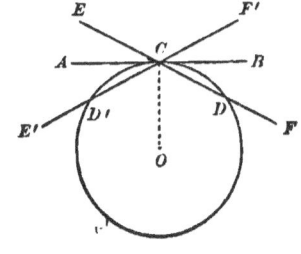

the revolving line again becomes a secant, as $E'F'$, and the second point of intersection reappears on the other side of C, as at D'.

If, then, our revolving line be required to be a *secant* in the strict sense imposed by our definition, that is a line meeting the circumference in *two* points, this condition can be satisfied only by keeping the second point of intersection, D, distinct from the first point, C, however near these points may be brought to each other; and, therefore, under this condition, the tangent is often called the *limit of the secants* drawn through the point of contact; that is to say, a limit toward which the secant continually approaches, as the second point of intersection (on either side of the first) continually approaches the first, but *a limit which is never reached by the secant as such*.

On the other hand, as the tangent is but one of the positions of our revolving line, it has properties in common with the secant; and in order to exhibit such common properties in the most striking manner, it is often expedient to regard *the tangent as a secant whose two points of intersection are coincident*. But it is to be observed that we then no longer consider the secant as a *cutting* line, but simply as a line drawn through two points of the curve; and we include the tangent as that special case of such a line in which the two points are coincident. In this, we generalize in the same way as in algebra, when we say that the expression $x = a - b$ signifies that x is the *difference* of a and b, even when $a = b$, and there is really *no difference* between a and b.

PROPOSITION XII.—THEOREM.

29. *Two parallels intercept equal arcs on a circumference.*

We may have three cases:

1st. When the parallels AB, CD, are both secants; then, the intercepted arcs AC and BD are equal. For, let OM be the radius drawn perpendicular to the parallels. By Prop. VII. the point M is at once the middle of the arc AMB and of the arc CMD, and hence we have

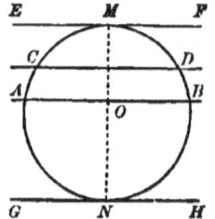

$$AM = BM \text{ and } CM = DM,$$

whence, by subtraction,

$$AM - CM = BM - DM;$$

that is,

$$AC = BD.$$

2d. When one of the parallels is a secant, as AB, and the other is a tangent, as EF at M, then, the intercepted arcs AM and BM are equal. For, the radius OM drawn to the point of contact is perpendicular to the tangent (27), and consequently perpendicular also to its parallel AB; therefore, by Prop. VII., $AM = BM$.

3d. When both the parallels are tangents, as EF at M, and GH

at N; then, the intercepted arcs MAN and MBN are equal. For, drawing any secant AB parallel to the tangents, we have by the second case,

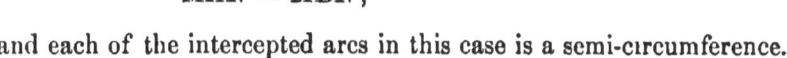

$$AM = BM \text{ and } AN = BN,$$

whence, by addition,

$$AM + AN = BM + BN,$$

that is,

$$MAN = MBN;$$

and each of the intercepted arcs in this case is a semi-circumference.

30. *Scholium* 1. The straight line joining the points of contact of two parallel tangents is a diameter.

31. *Scholium* 2. According to the principle of (28), the tangent being regarded as a secant whose two points of intersection are coincident, the demonstration of the first case in the preceding theorem embraces that of the other two cases.

RELATIVE POSITION OF TWO CIRCLES.

32. *Definition.* Two circles are *concentric*, when they have the same centre.

33. *Definition.* Two circumferences are *tangent* to each other, or *touch* each other, when they have but one point in common. The common point is called the *point of contact*, or the *point of tangency*.

Two kinds of contact are distinguished: *external contact*, when each circle is outside the other; *internal contact*, when one circle is within the other.

PROPOSITION XIII.—THEOREM.

34. *When two circumferences intersect, the straight line joining their centres bisects their common chord at right angles.*

Let O and O' be the centres of two circumferences which intersect in the points A, B; then, the straight line OO' bisects their common chord AB at right angles.

For, the perpendicular to AB erected

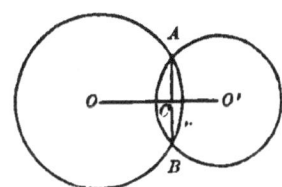

at its middle point C, passes through both centres (16); and there can be but one straight line drawn between the two points O and O'.

35. Corollary. *When two circumferences are tangent to each other, their point of contact is in the straight line joining their centres.* It has just been proved that when two circumferences intersect, the two points of intersection lie at equal distances from the line joining the centres and on opposite sides of this line. Now let the circles be supposed to be moved so as to cause the points of intersection to approach each other; these points will ultimately come together on the line joining the centres, and be blended in a single point C, common to the two circumferences, which will then be their point of contact. The perpendicular to OO' erected at C will then be a common

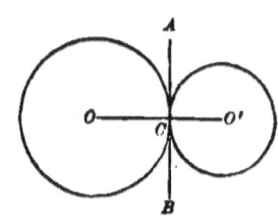

tangent to the two circumferences and take the place of the common chord.

PROPOSITION XIV.—THEOREM.

36. *When two circumferences are wholly exterior to each other, the distance of their centres is greater than the sum of their radii.*

Let O, O' be the centres. Their distance OO' is greater than the sum of the radii OA, $O'B$, by the portion AB interposed between the circles.

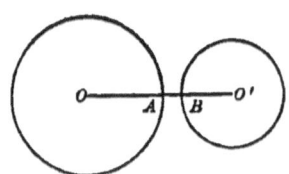

PROPOSITION XV.—THEOREM.

37. *When two circumferences are tangent to each other externally, the distance of their centres is equal to the sum of their radii.*

Let O, O', be the centres, and C the point of contact. The point C being in the line joining the centres (35), we have $OO' = OC + O'C$.

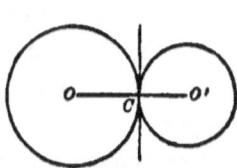

PROPOSITION XVI.—THEOREM.

38. *When two circumferences intersect, the distance of their centres is less than the sum of their radii and greater than the difference of their radii.*

Let O and O' be their centres, and A one of their points of intersection. The point A is not in the line joining the centres (34); and consequently there is formed the triangle AOO', in which we have $OO' < OA + O'A$, and also $OO' > OA - O'A$ (I. 67).

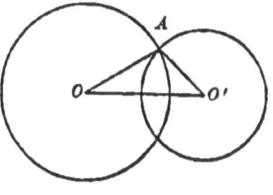

PROPOSITION XVII—THEOREM.

39. *When two circumferences are tangent to each other internally, the distance of their centres is equal to the difference of their radii.*

Let O, O', be the centres, and C the point of contact. The point C being in the line joining the centres (35), we have $OO' = OC - O'C$.

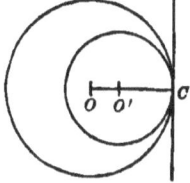

PROPOSITION XVIII.—THEOREM.

40. *When one circumference is wholly within another, the distance of their centres is less than the difference of their radii.*

Let O, O', be the centres. We have the difference of the radii $OA - O'B = OO' + AB$. Hence OO' is less than the difference of the radii by the distance AB.

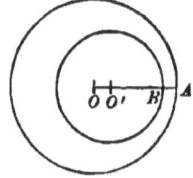

41. *Corollary.* The converse of each of the preceding five propositions is also true: namely—

1st. When the distance of the centres is greater than the sum of the radii, the circumferences are wholly exterior to each other.

2d. When the distance of the centres is equal to the sum of the radii, the circumferences touch each other externally.

3d. When the distance of the centres is less than the sum of the radii, but greater than their difference, the circumferences intersect.

4th. When the distance of the centres is equal to the difference of the radii, the circumferences touch each other internally.

5th. When the distance of the centres is less than the difference of the radii, one circumference is wholly within the other.

MEASURE OF ANGLES.

As the measurement of magnitude is one of the principal objects of geometry, it will be proper to premise here some principles in regard to the measurement of quantity in general.

42. *Definition.* To *measure* a quantity of any kind is to find how many times it contains another quantity of the same kind called the *unit*.

Thus, to measure a line is to find the number expressing how many times it contains another line called the *unit of length*, or the *linear unit*.

The number which expresses how many times a quantity contains the unit is called the *numerical measure* of that quantity.

43. *Definition.* The *ratio* of two quantities is the quotient arising from dividing one by the other; thus, the ratio of A to B is $\dfrac{A}{B}$.

To find the ratio of one quantity to another is, then, to find how many times the first contains the second; therefore, it is the same thing as to *measure* the first by the second taken as the unit (42). It is implied in the definition of ratio, that the quantities compared are of the *same kind*.

Hence, also, instead of the definition (42), we may say that to measure a quantity is to find its ratio to the unit.

The ratio of two quantities is the same as the ratio of their numerical measures. Thus, if P denotes the unit, and if P is contained m times in A and n times in B, then,

$$\frac{A}{B} = \frac{mP}{nP} = \frac{m}{n}.$$

44. *Definition.* Two quantities are *commensurable* when there is

some third quantity of the same kind which is contained a whole number of times in each. This third quantity is called the *common measure* of the proposed quantities.

Thus, the two lines, A and B, are commensurable, if there is some line, C, which is contained a whole number of times in each, as, for example, 7 times in A, and 4 times in B.

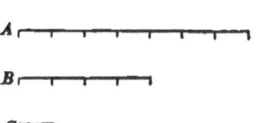

The ratio of two commensurable quantities can, therefore, be *exactly* expressed by a number whole or fractional (as in the preceding example by $\frac{7}{4}$), and is called a *commensurable ratio*.

45. *Definition.* Two quantities are *incommensurable* when they have no common measure. The ratio of two such quantities is called an *incommensurable ratio*.

If A and B are two incommensurable quantities, their ratio is still expressed by $\frac{A}{B}$.

46. *Problem. To find the greatest common measure of two quantities.* The well-known arithmetical process may be extended to quantities of all kinds. Thus, suppose AB and CD are two straight lines whose common measure is required. Their greatest common measure cannot be greater than the less line CD. Therefore, let CD be applied to AB as many times as possible, suppose 3 times, with a remainder EB less than CD. Any common measure of AB and CD must also be a common measure of CD and EB; for it will be contained a whole number of times in CD, and in AE, which is a multiple of CD, and therefore to measure AB it must also measure the part EB. Hence, the *greatest* common measure of AB and CD must also be the *greatest* common measure of CD and EB. This greatest common measure of CD and EB cannot be greater than the less line EB; therefore, let EB be applied as many times as possible to CD, suppose twice, with a remainder FD. Then, by the same reasoning, the greatest common measure of CD and EB, and consequently also that of AB and CD, is the greatest common measure of EB and FD. Therefore, let FD be applied to EB as many times as possible: suppose it is contained exactly twice in EB without remainder; the process is then completed, and we have found FD as the required greatest common measure.

The measure of each line, referred to FD as the unit, will then be as follows: we have

$$EB = 2FD,$$
$$CD = 2EB + FD = 4FD + FD = 5FD,$$
$$AB = 3CD + EB = 15FD + 2FD = 17FD.$$

The proposed lines are therefore numerically expressed, in terms of the unit FD, by the numbers 17 and 5; and their ratio is $\frac{17}{5}$.

47. When the preceding process is applied to two quantities and no remainder can be found which is exactly contained in a preceding remainder, however far the process be continued, the two quantities have no common measure; that is, they are incommensurable, and their ratio cannot be exactly expressed by any number whole or fractional.

48. But although an incommensurable ratio cannot be *exactly* expressed by a number, it may be *approximately* expressed by a number within any assigned measure of precision.

Suppose $\frac{A}{B}$ denotes the incommensurable ratio of two quantities A and B; and let it be proposed to obtain an approximate numerical expression of this ratio that shall be correct within an assigned measure of precision, say $\frac{1}{100}$. Let B be divided into 100 equal parts, and suppose A is found to contain 314 of these parts with a remainder less than one of the parts; then, evidently, we have

$$\frac{A}{B} = \frac{314}{100} \text{ within } \frac{1}{100},$$

that is, $\frac{314}{100}$ is an approximate value of the ratio $\frac{A}{B}$ within the assigned measure of precision.

To generalize this, $\frac{A}{B}$ denoting as before the incommensurable ratio of the two quantities A and B, let B be divided into n equal parts, and let A contain m of these parts with a remainder less than one of the parts; then we have

$$\frac{A}{B} = \frac{m}{n} \text{ within } \frac{1}{n};$$

and, since n may be taken as great as we please, $\frac{1}{n}$ may be made less than any assigned measure of precision, and $\frac{m}{n}$ will be the approximate value of the ratio $\frac{A}{B}$ within that assigned measure.

49. *Theorem. Two incommensurable ratios are equal, if their approximate numerical values are always equal, when both are expressed within the same measure of precision however small.*

Let $\frac{A}{B}$ and $\frac{A'}{B'}$ be two incommensurable ratios whose approximate numerical values are always the same when the same measure of precision is employed in expressing both; then, we say that

$$\frac{A}{B} = \frac{A'}{B'}.$$

For, let $\frac{1}{n}$ be any assumed measure of precision, and in accordance with the hypothesis of the theorem, suppose that for any value of $\frac{1}{n}$, the ratios $\frac{A}{B}$, $\frac{A'}{B'}$ have the same approximate numerical expression, say $\frac{m}{n}$, each ratio *exceeding* $\frac{m}{n}$ by a quantity less than $\frac{1}{n}$; then, these ratios cannot differ *from each other* by so much as $\frac{1}{n}$. But the measure $\frac{1}{n}$ may be assumed as small as we please, that is less than any assignable quantity however small; hence $\frac{A}{B}$ and $\frac{A'}{B'}$ cannot differ by any assignable quantity however small, and therefore they must be equal.

The student should study this demonstration in connection with that of Proposition XIX., which follows.

BOOK II.

50. Definition. *A proportion* is an equality of ratios. Thus, if the ratio $\dfrac{A}{B}$ is equal to the ratio $\dfrac{A'}{B'}$, the equality

$$\frac{A}{B} = \frac{A'}{B'}$$

is a proportion. It may be read: "Ratio of A to B equals ratio of A' to B'," or "A is to B as A' is to B'."

A proportion is often written as follows:

$$A : B = A' : B'$$

where the notation $A : B$ is equivalent to $A \div B$. When thus written, A and B' are called the *extremes*, B and A' the *means*, and B' is called a fourth proportional to A, B and A'; the first terms A and A', of the ratios are called the *antecedents*—the second terms, B and B', the *consequents*.

When the means are equal, as in the proportion

$$A : B = B : C,$$

the middle term B is called a *mean proportional* between A and C, and C is called a *third proportional* to A and B.

PROPOSITION XIX.—THEOREM.

51. *In the same circle, or in equal circles, two angles at the centre are in the same ratio as their intercepted arcs.*

Let AOB and AOC be two angles at the centre of the same, or at the centres of equal circles; AB and AC, their intercepted arcs; then,

$$\frac{AOB}{AOC} = \frac{AB}{AC}.$$

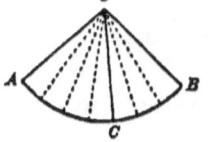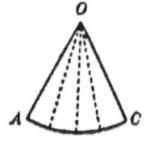

1st. Suppose the arcs to have a common measure which is contained, for example, 7 times in the arc AB and 4 times in the arc AC; so that if AB is divided into 7 parts, each equal to the common measure, AC will contain 4 of these parts. Then the ratio of the arcs AB and AC is 7 : 4; that is,

$$\frac{AB}{AC} = \frac{7}{4}$$

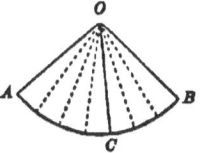 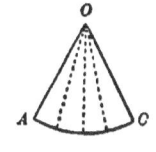

Drawing radii to the several points of division of the arcs, the partial angles at the centre subtended by the equal partial arcs will be equal (10); therefore the angle AOB will be divided into 7 equal parts, of which the angle AOC will contain 4; hence the ratio of the angles AOB and AOC is 7 : 4; that is,

$$\frac{AOB}{AOC} = \frac{7}{4}.$$

Therefore, we have

$$\frac{AOB}{AOC} = \frac{AB}{AC},$$

or,

$$AOB : AOC = AB : AC.$$

2d. If the arcs are incommensurable, suppose one of them, as AC, to be divided into any number n of equal parts; then AB will contain a certain number m of these parts, *plus* a remainder less than one of these parts. The numerical expression of the ratio $\frac{AB}{AC}$ will then be $\frac{m}{n}$, correct within $\frac{1}{n}$ (48). Drawing radii to the several points of division of the arcs, the angle AOC will be divided into n equal parts, and the angle AOB will contain m such parts, *plus* a remainder less than one of the parts. Therefore, the numerical expression of the ratio $\frac{AOB}{AOC}$ will also be $\frac{m}{n}$, correct within $\frac{1}{n}$; that is, the ratio $\frac{AOB}{AOC}$ has the same approximate numerical expression as the ratio $\frac{AB}{AC}$, however small the parts into which AC is divided; therefore these ratios must be absolutely equal (49), and we have for incommensurable, as well as for commensurable, arcs,

$$\frac{AOB}{AOC} = \frac{AB}{AC},$$

or,

$$AOB : AOC = AB : AC.$$

BOOK II. 71

PROPOSITION XX.—THEOREM.

52. *The numerical measure of an angle at the centre of a circle is the same as the numerical measure of its intercepted arc, if the adopted unit of angle is the angle at the centre which intercepts the adopted unit of arc.*

Let AOB be an angle at the centre O, and AB its intercepted arc. Let AOC be the angle which is adopted as the *unit of angle*, and let its intercepted arc AC be the arc which is adopted as the *unit of arc*. By Proposition XIX. we have

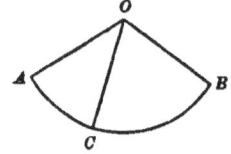

$$\frac{AOB}{AOC} = \frac{AB}{AC}.$$

But the first of these ratios is the measure (42) of the angle AOB referred to the unit AOC; and the second ratio is the measure of the arc AB referred to the unit AC. Therefore, with the adopted units, the numerical measure of the angle AOB is the same as that of the arc AB.

53. *Scholium* I. This theorem, being of frequent application, is usually more briefly, though inaccurately, expressed by saying that *an angle at the centre is measured by its intercepted arc*. In this conventional statement of the theorem, the condition that the adopted units of angle and arc correspond to each other is understood; and the expression "is measured by" is used for "has the same numerical measure as."

54. *Scholium* II. The right angle is, by its nature, the most simple unit of angle; nevertheless custom has sanctioned a different unit.

The unit of angle generally adopted is an angle equal to $\frac{1}{90}$th part of a right angle, called a *degree*, and denoted by the symbol °. The corresponding unit of arc is $\frac{1}{90}$th part of a quadrant (11), and is also called a degree.

A right angle and a quadrant are therefore both expressed by 90°. Two right angles and a semi-circumference are both expressed by 180°. Four right angles and a whole circumference are both expressed by 360°.

The degree (either of angle or arc) is subdivided into *minutes* and *seconds*, denoted by the symbols ′ and ″: a minute being $\frac{1}{60}$th part of a degree, and a second being $\frac{1}{60}$th part of a minute. Fractional parts of a degree less than one second are expressed by decimal parts of a second.

An angle, or an arc, of any magnitude is, then, numerically expressed by the unit degree and its subdivisions. Thus, for example, an angle equal to $\frac{1}{7}$th of a right angle, as well as its intercepted arc, will be expressed by 12° 51′ 25″.714 . . . .

55. *Definition.* When the sum of two arcs is a quadrant (that is, 90°), each is called the *complement* of the other.

When the sum of two arcs is a semi-circumference (that is, 180°), each is called the *supplement* of the other. See (I. 18, 19).

56. *Definitions.* An *inscribed angle* is one whose vertex is on the circumference and whose sides are chords; as BAC.

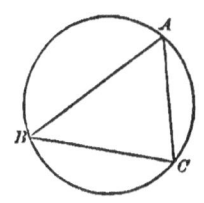

In general, any rectilinear figure, as ABC, is said to be inscribed in a circle, when its angular points are on the circumference; and the circle is then said to be *circumscribed* about the figure.

An angle is said to be *inscribed in a segment* when its vertex is in the arc of the segment, and its sides pass through the extremities of the subtending chord. Thus, the angle BAC is inscribed in the segment BAC.

PROPOSITION XXI.—THEOREM.

57. *An inscribed angle is measured by one-half its intercepted arc.*

There may be three cases:

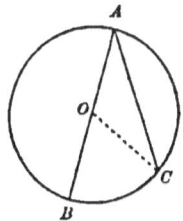

1st. Let one of the sides AB of the inscribed angle BAC be a diameter; then, the measure of the angle BAC is one-half the arc BC.

For, draw the radius OC. Then, AOC being an isosceles triangle, the angles OAC and OCA are equal (I. 86). The angle BOC, an exterior angle of the triangle AOC, is equal to the sum of the interior angles OAC and OCA (I. 69), and therefore double either of them. But the angle *BOC*, at the centre, is measured by the arc *BC*; therefore, the angle *OAC* is measured by one-half the arc *BC*.

2d. Let the centre of the circle fall within the inscribed angle *BAC*; then, the measure of the angle *BAC* is one-half of the arc *BC*.

For, draw the diameter *AD*. The measure of the angle *BAD* is, by the first case, one-half the arc *BD*; and the measure of the angle *CAD* is one-half the arc *CD*; therefore, the measure of the sum of the angles *BAD* and *CAD* is one-half the sum of the arcs *BD* and *CD*; that is, the measure of the angle *BAC* is one-half the arc *BC*.

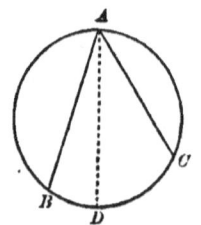

3d. Let the centre of the circle fall without the inscribed angle *BAC*; then, the measure of the angle *BAC* is one-half the arc *BC*.

For, draw the diameter *AD*. The measure of the angle *BAD* is, by the first case, one-half the arc *BD*; and the measure of the angle *CAD* is one-half the arc *CD*; therefore, the measure of the difference of the angles *BAD* and *CAD* is one-half the difference of the arcs *BD* and *CD*; that is, the measure of the angle *BAC* is one-half the arc *BC*.

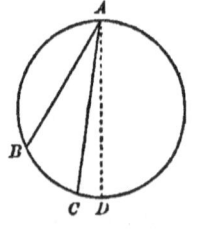

58. *Corollary* I. All the angles *BAC*, *BDC*, etc., inscribed in the same segment, are equal. For each is measured by one-half the same arc *BMC*.

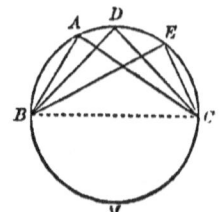

59. *Corollary* II. Any angle *BAC*, inscribed in a semicircle is a right angle. For it is measured by half a semi-circumference, or by a quadrant (54).

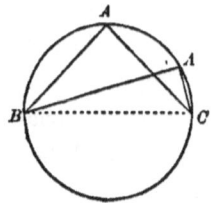

60. *Corollary* III. Any angle *BAC*, inscribed in a segment greater than a semicircle, is acute; for it is measured by half the arc *BDC*, which is less than a semi-circumference.

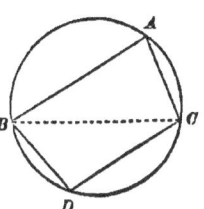

Any angle *BDC*, inscribed in a segment less than a semicircle, is obtuse; for it is measured by half the arc *BAC*, which is greater than a semi-circumference.

61. *Corollary* IV. The opposite angles of an inscribed quadrilateral *ABDC*, are supplements of each other. For the sum of two opposite angles, as *BAC* and *BDC*, is measured by one-half the circumference, which is the measure of two right angles, (54) and (I. 19).

PROPOSITION XXII.—THEOREM.

62. *An angle formed by a tangent and a chord is measured by one-half the intercepted arc.*

Let the angle *BAC* be formed by the tangent *AB* and the chord *AC*; then, it is measured by one-half the intercepted arc *AMC*.

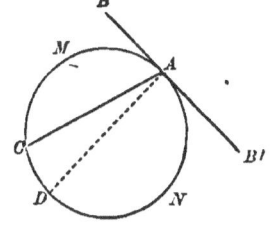

For, draw the diameter *AD*. The angle *BAD* being a right angle (27), is measured by one-half the semi-circumference *AMD*; and the angle *CAD* is measured by one-half the arc *CD*; therefore, the angle *BAC*, which is the difference of the angles *BAD* and *CAD*, is measured by one-half the difference of *AMD* and *CD*, that is, by one-half the arc *AMC*.

Also, the angle *B'AC* is measured by one-half the intercepted arc *ANC*. For, it is the sum of the right angle *B'AD* and the angle *CAD*, and is measured by one-half the sum of the semi-circumference *AND* and the arc *CD*; that is, by one-half the arc *ANC*.

63. *Scholium*. This proposition may be treated as a particular case of Prop. XXI. by an application of the principle of (28). For, consider the angle *CAD* which is measured by one-half the arc *CD*. Let the side *AC* remain fixed, while the side *AD*, regarded as a secant, revolves about *A* until it arrives at the position of the tangent

AB'. The point D will move along the circumference, and will ultimately coincide with A, when the line AD has become a tangent and the intercepted arc has become the arc CNA.

PROPOSITION XXIII.—THEOREM.

64. *An angle formed by two chords, intersecting within the circumference, is measured by one-half the sum of the arcs intercepted between its sides and between the sides of its vertical angle.*

Let the angle AEC be formed by the chords AB, CD, intersecting within the circumference; then will it be measured by one-half the sum of the arcs AC and BD, intercepted between the sides of AEC and the sides of its vertical angle BED.

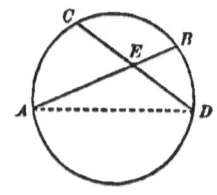

For, join AD. The angle AEC is equal to the sum of the angles EDA and EAD (I. 69), and these angles are measured by one-half of AC and one-half of BD, respectively; therefore, the angle AEC is measured by one-half the sum of the arcs AC and BD.

PROPOSITION XXIV.—THEOREM.

65. *An angle formed by two secants, intersecting without the circumference, is measured by one-half the difference of the intercepted arcs.*

Let the angle BAC be formed by the secants AB and AC; then, will it be measured by one-half the difference of the arcs BC and DE.

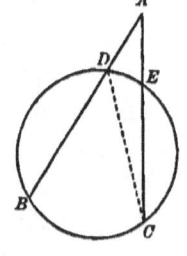

For, join CD. The angle BDC is equal to the sum of the angles DAC and ACD (I. 69); therefore, the angle A is equal to the difference of the angles BDC and ACD. But these angles are measured by one-half of BC and one-half of DE respectively; hence, the angle A is measured by one-half the difference of BC and DE.

66. *Corollary.* The angle *BAE*, formed by a tangent *AB* and a secant *AE*, is measured by one-half the difference of the intercepted arcs *BE* and *BC*. For, the tangent *AB* may be regarded as a secant whose two points of intersection are coincident at *B* (28).

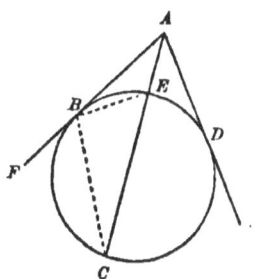

For, the same reason, the angle *BAD*, formed by two tangents *AB* and *AD*, is measured by one-half the difference of the intercepted arcs *BCD* and *BED*.

A proof may be given, without using the principle of (28), by drawing *EB* and *BC*.

PROBLEMS OF CONSTRUCTION.

Heretofore, our figures have been assumed to be constructed under certain conditions, although methods of constructing them have not been given. Indeed, the precise construction of the figures was not necessary, inasmuch as they were only required as *aids* in following the demonstration of *principles*. We now proceed, first, to apply these principles in the solution of the simple problems necessary for the construction of the plane figures already treated of, and then to apply these simple problems in the solution of more complex ones.

All the constructions of elementary geometry are effected solely by the straight line and the circumference, these being the only lines treated of in the elements; and these lines are practically *drawn*, or *described*, by the aid of the ruler and compasses, with the use of which the student is supposed to be familiar.

PROPOSITION XXV.—PROBLEM.

67. *To bisect a given straight line.*

Let *AB* be the given straight line.

With the points *A* and *B* as centres, and with a radius greater than the half of *AB*, describe arcs intersecting in the two points *D* and *E*. Through these points draw the straight line *DE*, which bisects *AB* at the point *C*. For, *D* and *E* being

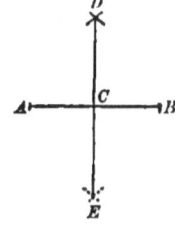

equally distant from *A* and *B*, the straight line *DE* is perpendicular to *AB* at its middle point (I. 41).

PROPOSITION XXVI.—PROBLEM.

68. *At a given point in a given straight line, to erect a perpendicular to that line.*

Let *AB* be the given line and *C* the given point.

Take two points, *D* and *E*, in the line and at equal distances from *C*. With *D* and *E* as centres and a radius greater than *DC* or *CE* describe two arcs intersecting in *F*. Then *CF* is the required perpendicular (I. 41).

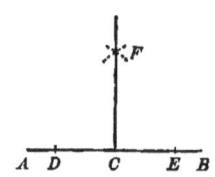

69. *Another solution.* Take any point *O*, without the given line, as a centre, and with a radius equal to the distance from *O* to *C* describe a circumference intersecting *AB* in *C* and in a second point *D*. Draw the diameter *DOE*, and join *EC*. Then *EC* will be the required perpendicular: for the angle *ECD*, inscribed in a semicircle, is a right angle (59).

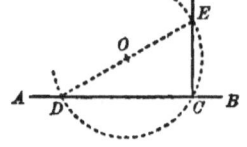

This construction is often preferable to the preceding, especially when the given point *C* is at, or near, one extremity of the given line, and it is not convenient to produce the line through that extremity. The point *O* must evidently be so chosen as not to lie in the required perpendicular.

PROPOSITION XXVII.—PROBLEM.

70. *From a given point without a given straight line, to let fall a perpendicular to that line.*

Let *AB* be the given line and *C* the given point.

With *C* as a centre, and with a radius sufficiently great, describe an arc intersecting *AB* in *D* and *E*. With *D* and *E* as centres and a radius greater than the half of *DE*,

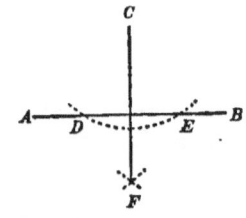

describe two arcs intersecting in *F*. The line *CF* is the required perpendicular (I. 41).

71. *Another solution.* With any point *O* in the line *AB* as a centre, and with the radius *OC,* describe an arc *CDE* intersecting *AB* in *D*. With *D* as a centre and a radius equal to the distance *DC* describe an arc

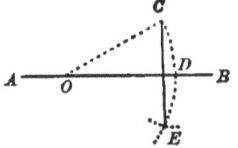

intersecting the arc *CDE* in *E*. The line *CE* is the required perpendicular. For, the point *D* is the middle of the arc *CDE*, and the radius *OD* drawn to this point is perpendicular to the chord *CE* (16).

PROPOSITION XXVIII.—PROBLEM.

72. *To bisect a given arc or a given angle.*

1st. Let *AB* be a given arc.
Bisect its chord *AB* by a perpendicular as in (67). This perpendicular also bisects the arc (16).

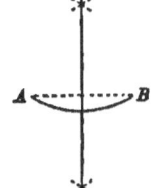

2d. Let *BAC* be a given angle. With *A* as a centre and with any radius, describe an arc intersecting the sides of the angle in *D* and *E*. With *D* and *E* as centres, and with equal radii, describe arcs intersecting in *F*. The straight line *AF* bisects the arc *DE*, and consequently also the angle *BAC* (12).

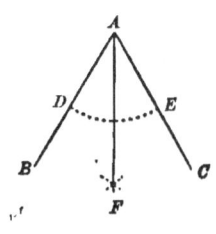

73. *Scholium.* By the same construction each of the halves of an arc, or an angle, may be bisected; and thus, by successive bisections, an arc, or an angle, may be divided into 4, 8, 16, 32, etc., equal parts.

PROPOSITION XXIX.—PROBLEM.

74. *At a given point in a given straight line, to construct an angle equal to a given angle.*

Let A be the given point in the straight line AB, and O the given angle.

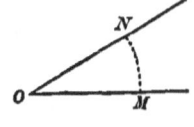

With O as a centre and with any radius describe an arc MN terminated by the sides of the angle. With A as a centre and with the same radius, OM, describe an indefinite arc BC. With B as a centre and with a radius equal to the chord of MN describe an arc intersecting the indefinite arc BC in D. Join AD. Then the angle BAD is equal to the angle O. For the chords of the arcs MN and BD are equal; therefore, these arcs are equal (12), and consequently also the angles O and A (10).

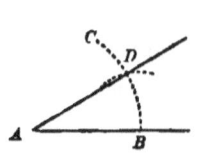

PROPOSITION XXX.—PROBLEM.

75. *Through a given point, to draw a parallel to a given straight line.*

Let A be the given point, and BC the given line.

From any point B in BC draw the straight line BAD through A. At the point A, by the preceding problem, construct the angle DAE equal to the angle ABC. Then AE is parallel to BC (I. 55).

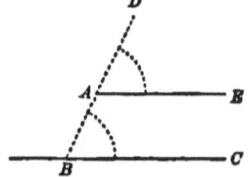

76. *Scholium.* This problem is, in practice, more accurately solved by the aid of a *triangle*, constructed of wood or metal. This triangle has one right angle, and its acute angles are usually made equal to 30° and 60°.

Let A be the given point, and BC the given line. Place the triangle, EFD, with one of its sides in coincidence with the given line BC. Then place the straight edge of a ruler MN

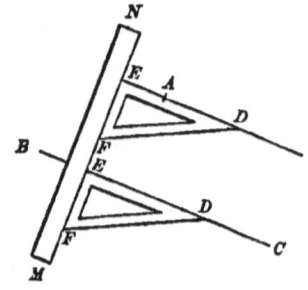

against the side *EF* of the triangle. Now, keeping the ruler firmly fixed, slide the triangle along its edge until the side *ED* passes through the given point *A*. Trace the line *EAD* along the edge *ED* of the triangle; then, it is evident that this line will be parallel to *BC*.

One angle of the triangle being made very precisely equal to a right angle, this instrument is also used in practice to construct perpendiculars, with more facility than by the methods of (68) and (70).

PROPOSITION XXXI.—PROBLEM.

77. *Two angles of a triangle being given, to find the third.*

Let *A* and *B* be the given angles.

Draw the indefinite line *QM*. From any point *O* in this line, draw *ON* making the angle *MON* = *A*, and the line *OP* making the angle *NOP* = *B*. Then *POQ* is the required third angle of the triangle (I. 72).

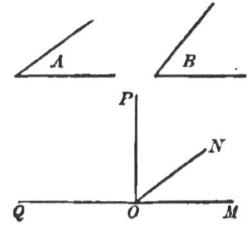

PROPOSITION XXXII.—PROBLEM.

78. *Two sides of a triangle and their included angle being given, to construct the triangle.*

Let *b* and *c* be the given sides and *A* their included angle.

Draw an indefinite line *AE*, and construct the angle *EAF* = *A*. On *AE* take *AC* = *b*, and on *AF* take *AB* = *c*; join *BC*. Then *ABC* is the triangle required; for it is formed with the data.

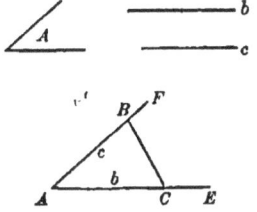

With the data, two sides and the included angle, only one triangle can be constructed; that is, *all* triangles constructed with these data are equal, and thus only repetitions of the same triangle (I. 76).

79. *Scholium.* It is evident that *one* triangle is always possible, whatever may be the magnitude of the proposed sides and their included angle.

PROPOSITION XXXIII.—PROBLEM.

80. *One side and two angles of a triangle being given, to construct the triangle.*

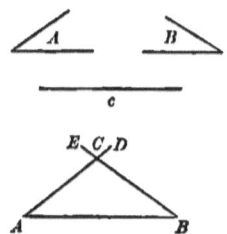

Two angles of the triangle being given, the third angle can be found by (77); and we shall therefore always have given the two angles adjacent to the given side. Let, then, c be the given side, A and B the angles adjacent to it.

Draw a line $AB = c$; at A make an angle $BAD = A$, and at B an angle $ABE = B$. The lines AD and BE intersecting in C, we have ABC as the required triangle.

With these data, but one triangle can be constructed (I. 78).

81. *Scholium.* If the two given angles are together equal to or greater than two right angles, the problem is *impossible;* that is, no triangle can be constructed with the data; for the lines AD and BC will not intersect on that side of AB on which the angles have been constructed.

PROPOSITION XXXIV.—PROBLEM.

82. *The three sides of a triangle being given, to construct the triangle.*

Let a, b and c be the three given sides.

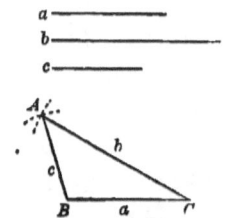

Draw $BC = a$; with C as a centre and a radius equal to b describe an arc; with B as a centre and a radius equal to c describe a second arc intersecting the first in A. Then, ABC is the required triangle.

With these data but one triangle can be constructed (I. 80).

83. *Scholium.* The problem is impossible when one of the given sides is equal to or greater than the sum of the other two (I. 66).

F

PROPOSITION XXXV.—PROBLEM.

84. *Two sides of a triangle and the angle opposite to one of them being given, to construct the triangle.*

We shall consider two cases.

1st. When the given angle A is acute, and the given side a, opposite to it in the triangle, is less than the other given side c.

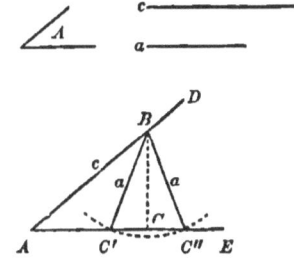

Construct an angle $DAE = A$. In one of its sides, as AD, take $AB = c$; with B as a centre and a radius equal to a, describe an arc which (since $a < c$) will intersect AE in two points, C' and C''', on the same side of A. Join BC' and BC'''. Then, either ABC' or ABC'' is the required triangle, since each is formed with the data; and the problem has two solutions.

There will, however, be but one solution, even with these data, when the side a is so much less than the side c as to be just equal to the perpendicular from B upon AE. For then the arc described from B as a centre and with the radius a, will touch AE in a single point C, and the required triangle will be ABC, right angled at C.

2d. When the given angle A is either acute, right or obtuse, and the side a opposite to it is greater than the other given side c.

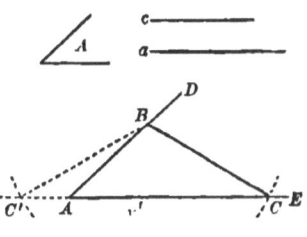

The same construction being made as in the first case, the arc described with B as a centre and with a radius equal to a, will intersect AE in only one point, C, on the same side of A. Then ABC will be the triangle required, and will be the only possible triangle with the data.

The second point of intersection, C', will fall in EA produced, and the triangle ABC' thus formed will not contain the given angle.

85. *Scholium.* The problem is impossible when the given angle A is acute and the proposed side opposite to it is less than the perpendicular from B upon AE; for then the arc described from B will not intersect AE.

The problem is also impossible when the given angle is right, or obtuse, if the given side opposite to the angle is less than the other given side; for either the arc described from *B* would not intersect *AE*, or it would intersect it only when produced through *A*. Moreover, a right or obtuse angle is the greatest angle of a triangle (I. 70), and the side opposite to it must be the greatest side (I. 92).

PROPOSITION XXXVI.—PROBLEM.

86. *The adjacent sides of a parallelogram and their included angle being given, to construct the parallelogram.*

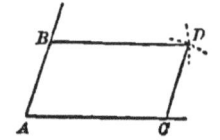

Construct an angle *A* equal to the given angle, and take *AC* and *AB* respectively equal to the given sides. With *B* as a centre and a radius equal to *AC*, describe an arc; with *C* as a centre and a radius equal to *AB*, describe another arc, intersecting the first in *D*. Draw *BD* and *CD*. Then *ABDC* is a parallelogram (I. 107), and it is the one required, since it is formed with the data.

Or thus: through *B* draw *BD* parallel to *AC*, and through *C* draw *CD* parallel to *AB*.

PROPOSITION XXXVII.—PROBLEM.

87. *To find the centre of a given circumference, or of a given arc.*

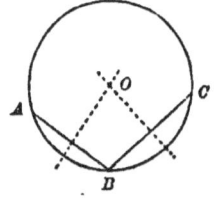

Take any three points, *A*, *B* and *C*, in the given circumference or arc. Bisect the arcs *AB*, *BC*, by perpendiculars to the chords *AB*, *BC* (72); these perpendiculars intersect in the required centre (16).

88. *Scholium.* The same construction serves to describe a circumference which shall pass through three given points *A*, *B*, *C*; or to *circumscribe* a circle about a given triangle *ABC*, that is, to describe a circumference in which the given triangle shall be *inscribed* (56).

PROPOSITION XXXVIII.—PROBLEM.

89. *At a given point in a given circumference, to draw a tangent to the circumference.*

Let A be the given point in the given circumference. Draw the radius OA, and at A draw BAC perpendicular to OA; BC will be the required tangent (26).

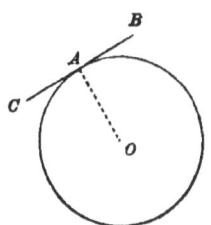

If the centre of the circumference is not given, it may first be found by the preceding problem, or we may proceed more directly as follows. Take two points D and E equidistant from A; draw the chord DE, and through A draw BAC parallel to DE. Since A is the middle point of the arc DE, the radius drawn to A will be perpendicular to DE (16), and consequently also to BC; therefore BC is a tangent at A.

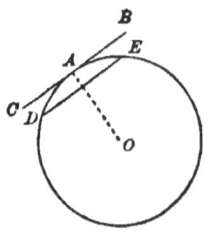

PROPOSITION XXXIX.—PROBLEM.

90. *Through a given point without a given circle to draw a tangent to the circle.*

Let O be the centre of the given circle and P the given point.

Upon OP, as a diameter, describe a circumference intersecting the circumference of the given circle in two points, A and A'. Draw PA and PA', both of which will be tangent to the given circle. For, drawing the radii OA and OA', the angles OAP and $OA'P$ are right angles (59); therefore PA and PA' are tangents (26).

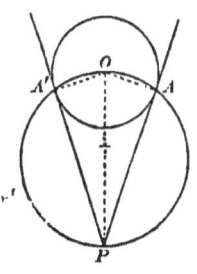

In practice, this problem is accurately solved by placing the straight edge of a ruler through the given point and tangent to the given circumference, and then tracing the tangent by the straight edge. The precise point of tangency is then determined by drawing a perpendicular to the tangent from the centre.

91. *Scholium.* This problem always admits of two solutions. Moreover, the portions of the two tangents intercepted between the given

BOOK II. 85

point and the points of tangency are equal, for the right triangles *POA* and *POA'* are equal (I. 83); therefore, $PA = PA'$.

PROPOSITION XL.—PROBLEM.

92. *To draw a common tangent to two given circles.*

Let O and O' be the centres of the given circles, and let the radius of the first be the greater.

1st. To draw an *exterior* common tangent. With the centre O, and a radius OM, equal to the difference of the given radii, describe a circumference; and from O' draw a tangent $O'M$ to this circumference (90). Join OM, and produce it to meet the given circumference in A. Draw $O'A'$ parallel to 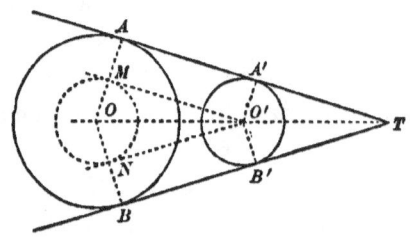 OA, and join AA'. Then AA' is a common tangent to the two given circles. For, by the construction, $OM = OA - O'A'$, and also $OM = OA - MA$, whence $MA = O'A'$, and $AMO'A'$ is a parallelogram (I. 108). But the angle M is a right angle; therefore, this parallelogram is a rectangle, and the angles at A and A' are right angles. Hence, AA' is a tangent to both circles.

Since two tangents can be drawn from O' to the circle OM, there are two exterior common tangents to the given circles, namely, AA' and BB', which meet in a point T in the line of centres OO' produced.

2d. To draw an *interior* common tangent. With the centre O and a radius OM equal to the sum of the given radii, describe a circumference, and from O' draw a tangent $O'M$ to this circumference. Join OM, intersecting the given circumference in A. Draw $O'A'$ parallel to OA. Then, since $OM = OA + O'A'$, we have $AM = O'A'$, and $AMO'A'$ is a rectangle. Therefore, AA' is a tangent to both the given circles.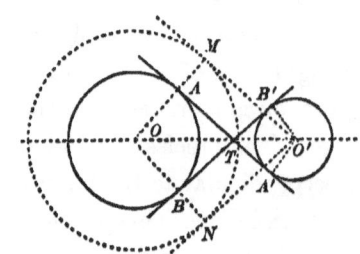

There are two interior common

8

tangents, AA' and BB', which intersect in a point T in the line of centres, between the two circles.

94. *Scholium.* If the given circles intersect each other, only the exterior tangents are possible. If they are tangent to each other externally, the two interior common tangents reduce to a single common tangent. If they are tangent internally, the two exterior tangents reduce to a single common tangent, and the interior tangents are not possible. If one circle is wholly within the other, there is no solution.

PROPOSITION XLI.—PROBLEM.

94. *To inscribe a circle in a given triangle.*

Let ABC be the given triangle. Bisect any two of its angles, as B and C, by straight lines meeting in O. From the point O let fall perpendiculars OD, OE, OF, upon the three sides of the triangle; these perpendiculars will be equal to each other (I. 129). Hence, the circumference of a circle, described with the centre O, and a radius $= OD$, will pass through the three points D, E, F, will be tangent to the three sides of the triangle at these points (26), and will therefore be inscribed in the triangle.

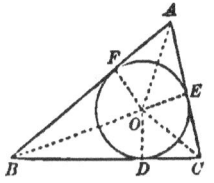

95. *Scholium.* If the sides of the triangle are produced and the exterior angles are bisected, the intersections O', O'', O''', of the

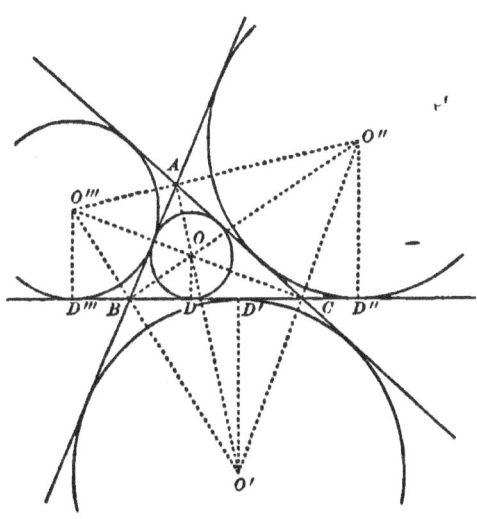

bisecting lines, will be the centres of three circles, each of which will touch one side of the triangle and the two other sides produced. In general, therefore, *four circles can be drawn tangent to three intersecting straight lines.* The three circles which lie without the triangle have been named *escribed* circles.

PROPOSITION XLII.—PROBLEM.

96. *Upon a given straight line, to describe a segment which shall contain a given angle.*

Let AB be the given line. At the point B construct the angle ABC equal to the given angle. Draw BO perpendicular to BC, and DO perpendicular to AB at its middle point D, intersecting BO in O. With O as a centre, and radius OB describe the circumference $AMBN$. The segment AMB is the required segment. For, the line BC, being perpendicular to the radius OB, is a tangent to the circle; therefore, the angle ABC is meas-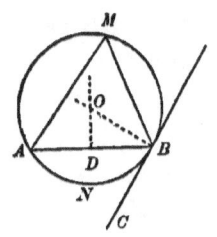
ured by one-half the arc ANB (62), which is also the measure of any angle AMB inscribed in the segment AMB (57). Therefore, any angle inscribed in this segment is equal to the given angle.

97. *Scholium.* If any point P is taken within the segment AMB, the angle APB is greater than the inscribed angle AMB (I. 74); and if any point Q is taken without this segment, but on the same side of the chord AB as the segment, the angle AQB is less than the inscribed angle AMB. Therefore, the angles whose vertices lie in the arc AMB are the only angles of the given magnitude whose sides pass through the two points A and B; hence, the arc AMB is the *locus* of the vertices of all the angles of the given magnitude whose sides pass through A and B.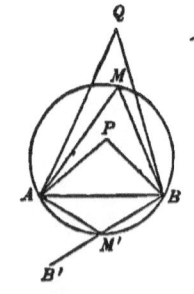

If any point M' be taken in the arc $AM'B$, the angle AMB is the supplement of the angle $AM'B$ (61); and if BM' be produced to B', the angle $AM'B'$ is also the supplement of $AM'B$; therefore $AM'B' = AMB$. Hence the vertices of all the angles of the given magnitude whose sides, *or sides produced*, pass through A and B, lie in the circumference $AMBM'$; that is, *the locus of the vertices of all the angles of a given magnitude whose sides, or sides produced, pass through two fixed points, is a circumference passing through these points*, and this locus may be constructed by the preceding problem.

It may here be remarked, that in order to establish a certain line as a locus of points subject to certain given conditions, it is necessary not only to show that every point in that line satisfies the conditions, but also that no other points satisfy them; for the asserted locus must be the assemblage of *all* the points satisfying the given conditions (I. 40).

INSCRIBED AND CIRCUMSCRIBED QUADRILATERALS.

98. *Definition.* An *inscriptible* quadrilateral is one which can be inscribed in a circle; that is, a circumference can be described passing through its four vertices.

PROPOSITION XLIII.—THEOREM.

99. *A quadrilateral is inscriptible if two opposite angles in it are supplements of each other.*

Let the angles A and C, of the quadrilateral $ABCD$, be supplements of each other. Describe a circumference passing through the three vertices B, C, D; and draw the chord BD. The angle A, being the supplement of C, is equal to any angle inscribed in the segment BMD (61); therefore the vertex A must

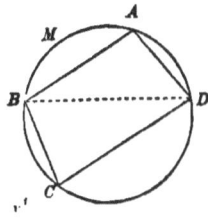

be on the arc AMD (97), and the quadrilateral is inscribed in the circle.

100. *Scholium.* This proposition is the converse of (61).

PROPOSITION XLIV.—THEOREM.

101. *In any circumscribed quadrilateral, the sum of two opposite sides is equal to the sum of the other two opposite sides.*

Let $ABCD$ be circumscribed about a circle; then,

$$AB + DC = AD + BC.$$

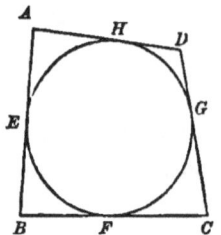

For, let E, F, G, H, be the points of contact of the sides; then we have (91),

$$AE = AH, \quad BE = BF, \quad CG = CF, \quad DG = DH.$$

Adding the corresponding members of these equalities, we have

$$AE + BE + CG + DG = AH + DH + BF + CF,$$

that is,

$$AB + DC = AD + BC.$$

PROPOSITION XLV.—THEOREM.

102. *Conversely, if the sum of two opposite sides of a quadrilateral is equal to the sum of the other two sides, the quadrilateral may be circumscribed about a circle.*

In the quadrilateral $ABCD$, let $AB + DC = AD + BC$; then, the quadrilateral can be circumscribed about a circle.

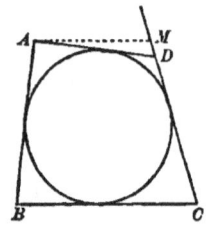

Since the sum of the four angles of the quadrilateral is equal to four right angles, there must be two consecutive angles in it whose sum is not greater than two right angles; let B and C be these angles. Let a circle be described tangent to the three sides AB, BC, CD, the centre of this circle being the intersection of the bisectors of the angles B and C; then it is to be proved that this circle is tangent also to the fourth side AD.

From the point A two tangents can be drawn to the circle (90). One of these tangents being AB, the other must be a line cutting CD (or CD produced); for, the sum of the angles B and C being not greater than two right angles, it is evident that no straight line can be drawn from *A*, falling on the same side of *BA* with *CD*, and not cutting the circle, which shall not cut *CD*. This second tangent, then, must be either *AD* or some other line, *AM*, cutting *CD* in a point *M* differing from *D*. If now *AM* is a tangent, *ABCM* is a circumscribed quadrilateral, and by the preceding proposition we shall have

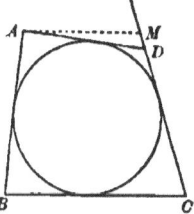

$$AB + CM = AM + BC.$$

But we also have, by the hypothesis of the present proposition,

$$AB + DC = AD + BC.$$

Taking the difference of these equalities, we have

$$DM = AM - AD;$$

that is, one side of a triangle is equal to the difference of the other two, which is absurd. Therefore, the hypothesis that the tangent drawn from *A* and cutting the line *CD*, cuts it in any other point than *D*, leads to an absurdity; therefore, that hypothesis must be false, and the tangent in question must cut *CD* in *D*, and consequently coincide with *AD*. Hence, a circle has been described which is tangent to the four sides of the quadrilateral; and the quadrilateral is circumscribed about the circle.

103. *Scholium.* The method of demonstration employed above is called the *indirect method*, or the *reductio ad absurdum*. At the outset of a demonstration, or at any stage of its progress, two or more hypotheses respecting the quantities under consideration may be admissible so far as has been proved up to that point. If, now, these hypotheses are such that one *must* be true,. and only one *can* be true, then, when all except one are shown to be absurd, that one must stand as the truth.

While admitting the validity of this method, geometers usually prefer the direct method whenever it is applicable. There are, however, propositions, such as the preceding, of which no direct proof is known, or at least no proof sufficiently simple to be admitted into elementary geometry. We have already employed the *reductio ad absurdum* in several cases without presenting the argument in full; see (I. 47), (I. 85), (27).

BOOK III.

PROPORTIONAL LINES. SIMILAR FIGURES.

THEORY OF PROPORTION.

1. DEFINITION. One quantity is said to be *proportional* to another when the ratio of any two values, A and B, of the first, is equal to the ratio of the two corresponding values, A' and B', of the second; so that the four values form the proportion

$$A : B = A' : B',$$

or
$$\frac{A}{B} = \frac{A'}{B'}.$$

This definition presupposes two quantities, each of which can have various values, so related to each other that each value of one corresponds to a value of the other. An example occurs in the case of an angle at the centre of a circle and its intercepted arc. The angle may *vary*, and with it also the arc; but to each value of the angle there corresponds a certain value of the arc. It has been proved (II. 51) that the ratio of any two values of the angle is equal to the ratio of the two corresponding values of the arc; and in accordance with the definition just given, this proposition would be briefly expressed as follows: "The angle at the centre of a circle is proportional to its intercepted arc."

2. *Definition.* One quantity is said to be *reciprocally proportional* to another when the ratio of two values, A and B, of the first, is equal to the reciprocal of the ratio of the two corresponding values, A' and B', of the second, so that the four values form the proportion

$$A : B = B' : A',$$

or
$$\frac{A}{B} = \frac{B'}{A'} = 1 \div \frac{A'}{B'};$$

For example, if the product p of two numbers, x and y, is given, so that we have
$$xy = p,$$
then, x and y may each have an indefinite number of values, but as x increases y diminishes. If, now, A and B are two values of x, while A' and B' are the two corresponding values of y, we must have
$$A \times A' = p,$$
$$B \times B' = p,$$
whence, by dividing one of these equations by the other,
$$\frac{A}{B} \times \frac{A'}{B'} = 1,$$
and therefore
$$\frac{A}{B} = \frac{1}{\frac{A'}{B'}} = \frac{B'}{A'};$$
that is, *two numbers whose product is constant are reciprocally proportional.*

3. Let the quantities in each of the couplets of the proportion
$$\frac{A}{B} = \frac{A'}{B'}, \quad \text{or } A : B = A' : B', \qquad [1]$$
be measured by a unit of their own kind, and thus expressed by numbers (II. 42); let a and b denote the numerical measures of A and B, a' and b' those of A' and B'; then (II. 43),
$$\frac{A}{B} = \frac{a}{b}, \qquad \frac{A'}{B'} = \frac{a'}{b'},$$
and the proportion [1] may be replaced by the *numerical* proportion,
$$\frac{a}{b} = \frac{a'}{b'}, \quad \text{or } a : b = a' : b'.$$

4. Conversely, if the numerical measures a, b, a', b', of four quantities A, B, A', B', are in proportion, these quantities themselves are in proportion, *provided* that A and B are quantities of the same kind, and A' and B' are quantities of the same kind (though not necessarily of the same kind as A and B); that is, if we have
$$a : b = a' : b',$$

we may, under these conditions, infer the proportion
$$A : B = A' : B'.$$

5. Let us now consider the numerical proportion
$$a : b = a' : b'.$$
Writing it in the form
$$\frac{a}{b} = \frac{a'}{b'},$$
and multiplying both members of this equality by bb', we obtain
$$ab' = a'b,$$
whence the theorem: *the product of the extremes of a (numerical) proportion is equal to the product of the means.*

Corollary. If the means are equal, as in the proportion $a : b = b : c$, we have $b^2 = ac$, whence $b = \sqrt{ac}$; that is, *a mean proportional between two numbers is equal to the square root of their product.*

6. Conversely, *if the product of two numbers is equal to the product of two others, either two may be made the extremes, and the other two the means, of a proportion.* For, if we have given
$$ab' = a'b,$$
then, dividing by bb', we obtain
$$\frac{a}{b} = \frac{a'}{b'}, \ \ \text{or } a : b = a' : b'.$$

Corollary. The terms of a proportion may be written in any order which will make the products of the extremes equal to the product of the means. Thus, any one of the following proportions may be inferred from the given equality $ab' = a'b$:
$$a : b = a' : b',$$
$$a : a' = b : b',$$
$$b : a = b' : a',$$
$$b : b' = a : a',$$
$$b' : a' = b : a, \text{ etc.}$$

Also, any one of these proportions may be inferred from any other.

7. *Definitions.* When we have given the proportion
$$a : b = a' : b',$$

and infer the proportion
$$a : a' = b : b',$$
the second proportion is said to be deduced *by alternation.*

When we infer the proportion
$$b : a = b' : a',$$
this proportion is said to be deduced *by inversion.*

8. It is important to observe, that when we speak of the products of the extremes and means of a proportion, it is implied that at least two of the terms are numbers. If, for example, the terms of the proportion
$$A : B = A' : B',$$
are all *lines*, no meaning can be directly attached to the products $A \times B'$, $B \times A'$, since in a product the multiplier at least must be a number.

But if we have a proportion such as
$$A : B = m : n,$$
in which m and n are numbers, while A and B are any two quantities of the same kind, then we may infer the equality $nA = mB$.

Nevertheless, we shall for the sake of brevity often speak of *the product of two lines*, meaning thereby *the product of the numbers which represent those lines when they are measured by a common unit.*

9. If A and B are any two quantities of the same kind, and m any number whole or fractional, we have, identically,
$$\frac{mA}{mB} = \frac{A}{B};$$
that is, *equimultiples of two quantities are in the same ratio as the quantities themselves.*

Similarly, if we have the proportion
$$A : B = A' : B',$$
and if m and n are any two numbers, we can infer the proportions
$$mA : mB = nA' : nB',$$
$$mA : nB = mA' : nB'.$$

10. *Composition and division.* Suppose we have given the proportion
$$\frac{A}{B} = \frac{A'}{B'}, \qquad [1]$$
in which A and B are any quantities of the same kind, and A' and B' quantities of the same kind. Let unity be added to both members of [1]; then
$$\frac{A}{B} + 1 = \frac{A'}{B'} + 1,$$
or, reducing,
$$\frac{A+B}{B} = \frac{A'+B'}{B'},$$
and dividing this by [1],
$$\frac{A+B}{A} = \frac{A'+B'}{A'}; \qquad [2]$$
results which are briefly expressed by the theorem, *if four quantities are in proportion, they are in proportion by composition;* the term *composition* being employed to express the *addition* of antecedent and consequent in each ratio.

If we had subtracted unity from both members of [1], we should have found
$$\frac{A-B}{B} = \frac{A'-B'}{B'},$$
$$\frac{A-B}{A} = \frac{A'-B'}{A'}; \qquad [3]$$
results which are briefly expressed by the theorem, *if four quantities are in proportion, they are in proportion by division;* where the term *division* is employed to express the *subtraction* of consequent from antecedent in each ratio, this subtraction being conceived to divide, or to separate, the antecedent into parts.

The quotient of [2] divided by [3] is
$$\frac{A+B}{A-B} = \frac{A'+B'}{A'-B'};$$
that is, *if four quantities are in proportion, they are in proportion by composition and division.*

11. *Definition.* A *continued proportion* is a series of equal ratios, as

$$A : B = A' : B' = A'' : B'' = A''' : B''' = \text{etc.}$$

12. Let r denote the common value of the ratio in the continued proportion of the preceding article; that is, let

$$r = \frac{A}{B} = \frac{A'}{B'} = \frac{A''}{B''} = \frac{A'''}{B'''} = \text{etc.};$$

then, we have

$$A = Br, \quad A' = B'r, \quad A'' = B''r, \quad A''' = B'''r, \text{ etc.},$$

and adding these equations,

$$A + A' + A'' + A''' + \text{etc.} = (B + B' + B'' + B''' + \text{etc.})\,r,$$

whence

$$\frac{A + A' + A'' + A''' + \text{etc.}}{B + B' + B'' + B''' + \text{etc.}} = r = \frac{A}{B} = \frac{A'}{B'} = \text{etc.};$$

that is, *the sum of any number of the antecedents of a continued proportion is to the sum of the corresponding consequents as any antecedent is to its consequent.*

If any antecedent and its corresponding consequent be taken with the negative sign, the theorem still holds, provided we read *algebraic sum* for *sum.*

In this theorem the quantities A, B, C, etc., must all be quantities of the same kind.

13. If we have any number of proportions, as

$$a : b = c : d,$$
$$a' : b' = c' : d',$$
$$a'' : b'' = c'' : d'', \text{ etc.};$$

then, writing them in the form,

$$\frac{a}{b} = \frac{c}{d}, \quad \frac{a'}{b'} = \frac{c'}{d'}, \quad \frac{a''}{b''} = \frac{c''}{d''}, \text{ etc.,}$$

and multiplying these equations together, we have

$$\frac{a\,a'\,a''\ldots}{b\,b'\,b''\ldots} = \frac{c\,c'\,c''\ldots}{d\,d'\,d''\ldots},$$

or

$$a\,a'\,a''\ldots : b\,b'\,b''\ldots = c\,c'\,c''\ldots : d\,d'\,d''\ldots;$$

that is, *if the corresponding terms of two or more proportions are multiplied together, the products are in proportion.*

If the corresponding terms of the several proportions are equal, that is, if $a = a' = a''$, $b = b' = b''$, etc., then the multiplication of two or more proportions gives

$$a^2 : b^2 = c^2 : d^2,$$
$$a^3 : b^3 = c^3 : d^3;$$

that is, *if four numbers are in proportion, like powers of these numbers are in proportion.*

14. If A, B and C are like quantities of any kind, and if

$$\frac{A}{B} = m, \text{ and } \frac{B}{C} = n,$$

then

$$\frac{A}{C} = mn.$$

If A, B and C were numbers, this would be proved, arithmetically, by simply omitting the common factor B in the multiplication of the two fractions; but when they are not numbers we cannot regard B as a factor, or multiplier, and therefore we should proceed more strictly as follows. By the nature of ratio we have

$$A = B \times m, \quad B = C \times n,$$

therefore, putting $C \times n$ for B, we have

$$A = C \times n \times m = C \times mn,$$

that is,

$$\frac{A}{C} = mn;$$

a result usually expressed as follows: *the ratio of the first of three quantities to the third is compounded of the ratio of the first to the second and the ratio of the second to the third.*

PROPORTIONAL LINES.

PROPOSITION I.—THEOREM.

15. *A parallel to the base of a triangle divides the other two sides proportionally.*

Let DE be a parallel to the base, BC, of the triangle ABC; then,

$$AB : AD = AC : AE.$$

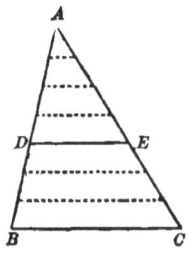

1st. Suppose the lines AB, AD, to have a common measure which is contained, for example, 7 times in AB, and 4 times in AD; so that if AB is divided into 7 parts each equal to the common measure, AD will contain 4 of these parts. Then the ratio of AB to AD is $7 : 4$ (II. 43); that is

$$\frac{AB}{AD} = \frac{7}{4}.$$

Through the several points of division of AB, draw parallels to the base; then AC will be divided into 7 equal parts (I. 125), of which AE will contain 4. Hence the ratio of AC to AE is $7 : 4$; that is,

$$\frac{AC}{AE} = \frac{7}{4}.$$

Therefore, we have

$$\frac{AB}{AD} = \frac{AC}{AE};$$

or $$AB : AD = AC : AE.$$

2d. If AB and AD are incommensurable, suppose one of them, as AD, to be divided into any number n of equal parts; then, AB will contain a certain number m of these parts *plus* a remainder less than one of these parts. The numerical expression of the ratio $\frac{AB}{AD}$ will then be $\frac{m}{n}$, correct within $\frac{1}{n}$ (II. 48). Drawing parallels to BC, through the several points of division of AB, the line AE will be divided into n equal parts, and the line AC will contain m such parts *plus* a remainder less than one of the parts. Therefore, the numerical expression of the ratio $\frac{AC}{AE}$ will also be $\frac{m}{n}$, correct within $\frac{1}{n}$. Since, then, the two ratios always have the same approximate numerical expression, however small the parts into which AD is divided, these ratios must be absolutely equal (II. 49), and we have, as before,

$$\frac{AB}{AD} = \frac{AC}{AE},$$

or $$AB : AD = AC : AE. \qquad [1]$$

16. *Corollary* I. By division (10), the proportion [1] gives

$$AB - AD : AB = AC - AE : AC,$$

or $$DB : AB = EC : AC.$$

Also, if the parallel DE intersect the sides BA and CA produced through A, we find, as in the preceding demonstration,

$$AB : AD = AC : AE,$$

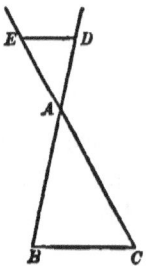

from which, by composition (10),

$$AB + AD : AB = AC + AE : AC,$$

or $$DB : AB = EC : AC.$$

17. *Corollary* II. By alternation (7), the preceding proportions give

$$AB : AC = AD : AE,$$

$$DB : EC = AB : AC,$$

which may both be expressed in one continued proportion,

$$\frac{AB}{AC} = \frac{AD}{AE} = \frac{DB}{EC}.$$

This proportion is indeed the most general statement of the proposition (15), which may also be expressed as follows: *if a straight line is drawn parallel to the base of a triangle, the corresponding segments on the two sides are in a constant ratio.*

18. Corollary III. *If two straight lines MN, M'N', are intersected by any number of parallels AA', BB', CC', etc., the corresponding segments of the two lines are proportional.*

For, let the two lines meet in O; then, by Corollary II.,

$$\frac{OA}{OA'}=\frac{AB}{A'B'}=\frac{OB}{OB'}=\frac{BC}{B'C'}=\frac{OC}{OC'}=\frac{CD}{C'D'}, \text{etc.},$$

whence, by (12),

$$\frac{AB}{A'B'}=\frac{BC}{B'C'}=\frac{CD}{C'D'}=\frac{AC}{A'C'}=\frac{BD}{B'D'}, \text{etc.}$$

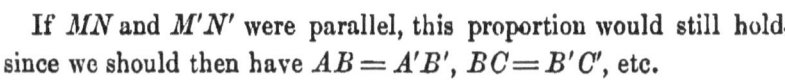

If MN and $M'N'$ were parallel, this proportion would still hold, since we should then have $AB=A'B'$, $BC=B'C'$, etc.

PROPOSITION II.—THEOREM.

19. Conversely, *if a straight line divides two sides of a triangle proportionally, it is parallel to the third side.*

Let DE divide the sides AB, AC, of the triangle ABC, proportionally; then, DE is parallel to BC.

For, if DE is not parallel to BC, let some other line DE', drawn through D, be parallel to BC. Then, by the preceding theorem,

$$AB:AD=AC:AE'.$$

But, by hypothesis, we have

$$AB:AD=AC:AE,$$

whence it follows that $AE' = AE$, which is impossible unless DE' coincides with DE. Therefore, DE is parallel to BC.

20. *Scholium.* The converse of (18) is not generally true.

PROPOSITION III.—THEOREM.

21. *In any triangle, the bisector of an angle, or the bisector of its exterior angle, divides the opposite side, internally or externally, into segments which are proportional to the adjacent sides.*

1st. Let AD bisect the angle A of the triangle ABC; then,

$$DB : DC = AB : AC.$$

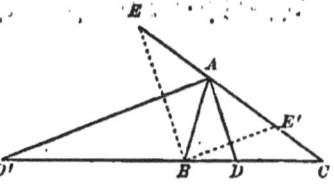

For, through B draw BE parallel to DA, meeting CA produced in E. The angle $ABE = BAD$ (I. 49), and the angle $AEB = CAD$ (I. 51); and, by hypothesis, the angle $BAD = CAD$; therefore, the angle $ABE = AEB$, and $AE = AB$ (I. 90).

Now, in the triangle CEB, AD being parallel to EB, we have (17),

$$DB : DC = AE : AC,$$
or
$$DB : DC = AB : AC;$$

that is, the side BC is divided by AD *internally* into segments proportional to the adjacent sides AB and AC.

2d. Let AD' bisect the exterior angle BAE; then,

$$D'B : D'C = AB : AC.$$

For, draw BE' parallel to $D'A$; then, ABE' is an isosceles triangle, and $AE' = AB$. In the triangle CAD', we have (17),

$$D'B : D'C = AE' : AC,$$
or
$$D'B : D'C = AB : AC;$$

that is, the side BC is divided by AD' *externally* into segments proportional to the adjacent sides AB and AC.

22. *Scholium.* When a point is taken on a given finite line, or on the line produced, the distances of the point from the extremities of the line are called the *segments*, internal or external, of the line. The given line is the *sum* of two *internal* segments, or the *difference* of two *external* segments.

23. *Corollary.* If a straight line, drawn from the vertex of any angle of a triangle to the opposite side, divides that side internally in the ratio of the other two sides, it is the bisector of the angle; if it divides the opposite side externally in that ratio, it is the bisector of the exterior angle. (To be proved).

SIMILAR POLYGONS.

24. *Definitions.* Two polygons are *similar*, when they are mutually equiangular and have their homologous sides proportional.

In similar polygons, any points, angles or lines, similarly situated in each, are called *homologous*.

The ratio of a side of one polygon to its homologous side in the other is called the *ratio of similitude* of the polygons.

PROPOSITION IV.—THEOREM.

25. *Two triangles are similar, when they are mutually equiangular.*

Let ABC, $A'B'C'$, be mutually equiangular triangles, in which $A = A'$, $B = B'$, $C = C'$; then, these triangles are similar.

For, place the angle A' upon its equal angle A, and let B' fall at b and C' at c. Since the angle Abc is equal to B, bc is parallel to BC (I. 55), and we have (15),

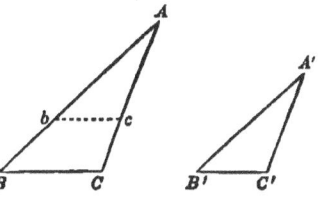

$$AB : Ab = AC : Ac,$$

or

$$AB : A'B' = AC : A'C'.$$

In the same manner, it is proved that

$$AB : A'B' = BC : B'C';$$

and, combining these proportions,

$$\frac{AB}{A'B'} = \frac{AC}{A'C'} = \frac{BC}{B'C'}. \qquad [1]$$

Therefore, the homologous sides are proportional, and the triangles are similar (24).

26. *Corollary.* Two triangles are similar when two angles of the one are respectively equal to two angles of the other (I. 73).

27. *Scholium* I. The homologous sides lie opposite to equal angles.

28. *Scholium* II. The ratio of similitude (24) of the two similar triangles, is any one of the equal ratios in the continued proportion [1].

29. *Scholium* III. In two similar triangles, any two homologous lines are in the ratio of similitude of the triangles. For example, the perpendiculars AD, $A'D'$, drawn from the homologous vertices A, A', to the opposite sides, are homologous lines of the two triangles; and the right triangles ABD, $A'B'D'$, being similar (25), we have

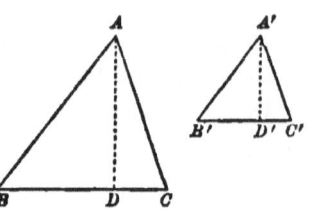

$$\frac{AD}{A'D'} = \frac{AB}{A'B'} = \frac{AC}{A'C'} = \frac{BC}{B'C'}.$$

In like manner, if the lines AD, $A'D'$, were drawn from A, A', to the middle points of the opposite sides, or to two points which divide the opposite sides in the same ratio in each triangle, these lines would still be to each other in the ratio of similitude of the two triangles.

PROPOSITION V.—THEOREM.

30. *Two triangles are similar, when their homologous sides are proportional.*

In the triangles ABC, $A'B'C'$, let

$$\frac{AB}{A'B'} = \frac{AC}{A'C'} = \frac{BC}{B'C'}; \qquad [1]$$

then, these triangles are similar.

For, on AB take $Ab = A'B'$, and draw bc parallel to BC. Then, the triangles Abc and ABC are mutually equiangular, and we have (25),

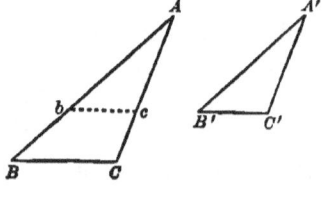

$$\frac{AB}{Ab} \text{ or } \frac{AB}{A'B'} = \frac{AC}{Ac} = \frac{BC}{Bc}.$$

Comparing this with the given proportion [1], we see that the first ratio is the same in both; hence the second and third ratios in each are equal respectively, and, the numerators being the same, the denominators are equal; that is, $A'C' = Ac$, and $B'C' = Bc$. Therefore, the triangles $A'B'C'$ and Abc are equal (I. 80); and since Abc is similar to ABC, $A'B'C'$ is also similar to ABC.

31. *Scholium.* In order to establish the similarity of two polygons according to the definition (24), it is necessary, in general, to show that they fulfill two conditions: 1st, they must be mutually equiangular, and 2d, their homologous sides must be proportional. In the case of triangles, however, either of these conditions involves the other; and to establish the similarity of two triangles it will be sufficient to show, *either* that they are mutually equiangular, *or* that their homologous sides are proportional.

PROPOSITION VI.—THEOREM.

32. *Two triangles are similar, when an angle of the one is equal to an angle of the other, and the sides including these angles are proportional.*

In the triangles ABC, $A'B'C'$, let $A = A'$, and

$$\frac{AB}{A'B'} = \frac{AC}{A'C'};$$

then, these triangles are similar.

For, place the angle A' upon its equal angle A; let B' fall at b, and C' at c. Then, by the hypothesis,

$$\frac{AB}{Ab} = \frac{AC}{Ac}.$$

Therefore, bc is parallel to BC (19), and the triangle Abc is similar to ABC (25). But Abc, is equal to $A'B'C'$; therefore, $A'B'C'$ is also similar to ABC.

PROPOSITION VII.—THEOREM.

33. *Two triangles are similar, when they have their sides parallel each to each, or perpendicular each to each.*

Let ABC, abc have their sides parallel each to each, or perpendicular each to each; then, these triangles are similar.

For, when the sides of two angles

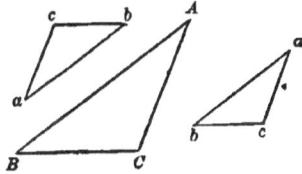

are parallel each to each, or perpendicular each to each, these angles are either equal, or supplements of each other, (I. 60, 62, 63). In the present case, therefore, three hypotheses may be made, namely, denoting a right angle by R,

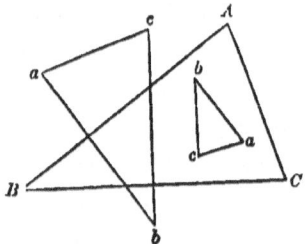

1st hyp. $A + a = 2R$, $B + b = 2R$, $C + c = 2R$;

2d " $A = a$, $B + b = 2R$, $C + c = 2R$;

3d " $A = a$, $B = b$, whence $C = c$.

The 1st and 2d hypotheses cannot be admitted, since the sum of all the angles of the two triangles would then exceed four right angles (I. 68). The 3d hypothesis is therefore the only admissible one; that is, the two triangles are mutually equiangular and consequently similar.

34. *Scholium.* Homologous sides in the two triangles are either two parallel sides, or two perpendicular sides; and homologous, or equal, angles, are angles included by homologous sides.

PROPOSITION VIII.—THEOREM.

35. *If three or more straight lines drawn through a common point intersect two parallels, the corresponding segments of the parallels are in proportion.*

Let OA, OB, OC, OD, drawn through the common point O, intersect the parallels AD and ad, in the points A, B, C, D and a, b, c, d, respectively; then,

$$\frac{AB}{ab} = \frac{BC}{bc} = \frac{CD}{cd}.$$

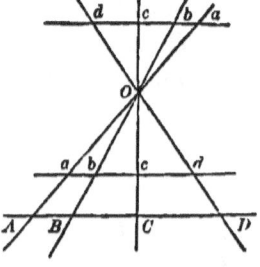

For, the triangle OAB is similar to the triangle Oab (25); OBC is similar to Obc; and OCD to Ocd; therefore, we have

$$\frac{AB}{ab} = \frac{OB}{Ob} = \frac{BC}{bc} = \frac{OC}{Oc} = \frac{CD}{cd},$$

which includes the proportion that was to be proved.

106   GEOMETRY.

36. *Scholium.* The demonstration is the same whether the parallels cut the system of diverging lines on the same side, or on opposite sides, of the point O. Moreover, the demonstration extends to *any* corresponding segments, as AC and ac, BD and bd, etc.; and the ratio of any two corresponding segments is equal to the ratio of the distances of the parallels from the point $\dot{O}$, measured on any one of the diverging lines.

PROPOSITION IX.—THEOREM.

37. Conversely, *if three or more straight lines divide two parallels proportionally, they pass through a common point.*

Let Aa, Bb, Cc, Dd, divide the parallels AD and ad proportionally; that is, so that

$$\frac{AB}{ab} = \frac{BC}{bc} = \frac{CD}{cd}; \qquad [1]$$

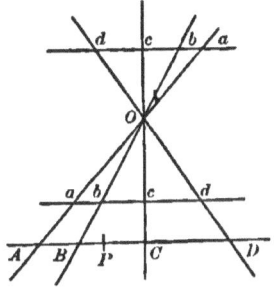

then, Aa, Bb, etc., meet in a common point.

For, let Aa and Cc meet in O; join Ob. Then, in order to prove that Bb passes through O, we have to prove that Ob and Bb are in the same straight line. Now, if they are not in the same straight line, Ob produced cuts AD in some point P differing from B; and by the preceding theorem, we have

$$\frac{AP}{ab} = \frac{AC}{ac}.$$

But, from the hypothesis [1], we have by (12),

$$\frac{AB}{ab} = \frac{AC}{ac},$$

whence, $AP = AB$, which is impossible unless P coincides with B, and Ob produced coincides with Bb. Therefore, Bb passes through O. In the same way, Dd is shown to pass through O.

PROPOSITION X.—THEOREM.

38. *If two polygons are composed of the same number of triangles similar each to each and similarly placed, the polygons are similar.*

Let the polygon $ABCD$, etc., be composed of the triangles ABC, ACD, etc.; and let the polygon $A'B'C'D'$, etc., be composed of the triangles $A'B'C'$, $A'C'D'$, etc., similar to ABC, ACD, etc., respectively, and similarly placed; then, the polygons are similar.

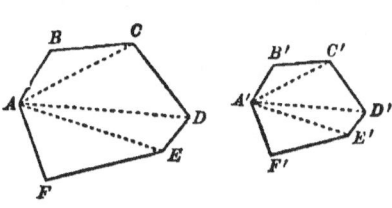

1st. The polygons are mutually equiangular. For, the homologous angles of the similar triangles are equal; and any two corresponding angles of the polygons are either homologous angles of two similar triangles, or sums of homologous angles of two or more similar triangles. Thus $B = B'$; $BCD = BCA + ACD = B'C'A' + A'C'D' = B'C'D'$; etc.

2d. Their homologous sides are proportional. For, from the similar triangles, we have

$$\frac{AB}{A'B'} = \frac{BC}{B'C'} = \frac{AC}{A'C'} = \frac{CD}{C'D'} = \frac{AD}{A'D'} = \frac{DE}{D'E'} = \text{etc.}$$

Therefore, the polygons fulfill the two conditions of similarity (24).

PROPOSITION XI.—THEOREM.

39. Conversely, *two similar polygons may be decomposed into the same number of triangles similar each to each and similarly placed.*

Let $ABCD$, etc., $A'B'C'D'$, etc., be two similar polygons. From two homologous vertices, A and A', let diagonals be drawn in each polygon; then, the polygons will be decomposed as required.

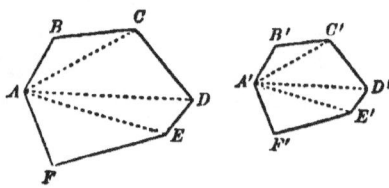

For, 1st. We have, by the definition of similar polygons,

Angle $B = B'$, and $\dfrac{AB}{A'B'} = \dfrac{BC}{B'C'}$;

therefore, the triangles ABC and $A'B'C'$ are similar (32).

2d. Since ABC and $A'B'C'$ are similar, the angles BCA and $B'C'A'$ are equal; subtracting these equals from the equals BCD

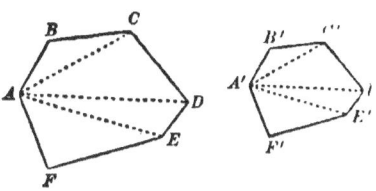

and $B'C'D'$, respectively, there remain the equals ACD and $A'C'D'$. Also, from the similarity of the triangles ABC and $A'B'C'$, and from that of the polygons, we have

$$\frac{AC}{A'C'} = \frac{BC}{B'C'} = \frac{CD}{C'D'};$$

therefore, the triangles ACD and $A'C'D'$ are similar (32).

Thus, successively, each triangle of one polygon may be shown to be similar to the triangle similarly situated in the other.

40. *Scholium.* Two similar polygons may be decomposed into similar triangles, not only by diagonals, but *by lines drawn from any two homologous points.* Thus, let O be any arbitrarily assumed point in the plane of the polygon $ABCD$, etc.; and draw OA, OB, OC, etc. In the similar polygon $A'B'C'D'$, etc., draw $A'O'$ making the angle $B'A'O'$ equal to BAO, and $B'O'$ making the angle

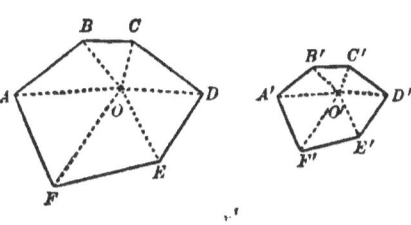

$A'B'O'$ equal to ABO. The intersection O' of these lines, regarded as a point belonging to the polygon $A'B'C'D'$, etc., is homologous to the point O of the polygon $ABCD$, etc.; and the lines $O'A'$, $O'B'$, $O'C'$, etc., being drawn, the triangles $O'A'B'$, $O'B'C'$, etc., are shown to be similar to OAB, OBC, etc., respectively, by the same method as was employed in the preceding demonstration.

If the point O is taken without the polygon, and its homologous

point O' found as before by constructing the triangle $OA'B'$ similar

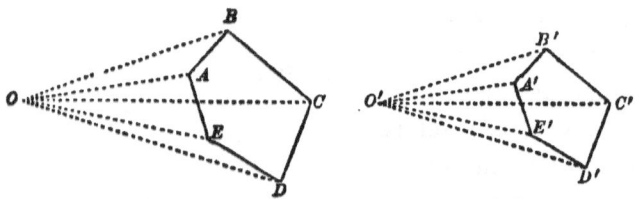

to OAB, the polygons will be decomposed into triangles partly additive and partly subtractive. Thus the polygon $ABCDE$ is equal to the sum of the two triangles OBC and OCD, diminished by the triangles OBA, OAE and OED; and the polygon $A'B'C'D'E'$ is similarly decomposed.

Homologous lines in the two polygons are lines joining pairs of homologous points, such as OA and $O'A'$, OB and $O'B'$, etc., the diagonals joining homologous vertices, etc.; and it is readily shown that any two such homologous lines are in the same ratio as any two homologous sides, that is, in the *ratio of similitude* of the polygons (24).

41. *Corollary.* Two similar polygons are equal when any line in one is equal to its homologous line in the other.

PROPOSITION XII.—THEOREM.

42. *The perimeters of two similar polygons are in the same ratio as any two homologous sides.*

For, we have (see preceding figures),

$$\frac{AB}{A'B'} = \frac{BC}{B'C'} = \frac{CD}{C'D'} = \text{etc.},$$

whence (12),

$$\frac{AB + BC + CD + \text{etc.}}{A'B' + B'C' + C'D' + \text{etc.}} = \frac{AB}{A'B'} = \frac{BC}{B'C'} = \text{etc.}$$

43. *Corollary.* The perimeters of two similar polygons are in the same ratio as any two homologous lines; that is, in the ratio of similitude of the polygons (40).

APPLICATIONS.

PROPOSITION XIII.—THEOREM.

44. *If a perpendicular is drawn from the vertex of the right angle to the hypotenuse of a right triangle:*

1st. *The two triangles thus formed are similar to each other and to the whole triangle;*

2d. *The perpendicular is a mean proportional between the segments of the hypotenuse;*

3d. *Each side about the right angle is a mean proportional between the hypotenuse and the adjacent segment.*

Let C be the right angle of the triangle ABC, and CD the perpendicular to the hypotenuse; then,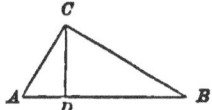

1st. The triangles ACD and CBD are similar to each other and to ABC. For, the triangles ACD and ABC have the angle A common, and the right angles, ADC, ACB, equal; therefore, they are similar (26). For a like reason CBD is similar to ABC, and consequently also to ACD.

2d. The perpendicular CD is a mean proportional between the segments AD and DB. For, the similar triangles, ACD, CBD, give

$$AD : CD = CD : BD.$$

3d. The side AC is a mean proportional between the hypotenuse AB and the adjacent segment AD. For, the similar triangles, ACD, ABC, give

$$AB : AC = AC : AD.$$

In the same way, the triangles CBD and ABC give,

$$AB : BC = BC : BD.$$

45. *Corollary* I. If all the lines of the figure are supposed to be expressed in numbers, being measured by any common unit, the preceding proportions give, by (5),

$$\overline{CD}^2 = AD \times BD,$$
$$\overline{AC}^2 = AB \times AD,$$
$$\overline{BC}^2 = AB \times BD;$$

BOOK III.   111

where we employ the notation $\overline{CD}$, as in algebra, to signify the product of CD multiplied by itself, or the *second power* of CD; observing, however, that this is but a conventional abbreviation for "second power of the *number* representing CD" (8). It may be read "the square of CD," for a reason that will appear hereafter.

46. *Corollary* II. By division, the last two equations of the preceding corollary give

$$\frac{\overline{AC}^2}{\overline{BC}^2} = \frac{AB \times AD}{AB \times BD} = \frac{AD}{BD};$$

that is, *the squares of the sides including the right angle are proportional to the segments of the hypotenuse.*

47. *Corollary* III. If from any point C in the circumference of a circle, a perpendicular CD is drawn to a diameter AB, and also the chords CA, CB; then, since ACB is a right angle (II. 59),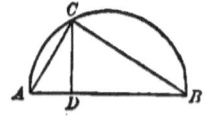
it follows that *the perpendicular is a mean proportional between the segments of the diameter; and each chord is a mean proportional between the diameter and the segment adjacent to that chord.*

PROPOSITION XIV.—THEOREM.

48. *The square of the hypotenuse of a right triangle is equal to the sum of the squares of the other two sides.*

Let ABC be right angled at C; then,

$$\overline{AB}^2 = \overline{AC}^2 + \overline{BC}^2.$$

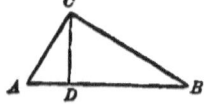

For, by the preceding proposition, we have

$$\overline{AC}^2 = AB \times AD, \text{ and } \overline{BC}^2 = AB \times BD,$$

the sum of which is

$$\overline{AC}^2 + \overline{BC}^2 = AB \times (AD + BD) = AB \times AB = \overline{AB}^2.$$

49. *Corollary* I. By this theorem, if the numerical measures of two sides of a right triangle are given, that of the third is found. For example, if $AC = 3$, $BC = 4$; then, $AB = \sqrt{[3^2 + 4^2]} = 5$.

If the hypotenuse, AB, and one side, AC, are given, we have $\overline{BC}^2 = \overline{AB}^2 - \overline{AC}^2$; thus, if there are given $AB = 5$, $AC = 3$, then, we find $BC = \sqrt{[5^2 - 3^2]} = 4$.

50. *Corollary* II. If AC is the diagonal of a square $ABCD$, we have, by the preceding theorem,

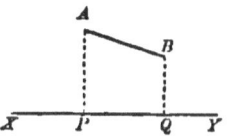

$$\overline{AC}^2 = \overline{AB}^2 + \overline{BC}^2 = 2\overline{AB}^2,$$

whence,

$$\frac{\overline{AC}^2}{\overline{AB}^2} = 2,$$

and extracting the square root,

$$\frac{AC}{AB} = \sqrt{2} = 1.41421 + ad\ inf.$$

Since the square root of 2 is an incommensurable number, it follows that *the diagonal of a square is incommensurable with its side.*

51. *Definition.* The *projection of a point A* upon an indefinite straight line XY is the foot P of the perpendicular let fall from the point upon the line.

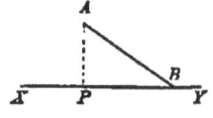

The *projection of a finite straight line AB* upon the line XY is the distance PQ between the projections of the extremities of AB.

If one extremity B of the line AB is in the line XY, the distance from B to P (the projection of A) is the projection of AB on XY; for the point B is in this case its own projection.

PROPOSITION XV.—THEOREM.

52. *In any triangle, the square of the side opposite to an acute angle is equal to the sum of the squares of the other two sides diminished by twice the product of one of these sides and the projection of the other upon that side.*

Let C be an acute angle of the triangle ABC, P the projection of A upon BC by the perpendicular AP, PC the projection of AC upon BC; then,

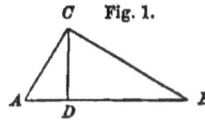

Fig. 1.

$$\overline{AB}^2 = \overline{BC}^2 + \overline{AC}^2 - 2BC \times PC.$$

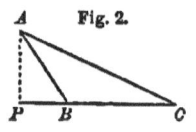

Fig. 2.

For, if P falls on the base, as in Fig. 1, we have

$$PB = BC - PC,$$

and if P falls upon the base produced, as in Fig. 2, we have

$$PB = PC - BC,$$

but in either case the square of PB is, by a theorem of algebra,*

$$\overline{PB}^2 = \overline{BC}^2 + \overline{PC}^2 - 2BC \times PC.$$

Adding $\overline{AP}^2$ to both members of this equality, and observing that by the preceding theorem, $\overline{PB}^2 + \overline{AP}^2 = \overline{AB}^2$, and $\overline{PC}^2 + \overline{AP}^2 = \overline{AC}^2$, we obtain

$$\overline{AB}^2 = \overline{BC}^2 + \overline{AC}^2 - 2BC \times PC.$$

PROPOSITION XVI.—THEOREM.

53. *In an obtuse angled triangle, the square of the side opposite to the obtuse angle is equal to the sum of the squares of the other two sides, increased by twice the product of one of these sides and the projection of the other upon that side.*

Let C be the obtuse angle of the triangle ABC, P the projection of A upon BC (produced); then,

$$\overline{AB}^2 = \overline{BC}^2 + \overline{AC}^2 + 2BC \times PC.$$

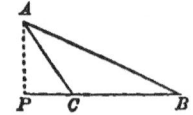

For, since P can only fall upon BC produced, ACB being an obtuse angle, we shall in all cases have

$$PB = BC + PC,$$

and the square of PB will be, by an algebraic theorem, †

$$\overline{PB}^2 = \overline{BC}^2 + \overline{PC}^2 + 2BC \times PC.$$

Adding $\overline{AP}^2$ to both members, we obtain

$$\overline{AB}^2 = \overline{BC}^2 + \overline{AC}^2 + 2BC \times PC.$$

---

* $(x - y)^2$ or $(y - x)^2 = x^2 + y^2 - 2xy$.
† $(x + y)^2 = x^2 + y^2 + 2xy$.

54. *Corollary.* From the preceding three theorems, it follows that an angle of a triangle is acute, right or obtuse, according as the square of the side opposite to it is less than, equal to, or greater than, the sum of the squares of the other two sides.

PROPOSITION XVII.—THEOREM.

55. *If through a fixed point within a circle any chord is drawn, the product of its two segments has the same value, in whatever direction the chord is drawn.*

Let P be any fixed point within the circle O, AB and $A'B'$ any two chords drawn through P; then,

$$PA \times PB = PA' \times PB'.$$

For, join AB' and $A'B$. The triangles APB', $A'PB$, are similar, having the angles at P equal, and also the angles A and A' equal (II. 58); therefore,

$$PA : PA' = PB' : PB,$$

whence (5),

$$PA \times PB = PA' \times PB'.$$

56. *Corollary.* If AB is the *least chord*, drawn through P (II. 20), then, since it is perpendicular to OP, we have $PA = PB$ (II. 15), and hence $\overline{PA}^2 = PA' \times PB'$; that is, *either segment of the least chord drawn through a fixed point is a mean proportional between the segments of any other chord drawn through that point.*

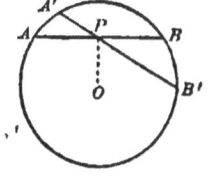

57. *Scholium.* If a chord constantly passing through a fixed point P, be conceived to revolve upon this point as upon a pivot, one segment of the chord increases while the other decreases, but their product being *constant* (being always equal to the square of half the least chord), the two segments are said to *vary reciprocally*, or to be *reciprocally proportional* (2).

PROPOSITION XVIII.—THEOREM.

58. *If through a fixed point without a circle a secant is drawn, the product of the whole secant and its external segment has the same value, in whatever direction the secant is drawn.*

Let P be any fixed point without the circle O, PAB and $PA'B'$ any two secants drawn through P; then,

$$PA \times PB = PA' \times PB'.$$

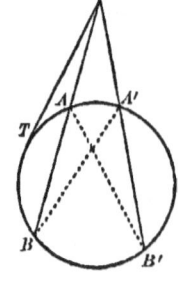

For, join AB' and $A'B$. The triangles APB', $A'PB$, are similar, having the angle at P common, and also the angles B and B' equal (II. 58); therefore,

$$PA : PA' = PB' : PB,$$

whence (5),

$$PA \times PB = PA' \times PB'.$$

59. *Corollary.* If the line PAB, constantly passing through the fixed point P, be conceived to revolve upon P, as upon a pivot, and to approach the tangent PT, the two points of intersection, A and B, will approach each other; and when the line has come into coincidence with the tangent, the two points of intersection will coincide in the point of tangency T. The whole secant and its external segment will then both become equal to the tangent PT; therefore, regarding the tangent as a secant whose two points of intersection are coincident (II. 28), we shall have

$$\overline{PT}^2 = PA' \times PB';$$

that is, *if through a fixed point without a circle a tangent to the circle is drawn, and also any secant, the tangent is a mean proportional between the whole secant and its external segment.*

60. *Scholium* I. When a secant, constantly passing through a fixed point, changes its direction, the whole secant and its external segment *vary reciprocally,* or they are *reciprocally proportional,* since their product is constant (2).

61. *Scholium* II. The analogy between the two preceding propositions is especially to be remarked. They may, indeed, be reduced to a single proposition in the following form: *If through any fixed* point in the plane of a circle a straight line is drawn intersecting the circumference, the product of the distances of the fixed point from the two points of intersection is constant.

PROPOSITION XIX.—THEOREM.

62. *In any triangle, if a medial line is drawn from the vertex to the base:*

1st. *The sum of the squares of the two sides is equal to twice the square of half the base increased by twice the square of the medial line;*

2d. *The difference of the squares of the two sides is equal to twice the product of the base by the projection of the medial line on the base.*

In the triangle ABC, let D be the middle point of the base BC, AD the medial line from A to the base, P the projection of A upon the base, DP the projection of AD upon the base; then,

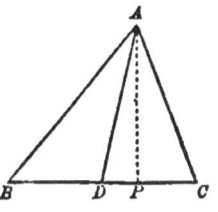

1st. $\overline{AB}^2 + \overline{AC}^2 = 2\overline{BD}^2 + 2\overline{AD}^2$;

2d. $\overline{AB}^2 - \overline{AC}^2 = 2BC \times DP$.

For, if $AB > AC$, the angle ADB will be obtuse and ADC will be acute, and in the triangles ABD, ADC, we shall have, by (53) and (52).

$$\overline{AB}^2 = \overline{BD}^2 + \overline{AD}^2 + 2BD \times DP,$$

$$\overline{AC}^2 = \overline{DC}^2 + \overline{AD}^2 - 2DC \times DP.$$

Adding these equations, and observing that $BD = DC$, we have

1st. $\overline{AB}^2 + \overline{AC}^2 = 2\overline{BD}^2 + 2\overline{AD}^2$.

Subtracting the second equation from the first, we have

$$\overline{AB}^2 - \overline{AC}^2 = 2(BD + DC) \times DP;$$

that is,

2d. $\overline{AB}^2 - \overline{AC}^2 = 2BC \times DP$.

63. Corollary I. In any quadrilateral, the sum of the squares of the four sides is equal to the sum of the squares of the diagonals *plus* four times the square of the line joining the middle points of the diagonals.

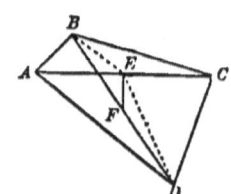

For, let E and F be the middle points of the diagonals of the quadrilateral $ABCD$; join EF, EB, ED. Then, by the preceding theorem, we have in the triangle ABC,

$$\overline{AB}^2 + \overline{BC}^2 = 2\overline{AE}^2 + 2\overline{BE}^2,$$

and in the triangle ADC,

$$\overline{CD}^2 + \overline{DA}^2 = 2\overline{AE}^2 + 2\overline{DE}^2,$$

whence, by addition,

$$\overline{AB}^2 + \overline{BC}^2 + \overline{CD}^2 + \overline{DA}^2 = 4\overline{AE}^2 + 2(\overline{BE}^2 + \overline{DE}^2).$$

Now, in the triangle BED, we have

$$\overline{BE}^2 + \overline{DE}^2 = 2\overline{BF}^2 + 2\overline{EF}^2;$$

therefore,

$$\overline{AB}^2 + \overline{BC}^2 + \overline{CD}^2 + \overline{DA}^2 = 4\overline{AE}^2 + 4\overline{BF}^2 + 4\overline{EF}^2.$$

But $4\overline{AE}^2 = (2AE)^2 = \overline{AC}^2$, and $4\overline{BF}^2 = (2BF)^2 = \overline{BD}^2$; hence, finally,

$$\overline{AB}^2 + \overline{BC}^2 + \overline{CD}^2 + \overline{DA}^2 = \overline{AC}^2 + \overline{BD}^2 + 4\overline{EF}^2.$$

64. Corollary II. In a parallelogram, the sum of the squares of the four sides is equal to the sum of the squares of the diagonals. For if the quadrilateral in the preceding corollary is a parallelogram, the diagonals bisect each other, and the distance EF is zero.

PROPOSITION XX.—THEOREM.

65. *In any triangle, the product of two sides is equal to the product of the diameter of the circumscribed circle by the perpendicular let fall upon the third side from the vertex of the opposite angle.*

Let AB, AC, be two sides of a triangle ABC, AD the perpendicular upon BC, AE the diameter of the circumscribed circle; then,

$$AB \times AC = AE \times AD.$$

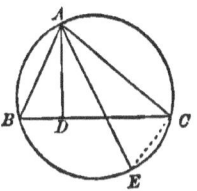

For, joining CE, the angle ACE is a right angle (II. 59), and the angles E and B are equal (II. 58); therefore, the right triangles AEC, ABD, are similar, and give

$$AB : AE = AD : AC,$$

whence, $AB \times AC = AE \times AD$.

PROPOSITION XXI.—THEOREM.

66. *In any triangle, the product of two sides is equal to the product of the segments of the third side formed by the bisector of the opposite angle* plus *the square of the bisector.*

Let AD bisect the angle A of the triangle ABC; then,

$$AB \times AC = DB \times DC + \overline{DA}^2.$$

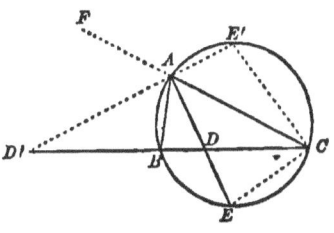

For, circumscribe a circle about ABC, produce AD to meet the circumference in E, and join CE. The triangles ABD, AEC, are similar, and give

$$AB : AE = DA : AC,$$

whence $AB \times AC = AE \times DA = (DE + DA) \times DA$
$$= DE \times DA + \overline{DA}^2.$$

Now, by (55), we have $DE \times DA = DB \times DC$, and hence

$$AB \times AC = DB \times DC + \overline{DA}^2.$$

67. *Corollary.* If the exterior angle BAF is bisected by AD', the same theorem holds, except that *plus* is to be changed to *minus*.

For, producing $D'A$ to meet the circumference in E', and joining CE', the triangles ABD', $AE'C$, are similar, and give

$$AB : AE' = AD' : AC,$$

whence $AB \times AC = AE' \times AD' = (D'E' - D'A) \times D'A$
$$= D'E' \times D'A - \overline{D'A}^2,$$

or, by (58), $AB \times AC = D'B \times D'C - \overline{D'A}^2$.

PROBLEMS OF CONSTRUCTION.

PROPOSITION XXII.—PROBLEM.

68. *To divide a given straight line into parts proportional to given straight lines.*

Let it be required to divide AB into parts proportional to M, N and P. From A draw an indefinite straight line AX, upon which lay off $AC = M$, $CD = N$, $DE = P$, join EB, and draw CF, DG, parallel to EB; then AF, FG, GB, are proportional to M, N, P (18).

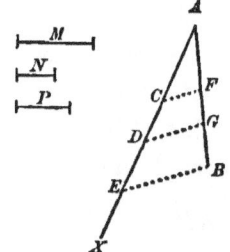

69. *Corollary.* To divide a given straight line AB into any number of *equal* parts, draw an indefinite line AX, upon which lay off the same number of equal distances, each distance being of any convenient length; through M the last point of division on AX draw MB, and through the other points of division of AX draw parallels to MB, which will divide AB into the required number of equal parts. This follows both from the theory of proportional lines and from (I. 125).

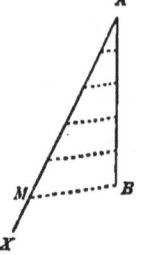

PROPOSITION XXIII.—PROBLEM.

70. *To find a fourth proportional to three given straight lines.*

Let it be required to find a fourth proportional to M, N and P. Draw the indefinite lines AX, AY, making any angle with each other. Upon AX lay off $AB = M$, $AD = N$; and upon AY lay off $AC = P$; join BC, and draw DE parallel to BC; then AE is the required fourth proportional.

For, we have (15),

$$AB : AD = AC : AE, \text{ or } M : N = P : AE.$$

71. *Corollary.* If $AB = M$, and both AD and AC are made equal to N, AE will be a third proportional to M and N; for we shall have $M : N = N : AE$.

PROPOSITION XXIV.—PROBLEM.

72. *To find a mean proportional between two given straight lines.*

Let it be required to find a mean proportional between M and N. Upon an indefinite line lay off $AB = M$, $BC = N$; upon AC describe a semi-circumference, and at B erect a perpendicular, BD, to AC. Then BD is the required mean proportional (47).

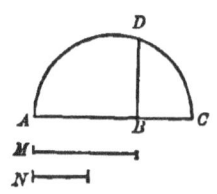

Second method. Take AB equal to the greater line M, and upon it lay off $BC = N$. Upon AB describe a semi-circumference, erect CD perpendicular to AB and join BD. Then BD is the required mean proportional (47).

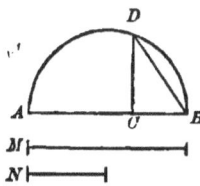

73. *Definition.* When a given straight line is divided into two segments such that one of the segments is a mean proportional between the given line and the other segment, it is said to be divided *in extreme and mean ratio.*

Thus AB is divided in extreme and mean ratio at C, if $AB : AC = AC : CB$.

If C' is taken in BA produced so that $AB : AC' = AC' : C'B$, then AB is divided at C', *externally*, in extreme and mean ratio.

PROPOSITION XXV.—PROBLEM.

74. *To divide a given straight line in extreme and mean ratio.*

Let AB be the given straight line. At B erect the perpendicular BO equal to one half of AB. With the centre O and radius OB, describe a circumference, and through A and O draw AO cutting the circumference first in D and a second time in D'.

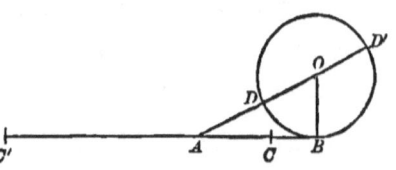

Upon AB lay off $AC = AD$, and upon BA produced lay off $AC' = AD'$. Then AB is divided at C internally, and at C' externally, in extreme and mean ratio.

For, 1st, we have (59),

$$AD' : AB = AB : AD \text{ or } AC, \qquad [1]$$

whence, by division (10),

$$AD' - AB : AB = AB - AC : AC,$$

or, since $DD' = 2OB = AB$, and therefore $AD' - AB = AD' - DD' = AD = AC$,

$$AC : AB = CB : AC,$$

and, by inversion (7),

$$AB : AC = AC : CB;$$

that is, AB is divided at C, internally, in extreme and mean ratio.

2d. The proportion [1] gives by composition (10),

$$AD' + AB : AD' = AB + AD : AB,$$

or, since $AD' = AC'$, $AD' + AB = C'B$, $AB + AD = DD' + AD = AD' = AC'$,

$$C'B : AC' = AC' : AB,$$

and, by inversion,

$$AB : AC' = AC' : C'B;$$

that is, AB is divided at C', externally, in extreme and mean ratio.

75. *Scholium.* Since $OD = OD' = \dfrac{AB}{2}$, we have

$$AC = AO - \dfrac{AB}{2}, \quad AC' = AO + \dfrac{AB}{2}.$$

But the right triangle AOB gives

$$\overline{AO}^2 = \overline{AB}^2 + \left(\dfrac{AB}{2}\right)^2 = \overline{AB}^2 \cdot \dfrac{5}{4},$$

whence, extracting the square root,

$$AO = AB \cdot \dfrac{\sqrt{5}}{2}.$$

Therefore,

$$AC = AB \cdot \dfrac{\sqrt{5}-1}{2}, \quad AC' = AB \cdot \dfrac{\sqrt{5}+1}{2}.$$

76. *Definitions.* When a straight line is divided internally and externally in the same ratio, it is said to be divided *harmonically*.

Thus, AB is divided harmonically at C and D, if $CA : CB = DA : DB$; 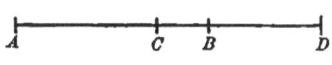 that is, if the ratio of the distances of C from A and B is equal to the ratio of the distances of D from A and B.

Since this proportion may also be written in the form

$$AC : AD = BC : BD,$$

the ratio of the distances of A from C and D is equal to the ratio of the distances of B from C and D; consequently the line CD is divided harmonically at A and B.

The four points A, B, C, D, thus related, are called *harmonic points*, and A and B are called *conjugate* points, as also C and D.

PROPOSITION XXVI.—PROBLEM.

77. *To divide a given straight line harmonically in a given ratio.*

Let it be required to divide AB harmonically in the ratio of M to N.

Upon the indefinite line AX, lay off $AE = M$, and from E lay off EF and EG, each equal to N; join FB, GB; and draw EC parallel to FB, ED parallel to GB.

Then, by the construction we have (17),

$$\frac{M}{N} = \frac{CA}{CB} = \frac{DA}{DB};$$

therefore, by the definition (76), AB is divided harmonically at C and D, and in the given ratio.

78. *Scholium.* If the extreme points A and D are given, and it is required to insert their conjugate harmonic points B and C, the harmonic ratio being given $= M : N$, we take on AX, as before, $AE = M$ and $EF = EG = N$, join ED, and draw GB parallel to ED, which determines B; then, join FB and draw EC parallel to FB, which determines C.

Also if, of four harmonic points A, B, C, D, any three are given, the fourth can be found.

PROPOSITION XXVII.—PROBLEM.

79. *To find the locus of all the points whose distances from two given points are in a given ratio.*

Let A and B be the given points, and let the given ratio be $M : N$. Suppose the problem solved, and that P is a point of the required locus. Divide AB internally at C and externally at D, in the ratio $M : N$, and join PA, PB, PC, PD. By the condition imposed upon P we must have

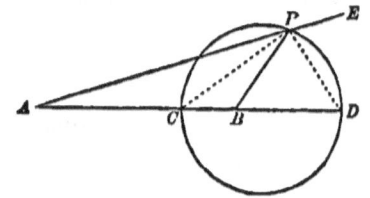

$$PA : PB = M : N = CA : CB = DA : DB;$$

therefore, PC bisects the angle APB, and PD bisects the exterior angle BPE (23). But the bisectors PC and PD are perpendicular to each other (I. 25); therefore, the point P is the vertex of a right angle whose sides pass through the fixed points C and D, and the locus of P is the circumference of a circle described upon CD as a diameter (II. 59, 97). Hence, we derive the following

Construction. Divide AB harmonically, at C and D, in the given ratio (77), and upon CD as a diameter describe a circumference. This circumference is the required locus.

PROPOSITION XXVIII.—PROBLEM.

80. *On a given straight line, to construct a polygon similar to a given polygon.*

Let it be required to construct upon $A'B'$ a polygon similar to $ABCDEF$.

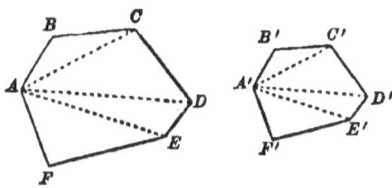

Divide $ABCDEF$ into triangles by diagonals drawn from A. Make the angles $B'A'C'$ and $A'B'C'$ equal to BAC and ABC respectively; then, the triangle $A'B'C'$ will be similar to ABC (25). In the same manner construct the triangle $A'D'C'$ similar to ADC, $A'E'D'$ similar to AED, and $A'E'F'$ similar to AEF. Then, $A'B'C'D'E'F'$ is the required polygon (38).

PROPOSITION XXIX.—PROBLEM.

81. *To construct a polygon similar to a given polygon, the ratio of similitude of the two polygons being given.*

Let $ABCDE$ be the given polygon, and let the given ratio of similitude be $M : N$.

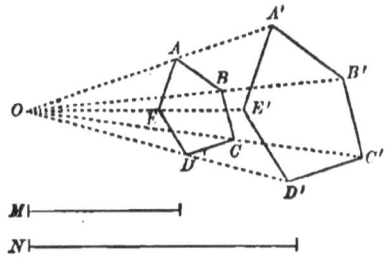

Take any point O, either within or without the given polygon, and draw straight lines from O through each of the vertices of the polygon. Upon any one of these lines, as OA, take OA' a fourth proportional to M, N, and OA, that is, so that

$$M : N = OA : OA'.$$

In the angle AOB draw $A'B'$ parallel to AB; then, in the angle BOC, $B'C'$ parallel to BC, and so on. The polygon $A'B'C'D'E'$ will be similar to $ABCDE$; for the two polygons will be composed of the same number of triangles, additive or subtractive, similarly placed; and their ratio of similitude will evidently be the given ratio $M : N$. (40).

82. *Scholium.* The point O in the preceding construction is called the *centre of similitude* of the two polygons.

BOOK IV.

COMPARISON AND MEASUREMENT OF THE SURFACES OF RECTILINEAR FIGURES.

1. DEFINITION. The *area* of a surface is its numerical measure, referred to some other surface as the unit; in other words, it is the *ratio* of the surface to the *unit of surface* (II. 43).

The unit of surface is called the *superficial unit*. The most convenient superficial unit is the square whose side is the linear unit.

2. *Definition. Equivalent figures* are those whose areas are equal.

PROPOSITION I.—THEOREM.

3. *Two rectangles having equal altitudes are to each other as their bases.*

Let $ABCD$, $AEFD$, be two rectangles having equal altitudes, AB and AE their bases; then,

$$\frac{ABCD}{AEFD} = \frac{AB}{AE}.$$

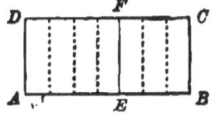

Suppose the bases to have a common measure which is contained, for example, 7 times in AB, and 4 times in AE; so that if AB is divided into 7 equal parts, AE will contain 4 of these parts; then, we have

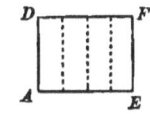

$$\frac{AB}{AE} = \frac{7}{4}.$$

If, now, at the several points of division of the bases, we erect perpendiculars to them, the rectangle $ABCD$ will be divided into 7 equal rectangles (I. 120), of which $AEFD$ will contain 4; consequently, we have

$$\frac{ABCD}{AEFD} = \frac{7}{4},$$

and therefore

$$\frac{ABCD}{AEFD} = \frac{AB}{AE}.$$

The demonstration is extended to the case in which the bases are incommensurable, by the process already exemplified in (II. 51) and (III. 15).

4. *Corollary.* Since AD may be called the base, and AB and AE the altitudes, it follows that *two rectangles having equal bases are to each other as their altitudes.*

Note. In these propositions, by "rectangle" is to be understood "surface of the rectangle."

PROPOSITION II.—THEOREM.

5. *Any two rectangles are to each other as the products of their bases by their altitudes.*

Let R and R' be two rectangles, k and k' their bases, h and h' their altitudes; then,

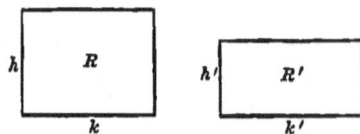

$$\frac{R}{R'} = \frac{k \times h}{k' \times h'}.$$

For, let S be a third rectangle having the same base k as the rectangle R, and the same altitude h' as the rectangle R'; then we have, by (4) and (3),

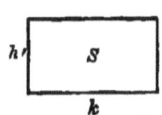

$$\frac{R}{S} = \frac{h}{h'}, \quad \frac{S}{R'} = \frac{k}{k'},$$

and multiplying these ratios, we find (III. 14),

$$\frac{R}{R'} = \frac{k \times h}{k' \times h'}.$$

6. *Scholium.* It must be remembered that by the product of two

lines, is to be understood the product of the numbers which represent them when they are measured by the linear unit (III. 8).

PROPOSITION III.—THEOREM.

7. *The area of a rectangle is equal to the product of its base and altitude.*

Let R be any rectangle, k its base and h its altitude numerically expressed in terms of the linear unit; and let Q be the square whose side is the linear unit; then, by the preceding theorem,

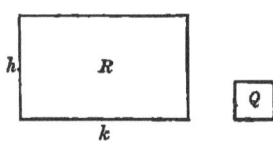

$$\frac{R}{Q} = \frac{k \times h}{1 \times 1} = k \times h.$$

But since Q is the unit of surface, $\dfrac{R}{Q}$ = the numerical measure, or area, of the rectangle R (1); therefore,

$$\text{Area of } R = k \times h.$$

8. *Scholium* I. When the base and altitude are exactly divisible by the linear unit, this proposition is rendered evident by dividing the rectangle into squares each equal to the superficial unit. Thus, if the base contains 7 linear units and the altitude 5, the rectangle can obviously be divided into 35 squares each equal to the superficial unit; that is, its area $= 5 \times 7$. The proposition, as above demonstrated, is, however, more general, and includes also the cases in which either the base, or the altitude, or both, are incommensurable with the unit of length.

9. *Scholium* II. The area of a square being the product of two equal sides, is the *second power* of a side. Hence it is, that in arithmetic and algebra, the expression "square of a number" has been adopted to signify "second power of a number."

We may also here observe that many writers employ the expression "rectangle of two lines" in the sense of "product of two lines," because the rectangle constructed upon two lines is measured by the product of the numerical measures of the lines.

PROPOSITION IV.—THEOREM.

10. *The area of a parallelogram is equal to the product of its base and altitude.*

Let $ABCD$ be a parallelogram, k the numerical measure of its base AB, h that of its altitude AF; and denote its area by S; then,

$$S = k \times h.$$

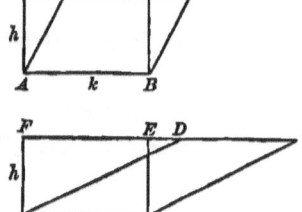

For, let the rectangle $ABEF$ be constructed having the same base and altitude as the parallelogram; the upper bases of the two figures will be in the same straight line FC (I. 58). The right triangles AFD and BEC are equal, having $AF = BE$, and $AD = BC$ (I. 83). If from the whole figure $ABCF$ we take away the triangle AFD, there remains the parallelogram $ABCD$; and if from the whole figure we take away the triangle BEC, there remains the rectangle $ABEF$; therefore the surface of the parallelogram is equal to that of the rectangle. But the area of the rectangle is $k \times h$ (7); therefore that of the parallelogram is also $k \times h$; that is $S = k \times h$.

11. *Corollary* I. Parallelograms having equal bases and equal altitudes are equivalent.

12. *Corollary* II. Parallelograms having equal altitudes are to each other as their bases; parallelograms having equal bases are to each other as their altitudes; and any two parallelograms are to each other as the products of their bases by their altitudes.

PROPOSITION V.—THEOREM.

13. *The area of a triangle is equal to half the product of its base and altitude.*

Let ABC be a triangle, k the numerical measure of its base BC, h that of its altitude AD; and S its area; then,

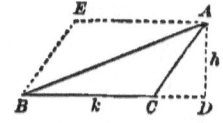

$$S = \tfrac{1}{2} k \times h.$$

For, through A draw AE parallel to CB, and through B draw BE

parallel to CA. The triangle ABC is one-half the parallelogram $AEBC$ (I. 105); but the area of the parallelogram $= k \times h$; therefore, for the triangle, we have $S = \frac{1}{2} k \times h$.

14. *Corollary* I. A triangle is equivalent to one-half of any parallelogram having the same base and the same altitude.

15. *Corollary* II. Triangles having equal bases and equal altitudes are equivalent.

16. *Corollary* III. Triangles having equal altitudes are to each other as their bases; triangles having equal bases are to each other as their altitudes; and any two triangles are to each other as the products of their bases by their altitudes.

PROPOSITION VI.—THEOREM.

17. *The area of a trapezoid is equal to the product of its altitude by half the sum of its parallel bases.*

Let $ABCD$ be a trapezoid; $MN = h$, its altitude; $AD = a$, $BC = b$, its parallel bases; and let S denote its area; then,

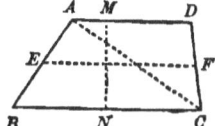

$$S = \tfrac{1}{2}(a + b) \times h.$$

For, draw the diagonal AC. The altitude of each of the triangles ADC and ABC is equal to h, and their bases are respectively a and b; the area of the first is $\frac{1}{2} a \times h$, that of the second is $\frac{1}{2} b \times h$; and the trapezoid being the sum of the two triangles, we have

$$S = \tfrac{1}{2} a \times h + \tfrac{1}{2} b \times h = \tfrac{1}{2}(a + b) \times h.$$

18. *Corollary.* The straight line EF, joining the middle points of AB and DC, being equal to half the sum of AD and BC (I. 124), the area of the trapezoid is equal to the product $MN \times EF$.

19. *Scholium.* The area of any polygon may be found by finding the areas of the several triangles into which it may be decomposed by drawing diagonals from any vertex.

The following method, however, is usually preferred, especially in surveying. Draw the longest diagonal AD of the proposed polygon

ABCDEF; and upon *AD* let fall the perpendiculars *BM, CN, EP, FQ*. The polygon is thus decomposed into right triangles and right trapezoids, and by measuring the lengths of the perpendiculars and also of the distances *AM, MN, ND, AQ, QP, PD*, the bases and altitudes of these triangles and

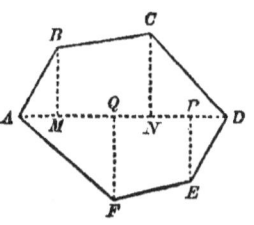

trapezoids are known. Hence their areas can be computed by the preceding theorems, and the sum of these areas will be the area of the polygon.

PROPOSITION VII.—THEOREM.

20. *Similar triangles are to each other as the squares of their homologous sides.*

Let *ABC, A'B'C'* be similar triangles; then,

$$\frac{ABC}{A'B'C'} = \frac{\overline{BC}^2}{\overline{B'C'}^2}.$$

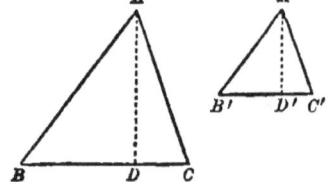

Let *AD, A'D'*, be the altitudes. By (16), we have

$$\frac{ABC}{A'B'C'} = \frac{BC \times AD}{B'C' \times A'D'} = \frac{BC}{B'C'} \times \frac{AD}{A'D'}.$$

But the homologous lines *AD, A'D'*, are in the ratio of similitude of the triangles (III. 29); that is,

$$\frac{AD}{A'D'} = \frac{BC}{B'C'},$$

therefore,

$$\frac{ABC}{A'B'C'} = \frac{BC}{B'C'} \times \frac{BC}{B'C'} = \frac{\overline{BC}^2}{\overline{B'C'}^2}.$$

21. *Corollary.* If we had put the ratio $AD : A'D'$ in the place of the ratio $BC : B'C'$, we should have found

$$\frac{ABC}{A'B'C'} = \frac{\overline{AD}^2}{\overline{A'D'}^2};$$

and in general, we may conclude that *the surfaces of two similar triangles are as the squares of any two homologous lines*; or, again, *the ratio of the surfaces of two similar triangles is the square of the ratio of similitude of the triangles.*

PROPOSITION VIII.—THEOREM.

22. *Two triangles having an angle of the one equal to an angle of the other are to each other as the products of the sides including the equal angles.*

Two triangles which have an angle of the one equal to an angle of the other may be placed with their equal angles in coincidence. Let ABC, ADE, be the two triangles having the common angle A; then,

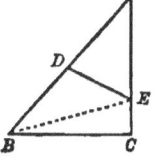

$$\frac{ABC}{ADE} = \frac{AB \times AC}{AD \times AE}.$$

For, join BE. The triangles ABC, ABE, having the common vertex B, and their bases AC, AE, in the same straight line, have the same altitude; therefore (16),

$$\frac{ABC}{ABE} = \frac{AC}{AE}.$$

The triangles ABE, ADE, having the common vertex E, and their bases AB, AD, in the same straight line, have the same altitude; therefore,

$$\frac{ABE}{ADE} = \frac{AB}{AD}.$$

Multiplying these ratios, we have (III. 14),

$$\frac{ABC}{ADE} = \frac{AB \times AC}{AD \times AE}.$$

PROPOSITION IX.—THEOREM.

23. *Similar polygons are to each other as the squares of their homologous sides.*

Let $ABCDEF$, $A'B'C'D'E'F'$, be two similar polygons; and denote their surfaces by S and S'; then,

$$\frac{S}{S'} = \frac{\overline{AB}^2}{\overline{A'B'}^2}.$$

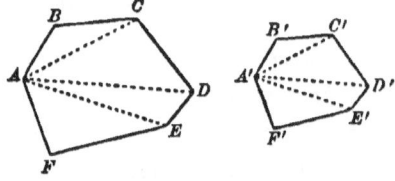

For, let the polygons be decomposed into homologous triangles (III. 39). The ratio of the surfaces of any pair of homologous triangles, as ABC and $A'B'C'$, ACD and $A'C'D'$, etc., will be the square of the ratio of two homologous sides of the polygons (20); therefore, we shall have

$$\frac{ABC}{A'B'C'} = \frac{ACD}{A'C'D'} = \frac{ADE}{A'D'E'} = \frac{AEF}{A'E'F'} = \frac{\overline{AB}^2}{\overline{A'B'}^2}.$$

Therefore, by addition of antecedents and consequents (III. 12),

$$\frac{ABC + ACD + ADE + AEF}{A'B'C' + A'C'D' + A'D'E' + A'E'F'} = \frac{S}{S'} = \frac{\overline{AB}^2}{\overline{A'B'}^2}.$$

24. *Corollary.* The ratio of the surfaces of two similar polygons is the square of the ratio of similitude of the polygons; that is, the square of the ratio of any two homologous lines of the polygons.

PROPOSITION X.—THEOREM.

25. *The square described upon the hypotenuse of a right triangle is equivalent to the sum of the squares described on the other two sides.*

Let the triangle ABC be right angled at C; then, the square AH, described upon the hypotenuse, is equal in area to the sum of the squares AF and BD, described on the other two sides.

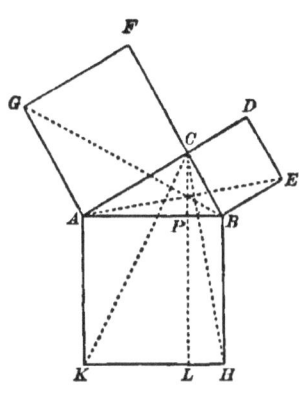

For, from C draw CP perpendicular to AB and produce it to meet KH in L. Join CK, BG. Since ACF and ACB are right angles, CF and CB are in the same straight line (I. 21); and for a similar reason AC and CD are in the same straight line.

In the triangles CAK, GAB, we have AK equal to AB, being sides of the same square; AC equal to AG, for the same reason; and the angles CAK, GAB, equal, being each equal to the sum of the angle CAB and a right angle; therefore, these triangles are equal (I. 76).

The triangle CAK and the rectangle AL have the same base AK; and since the vertex C is upon LP produced, they also have the same altitude; therefore, the triangle CAK is equivalent to one-half the rectangle AL (14).

The triangle GAB and the square AF have the same base AG; and, since the vertex B is upon FC produced, they also have the same altitude; therefore, the triangle GAB is equivalent to one-half the square AF (14).

But the triangles CAK, GAB, have been shown to be equal; therefore, the rectangle AL is equivalent to the square AF.

In the same way, it is proved that the rectangle BL is equivalent to the square BD.

Therefore, the square AH, which is the sum of the rectangles AL and BL, is equivalent to the sum of the squares AF and BD.

26. *Scholium.* This theorem is ascribed to Pythagoras (born about 600 B.C.), and is commonly called the *Pythagorean Theorem*. The preceding demonstration of it is that which was given by Euclid in his Elements (about 300 B.C.).

It is important to observe, that we may deduce the same result from the *numerical* relation $\overline{AB}^2 = \overline{AC}^2 + \overline{BC}^2$, already established in (III. 48). For, since the measure of the area of a square is the second power of the number which represents its side, it follows directly from this numerical relation that the area of which $\overline{AB}^2$ is the measure is equal to the sum of the areas of which $\overline{AC}^2$ and $\overline{BC}^2$ are the measures. In the same manner, most of the numerical relations demonstrated in the articles (III. 48) to (III. 67) give rise to theorems respecting areas by merely substituting, for a product, the area represented by that product. This may be called a transition from the *abstract* (pure number) to the *concrete* (actual space).

On the other hand, we may pass from the concrete to the abstract. For example, in the above figure it has been proved that the areas of the rectangles AL, BL, are respectively equal to the areas of the squares AF, BD. But the rectangles, having the same altitude, are to each other as their bases AP, PB; and the squares are to each other as their numerical measures $\overline{AC}^2$, $\overline{BC}^2$; hence, we infer the numerical relation

$$\overline{AC}^2 : \overline{BC}^2 = AP : PB,$$

which was otherwise proved in (III. 46).

Henceforth, we shall employ the equation $\overline{AB}^2 = \overline{AC}^2 + \overline{BC}^2$, as the expression of either one of the theorems (III. 48) and (IV. 25).

27. *Corollary. If the three sides of a right triangle be taken as the homologous sides of three similar polygons constructed upon them, then the polygon constructed upon the hypotenuse is equivalent to the sum of the polygons constructed upon the other two sides.*

For, let P, Q, R, denote the areas of the polygons constructed upon the sides AC, BC, and upon the hypotenuse AB, respectively. Then, the polygons being similar, we have

$$\frac{P}{Q} = \frac{\overline{AC}^2}{\overline{BC}^2}, \qquad \frac{R}{Q} = \frac{\overline{AB}^2}{\overline{BC}^2},$$

from the first of which we derive, by composition,

$$\frac{P+Q}{Q} = \frac{\overline{AC}^2 + \overline{BC}^2}{\overline{BC}^2} = \frac{\overline{AB}^2}{\overline{BC}^2},$$

which compared with the second gives at once

$$R = P + Q.$$

PROBLEMS OF CONSTRUCTION.

PROPOSITION XI.—PROBLEM.

28. *To construct a triangle equivalent to a given polygon.*
Let $ABCDEF$ be the given polygon.

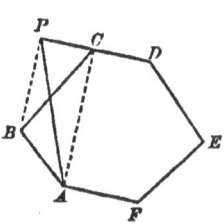

Take any three consecutive vertices, as $A, B, C,$ and draw the diagonal AC. Through B draw BP parallel to AC meeting DC produced in P; join AP.

The triangles $APC, ABC,$ have the same base AC; and since their vertices, P and B, lie on the same straight line BP parallel to AC, they also have the same altitude; therefore they are equivalent. Therefore, the pentagon $APDEF$ is equivalent to the hexagon $ABCDEF$. Now, taking any three consecutive vertices of this pentagon, we shall, by a precisely similar construction, find a quadrilateral of the same area; and, finally, by a similar operation upon the quadrilateral, we shall find a triangle of the same area.

Thus, whatever the number of the sides of the given polygon, a series of successive steps, each step reducing the number of sides by one, will give a series of polygons of equal areas, terminating in a triangle.

PROPOSITION XII.—PROBLEM.

29. *To construct a square equivalent to a given parallelogram or to a given triangle.*

1st. Let AC be a given parallelogram; k its base, and h its altitude.

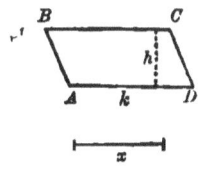

Find a mean proportional x between h and k, by (III. 72). The square constructed upon x will be equivalent to the parallelogram, since $x^2 = h \times k$.

2d. Let ABC be a given triangle; a its base and h its altitude.

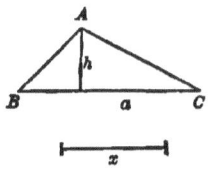

Find a mean proportional x between a and $\tfrac{1}{2} h$; the square constructed upon x will be equivalent to the triangle, since $x^2 = a \times \tfrac{1}{2} h = \tfrac{1}{2} ah$.

BOOK IV.   137

30. *Scholium.* By means of this problem and the preceding, a square can be found equivalent to any given polygon.

PROPOSITION XIII.—PROBLEM.

31. *To construct a square equivalent to the sum of two or more given squares, or to the difference of two given squares.*

1st. Let m, n, p, q, be the sides of given squares.

Draw $AB = m$, and $BC = n$, perpendicular to each other at B; join AC. Then (25), $\overline{AC}^2 = m^2 + n^2$.

Draw $CD = p$, perpendicular to AC, and join AD. Then $\overline{AD}^2 = \overline{AC}^2 + p^2 = m^2 + n^2 + p^2$.

Draw $DE = q$ perpendicular to AD, and join AE. Then, $\overline{AE}^2 = \overline{AD}^2 + q^2 = m^2 + n^2 + p^2 + q^2$; therefore, the square constructed upon AE will be equivalent to the sum of the squares constructed upon m, n, p, q.

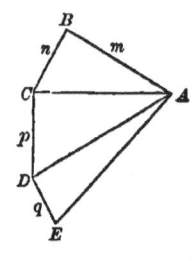

In this manner may the areas of any number of given squares be added.

2d. Construct a right angle ABC, and lay off $BA = n$. With the centre A and a radius $= m$, describe an arc cutting BC in C. Then $\overline{BC}^2 = \overline{AC}^2 - \overline{AB}^2 = m^2 - n^2$; therefore, the square constructed upon BC will be equivalent to the difference of the squares constructed upon m and n.

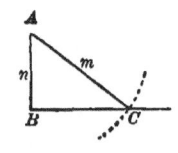

32. *Scholium* I. By means of this problem, together with the preceding ones, a square can be found equivalent to the sum of any number of given polygons; or to the difference of any two given polygons.

33. *Scholium* II. If m, n, p, q, in the preceding problem are homologous sides of given similar polygons, the line AE in the first figure is the homologous side of a similar polygon equivalent to the sum of the given polygons (27).

And the line BC, in the second figure, is the homologous side of a similar polygon, equivalent to the difference of two given similar polygons.

One side of a polygon, similar to a given polygon, being known, the polygon may be constructed by (III. 80).

PROPOSITION XIV.—PROBLEM.

34. *Upon a given straight line to construct a rectangle equivalent to a given rectangle.*

Let k' be the given straight line, and AC the given rectangle whose base is k and altitude h.

Find a fourth proportional h', to k', k and h, by (III. 70). Then, the rectangle constructed upon the base k' with the altitude h' is equivalent to AC; for, by the construction, $k' : k = h : h'$, whence, $k' \times h' = k \times h$ (7).

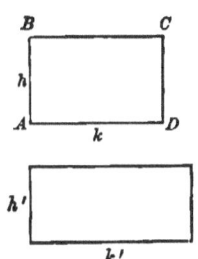

PROPOSITION XV.—PROBLEM.

35. *To construct a rectangle, having given its area and the sum of two adjacent sides.*

Let MN be equal to the given sum of the adjacent sides of the required rectangle; and let the given area be that of the square whose side is AB.

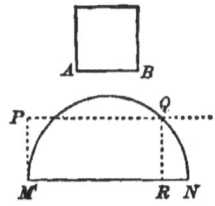

Upon MN as a diameter describe a semicircle. At M erect $MP = AB$ perpendicular to MN, and draw PQ parallel to MN, intersecting the circumference in Q. From Q let fall QR perpendicular to MN; then, MR and RN are the base and altitude of the required rectangle. For, by (III. 47), $MR \times RN = \overline{QR}^2 = \overline{PM}^2 = \overline{AB}^2$

PROPOSITION XVI.—PROBLEM.

36. *To construct a rectangle, having given its area and the difference of two adjacent sides.*

Let MN be equal to the given difference of the adjacent sides of the required rectangle; and let the given area be that of the square described on AB.

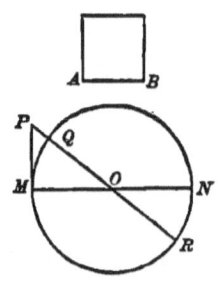

Upon MN as a diameter describe a circle. At M draw the tangent $MP = AB$, and from P, draw the secant PQR through the centre of the circle; then, PR and PQ are the base and altitude of the required rectangle. For, by (III. 59), $PR \times PQ = \overline{PM}^2 = \overline{AB}^2$, and the difference of PR and PQ is $QR = MN$.

PROPOSITION XVII.—PROBLEM.

37. *To find two straight lines in the ratio of the areas of two given polygons.*

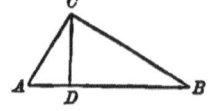

Let squares be found equal in area to the given polygons, respectively (30). Upon the sides of the right angle ACB, take CA and CB equal to the sides of these squares, join AB and let fall CD perpendicular to AB. Then, by (III. 46), we have $AD : DB = \overline{CA}^2 : \overline{CB}^2$; therefore, AD, DB, are in the ratio of the areas of the given polygons.

PROPOSITION XVIII.—PROBLEM.

38. *To find a square which shall be to a given square in the ratio of two given straight lines.*

Let $\overline{AB}^2$ be the given square, and $M : N$ the given ratio.

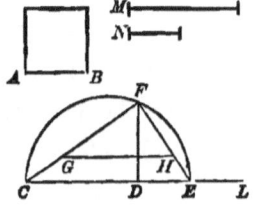

Upon an indefinite straight line CL, lay off $CD = M$, $DE = N$; upon CE as a diameter describe a semicircle; at D erect the perpendicular DF cutting the circumference in F; join FC, FE; lay off $FH = AB$, and through H draw HG parallel to EC; then, FG is the side of the required square. For, by (III. 15), we have

$$FG : FH = FC : FE,$$

whence (III. 13),
$$\overline{FG}^2 : \overline{FH}^2 = \overline{FC}^2 : \overline{FE}^2.$$
Also, by (III. 46),
$$\overline{FC}^2 : \overline{FE}^2 = CD : DE = M : N.$$
Hence,
$$\overline{FG}^2 : \overline{FH}^2 = M : N.$$

But $FH = AB$, therefore the square constructed upon FG is to the square upon AB in the ratio $M : N$.

PROPOSITION XIX.—PROBLEM.

39. *To construct a polygon similar to a given polygon and whose area shall be in a given ratio to that of the given polygon.*

Let P be the given polygon, and let a be one of its sides; let $M : N$ be the given ratio.

Find, by the preceding problem, the side a' of a square which shall be to a^2 in the ratio $M : N$; upon a', as a homologous side to a, construct the polygon P' similar to P (III. 80); this will be the polygon required.

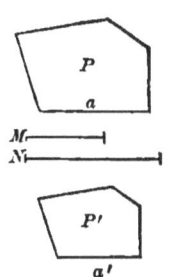

For, the polygons being similar, their areas are in the ratio $a'^2 : a^2$, or $M : N$, as required.

PROPOSITION XX.—PROBLEM.

40. *To construct a polygon similar to a given polygon P and equivalent to a given polygon Q.*

Find M and N, the sides of squares respectively equal in area to P and Q, (30).

Let a be any side of P, and find a fourth proportional a' to M, N and a: upon a', as a homologous side to a, construct the polygon P' similar to P; this will be the required polygon. For, by construction,

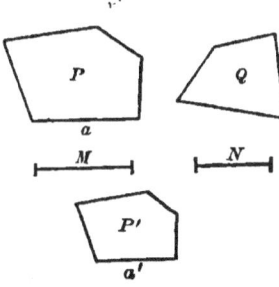

$$\frac{M}{N} = \frac{a}{a'};$$

therefore, taking the letters P, Q and P', to denote the areas of the polygons,
$$\frac{P}{Q} = \frac{M^2}{N^2} = \frac{a^2}{a'^2};$$
but, the polygons P and P' being similar, we have, by (23),
$$\frac{P}{P'} = \frac{a^2}{a'^2};$$
and comparing these equations, we have $P' = Q$.

Therefore, the polygon P' is similar to the polygon P and equivalent to the polygon Q, as required.

BOOK V.

REGULAR POLYGONS. MEASUREMENT OF THE CIRCLE. MAXIMA AND MINIMA OF PLANE FIGURES.

REGULAR POLYGONS.

1. DEFINITION. A *regular polygon* is a polygon which is at once equilateral and equiangular.

The equilateral triangle and the square are simple examples of regular polygons. The following theorem establishes the possibility of regular polygons of any number of sides.

PROPOSITION I.—THEOREM.

2. *If the circumference of a circle be divided into any number of equal parts, the chords joining the successive points of division form a regular polygon inscribed in the circle; and the tangents drawn at the points of division form a regular polygon circumscribed about the circle.*

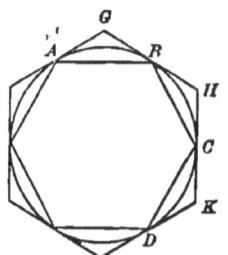

Let the circumference be divided into the equal arcs AB, BC, CD, etc.; then, 1st, drawing the chords AB, BC, CD, etc., $ABCD$, etc., is a regular inscribed polygon. For, its sides are equal, being chords of equal arcs; and its angles are equal, being inscribed in equal segments.

2d. Drawing tangents at A, B, C, etc., the polygon GHK, etc., is a regular circumscribed polygon. For, in the triangles AGB, BHC, CKD, etc., we have $AB = BC = CD$, etc., and the angles GAB, GBA, HBC, HCB, etc., are equal, since each is formed by a tangent and chord and is measured by half of one of the equal parts of the circumference

(II. 62); therefore, these triangles are all isosceles and equal to each other. Hence, we have the angles $G = H = K$, etc., and $AG = GB = BH = HC = CK$, etc., from which, by the addition of equals, it follows that $GH = HK$, etc.

3. *Corollary* I. Hence, if an inscribed polygon is given, a circumscribed polygon of the same number of sides can be formed by drawing tangents at the vertices of the given polygon. And if a circumscribed polygon is given, an inscribed polygon of the same number of sides can be formed by joining the points at which the sides of the given polygon touch the circle.

It is often preferable, however, to obtain the circumscribed polygon from the inscribed, and reciprocally, by the following methods:

1st. Let $ABCD\ldots$ be a given inscribed polygon. Bisect the arcs AB, BC, CD, etc., in the points E, F, G, etc., and draw tangents, $A'B'$, $B'C'$, $C'D'$, etc., at these points; then, since the arcs EF, FG, etc., are equal, the polygon $A'B'C'D'\ldots$ is, by the preceding proposition, a regular circumscribed polygon of the same number of sides as $ABCD\ldots$ Since the radius OE is perpendicular to 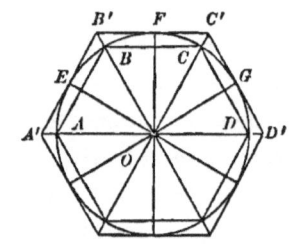 AB (II. 16) as well as to $A'B'$, the sides $A'B'$, AB, are parallel; and, for the same reason, all the sides of $A'B'C'D'\ldots$ are parallel to the sides of $ABCD\ldots$ respectively. Moreover, the radii OA, OB, OC, etc., when produced, pass through the vertices A', B', C', etc.; for since $B'E = B'F$, the point B' must lie on the line OB which bisects the angle EOF (I. 127).

2d. If the circumscribed polygon $A'B'C'D'\ldots$ is given, we have only to draw OA', OB', OC', etc., intersecting the circumference in A, B, C, etc., and then to join AB, BC, CD, etc., to obtain the inscribed polygon of the same number of sides.

4. *Corollary* II. If the chords AE, EB, BF, FC, etc., be drawn, a regular inscribed polygon will be formed of double the number of sides of $ABCD\ldots$

If tangents are drawn at A, B, C, etc., intersecting the tangents $A'B'$, $B'C'$, $C'D'$, etc., a regular circumscribed polygon will be formed of double the number of sides of $A'B'C'D'\ldots$

It is evident that the area of an inscribed polygon is less than that of the inscribed polygon of double the number of sides; and the area of a circumscribed polygon is greater than that of the circumscribed polygon of double the number of sides.

PROPOSITION II.—THEOREM.

5. *A circle may be circumscribed about any regular polygon; and a circle may also be inscribed in it.*

Let $ABCD...$ be a regular polygon; then,

1st. A circle may be circumscribed about it. For, describe a circumference passing through three consecutive vertices A, B, C (II. 88); let O be its centre, draw OH perpendicular to BC and bisecting it at H, and join OA, OD. Conceive the quadrilateral 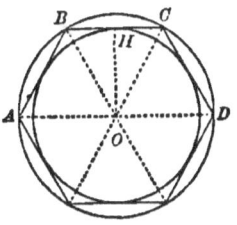 $AOHB$ to be revolved upon the line OH (*i. e.*, *folded* over), until HB falls upon its equal HC. The polygon being regular, the angle $HBA = HCD$, and the side $BA = CD$; therefore the side BA will take the direction of CD and the point A will fall upon D. Hence $OD = OA$, and the circumference described with the radius OA and passing through the three consecutive vertices A, B, C, also passes through the fourth vertex D. It follows that the circumference which passes through the three vertices B, C, D, also passes through the next vertex E, and thus through all the vertices of the polygon. The circle is therefore circumscribed about the polygon.

2d. A circle may be inscribed in it. For, the sides of the polygon being equal chords of the circumscribed circle, are equally distant from the centre; therefore, a circle described with the centre O and the radius OH will touch all the sides, and will consequently be inscribed in the polygon.

6. *Definitions.* The *centre of a regular polygon* is the common centre, O, of the circumscribed and inscribed circles.

The *radius* of a regular polygon is the radius, OA, of the circumscribed circle.

The *apothem* is the radius, OH, of the inscribed circle.

The *angle at the centre* is the angle, AOB, formed by radii drawn to the extremities of any side.

BOOK V.   145

7. The angle at the centre is equal to four right angles divided by the number of sides of the polygon.

8. Since the angle ABC is equal to twice ABO, or to $ABO + BAO$, it follows that the angle ABC of the polygon is the supplement of the angle at the centre (I. 68).

PROPOSITION III.—THEOREM.

9. *Regular polygons of the same number of sides are similar.*

Let $ABCDE$, $A'B'C'D'E'$, be regular polygons of the same number of sides; then, they are similar.

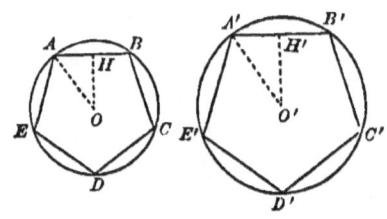

For, 1st, they are mutually equiangular, since the magnitude of an angle of either polygon depends only on the number of the sides (7 and 8), which is the same in both.

2d. The homologous sides are proportional, since the ratio $AB : A'B'$ is the same as the ratio $BC : B'C'$, or $CD : C'D'$, etc.

Therefore the polygons fulfill the two conditions of similarity.

10. *Corollary.* The perimeters of regular polygons of the same number of sides are to each other as the radii of the circumscribed circles, or as the radii of the inscribed circles; and their areas are to each other as the squares of these radii. For, these radii are homologous lines of the similar polygons (III. 43), (IV. 24).

PROPOSITION IV.—PROBLEM.

11. *To inscribe a square in a given circle.*

Draw any two diameters AC, BD, perpendicular to each other, and join their extremities by the chords AB, BC, CD, DA; then, $ABCD$ is an inscribed square (II. 12), (II. 59).

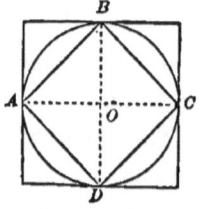

12. *Corollary.* To circumscribe a square about the circle, draw tangents at the extremities of two perpendicular diameters AC, BD.

13   K

13. *Scholium.* In the right triangle ABO, we have $\overline{AB}^2 = \overline{OA}^2 + \overline{OB}^2 = 2\overline{OA}^2$, whence $AB = OA \cdot \sqrt{2}$, by which the side of the inscribed square can be computed, the radius being given.

PROPOSITION V.—PROBLEM.

14. *To inscribe a regular hexagon in a given circle.*

Suppose the problem solved, and let $ABCDEF$ be a regular inscribed hexagon.

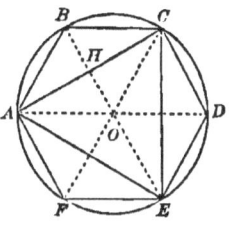

Join BE and AD; since the arcs AB, BC, CD, etc., are equal, the lines BE, AD, bisect the circumference and are diameters intersecting in the centre O. The inscribed angle ABO is measured by one-half the arc AFE, that is, by AF, or one of the equal divisions of the circumference; the angle AOB at the centre is also measured by one division, that is, by AB; and the angle $BAO = ABO$; therefore the triangle ABO is equiangular, and $AB = OA$. Therefore the side of the inscribed regular hexagon is equal to the radius of the circle.

Consequently, to inscribe a regular hexagon, apply the radius six times as a chord.

15. *Corollary.* To inscribe an equilateral triangle, ACE, join the alternate vertices of the regular hexagon.

16. *Scholium.* In the right triangle ACD, we have $\overline{AC}^2 = \overline{AD}^2 - \overline{DC}^2 = (2AO)^2 - \overline{AO}^2 = 3\overline{AO}^2$; whence, $AC = AO \cdot \sqrt{3}$, by which the side of the inscribed equilateral triangle can be computed, the radius being given.

The apothem, OH, of the inscribed equilateral triangle is equal to one-half the radius OB; for the figure $AOCB$ is a rhombus and its diagonals bisect each other at right angles (I. 110).

The apothem of the inscribed regular hexagon is equal to one-half the side of the inscribed equilateral triangle, that is, to $\dfrac{AO}{2}\sqrt{3}$; for the perpendicular from O upon AB is equal to the perpendicular from A upon OB, that is, to AH.

The *angle at the centre* of the regular inscribed hexagon is $\frac{1}{6}$ of 4 right angles, that is, $\frac{2}{3}$ of one right angle $= 60°$.

The angle of the hexagon, or ABC, is $\frac{4}{3}$ of a right angle $= 120°$.

The angle at the centre of the inscribed equilateral triangle is $\frac{4}{3}$ of one right angle $= 120°$.

PROPOSITION VI.—PROBLEM.

17. *To inscribe a regular decagon in a given circle.*

Suppose the problem solved, and let $ABC \ldots L$, be a regular inscribed decagon.

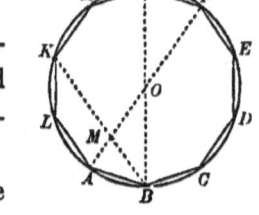

Join AF, BG; since each of these lines bisects the circumference, they are diameters and intersect in the centre O. Draw BK intersecting OA in M.

The angle AMB is measured by half the sum of the arcs KF and AB (II. 64), that is, by two divisions of the circumference; the inscribed angle MAB is measured by half the arc BF, that is, also by two divisions; therefore AMB is an isosceles triangle, and $MB = AB$.

Again, the inscribed angle MBO is measured by half the arc KG, that is, by one division, and the angle MOB at the centre has the same measure; therefore OMB is an isosceles triangle, and $OM = MB = AB$.

The inscribed angle MBA, being measured by half the arc AK, that is, by one division, is equal to the angle AOB. Therefore the isosceles triangles AMB and AOB are mutually equiangular and similar, and give the proportion

$$OA : AB = AB : AM,$$

whence

$$OA \times AM = \overline{AB}^2 = \overline{OM}^2;$$

that is, the radius OA is divided in extreme and mean ratio at M (III. 73); and the greater segment OM is equal to the side AB of the inscribed regular decagon.

Consequently, to inscribe a regular decagon, divide the radius in extreme and mean ratio (III. 74), and apply the greater segment ten times as a chord.

148 GEOMETRY.

18. *Corollary.* To inscribe a regular pentagon, $ACEGK$, join the alternate vertices of the regular inscribed decagon.

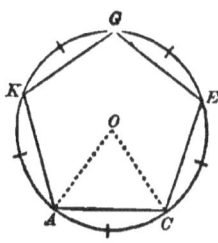

19. *Scholium.* By (III. 75), we have

$$AB = OA \cdot \frac{\sqrt{5}-1}{2},$$

by which the side of the regular decagon may be computed from the radius.

The angle at the centre of the regular decagon is $\frac{2}{5}$ of one right angle $= 36°$; the angle at the centre of the regular pentagon is $\frac{4}{5}$ of one right angle $= 72°$.

The angle ABC of the regular decagon is $\frac{8}{5}$ of one right angle $= 144°$; the angle ACE of the regular pentagon is $\frac{6}{5}$ of one right angle $= 108°$.

PROPOSITION VII.—PROBLEM.

20. *To inscribe a regular pentedecagon in a given circle.*

Suppose AB is the side of a regular inscribed pentedecagon, or that the arc AB is $\frac{1}{15}$ of the circumference.

Now the fraction $\frac{1}{15} = \frac{1}{6} - \frac{1}{10}$; therefore the arc AB is the difference between $\frac{1}{6}$ and $\frac{1}{10}$ of the circumference. Hence, if we inscribe the chord AC equal to the side of the regular inscribed hexagon, and then CB equal to that of the regular inscribed decagon, the chord AB will be the side of the regular inscribed pentedecagon required.

21. *Scholium.* Any regular inscribed polygon being given, a regular inscribed polygon of double the number of sides can be formed by bisecting the arcs subtended by its sides and drawing the chords of the semi-arcs (4). Also, any regular inscribed polygon being given, a regular circumscribed polygon of the same number of sides can be formed (3). Therefore, by means of the inscribed square, we can inscribe and circumscribe, successively, regular polygons of 8, 16, 32, etc., sides; by means of the hexagon, those of 12, 24, 48, etc., sides; by means of the decagon, those of 20, 40, 80, etc., sides; and,

finally, by means of the pentedecagon, those of 30, 60, 120, etc., sides.

Until the beginning of the present century, it was supposed that these were the only polygons that could be constructed by elementary geometry, that is, by the use of the straight line and circle only. GAUSS, however, in his *Disquisitiones Arithmeticæ, Lipsiæ*, 1801, proved that it is possible, by the use of the straight line and circle only, to construct regular polygons of 17 sides, of 257 sides, and in general of any number of sides which can be expressed by $2^n + 1$, n being an integer, provided that $2^n + 1$ is a prime number.

PROPOSITION VIII.—THEOREM.

22. *The area of a regular polygon is equal to half the product of its perimeter and apothem.*

For, straight lines drawn from the centre to the vertices of the polygon divide it into equal triangles whose bases are the sides of the polygon and whose common altitude is the apothem. The area of one of these triangles is equal to half the product of its base and altitude (IV. 13); therefore, the sum of their areas, or the area of the polygon, is half the product of the sum of the bases by the common altitude, that is, half the product of the perimeter and apothem.

PROPOSITION IX.—THEOREM.

23. *The area of a regular inscribed dodecagon is equal to three times the square of the radius.*

Let AB, BC, CD, DE, be four consecutive sides of a regular inscribed dodecagon, and draw the radii OA, OE; then, the figure $OABCDE$ is one-third of the dodecagon, and we have only to prove that the area of this figure is equal to the square of the radius.

Draw the radius OD; at A and D draw the tangents AF and GDF meeting in F; join AC and CE, and let AC and OE be produced to meet the tangent GF in H and G. The arc AD, containing three of the sides of the dodecagon, is one fourth of the circumference; therefore the angle AOD is a right angle, and OF is a square described on the radius.

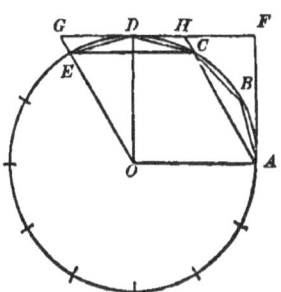

Since AC and CE are sides of the regular inscribed hexagon, each is equal to the radius; therefore $OACE$ is a parallelogram. Hence also $GOAH$ and $GECH$ are parallelograms.

The triangles DEC and BCA are equal (I. 80). The area of the triangle DEC is one-half that of the parallelogram EH (IV. 14); therefore the two triangles DEC and BCA are together equivalent to the parallelogram EH. Adding the parallelogram OC to these equals, we have the figure $OABCDE$ equivalent to the parallelogram OH. But the parallelogram OH is equivalent to the square OF (IV. 11); therefore the figure $OABCDE$, or one-third the dodecagon, is equivalent to the square OF, that is, to the square of the radius. Therefore, the area of the whole dodecagon is equal to three times the square of the radius.

24. *Scholium.* The area of the circumscribed square is evidently equal to four times the square of the radius. The area of the circle is greater than that of the inscribed regular dodecagon, and less than that of the circumscribed square; therefore, if the square of the radius is taken as the unit of surface, the area of a circle is greater than 3 and less than 4.

PROPOSITION X.—PROBLEM.

25. *Given the perimeters of a regular inscribed and a similar circumscribed polygon, to compute the perimeters of the regular inscribed and circumscribed polygons of double the number of sides.*

Let AB be a side of the given inscribed polygon, CD a side of the similar circumscribed polygon, tangent to the arc AB at its middle point E. Join AE, and at A and B draw the tangents AF and BG; then AE is a side of the regular inscribed polygon of double the number of sides, and FG is a side of the circumscribed polygon of double the number of sides (4).

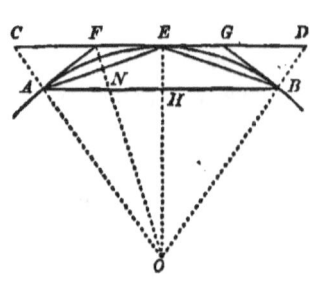

Denote the perimeters of the given inscribed and circumscribed polygons by p and P, respectively; and the perimeters of the required inscribed and circumscribed polygons of double the number of sides by p' and P', respectively.

Since OC is the radius of the circle circumscribed about the polygon whose perimeter is P, we have (10),

$$\frac{P}{p} = \frac{OC}{OA} \text{ or } \frac{OC}{OE};$$

and since OF bisects the angle COE, we have (III. 21),

$$\frac{OC}{OE} = \frac{CF}{FE};$$

therefore,

$$\frac{P}{p} = \frac{CF}{FE},$$

whence, by composition,

$$\frac{P+p}{2p} = \frac{CF+FE}{2FE} = \frac{CE}{FG}.$$

Now FG is a side of the polygon whose perimeter is P', and is contained as many times in P' as CE is contained in P, hence (III. 9),

$$\frac{CE}{FG} = \frac{P}{P'},$$

and therefore,

$$\frac{P+p}{2p} = \frac{P}{P'},$$

whence

$$P' = \frac{2pP}{P+p}. \qquad [1]$$

Again, the right triangles AEH and EFN are similar, since their

152 GEOMETRY.

acute angles EAH and FEN are equal, and give

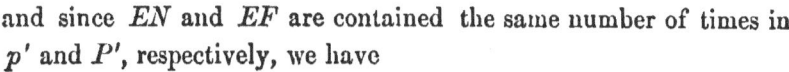

$$\frac{AH}{AE} = \frac{EN}{EF}.$$

Since AH and AE are contained the same number of times in p and p', respectively, we have

$$\frac{AH}{AE} = \frac{p}{p'},$$

and since EN and EF are contained the same number of times in p' and P', respectively, we have

$$\frac{EN}{EF} = \frac{p'}{P'};$$

therefore, we have

$$\frac{p}{p'} = \frac{p'}{P'},$$

whence

$$p' = \sqrt{p \times P'}. \qquad [2]$$

Therefore, from the given perimeters p and P, we compute P' by the equation [1], and then with p and P' we compute p' by the equation [2].

26. *Definition.* Two polygons are *isoperimetric* when their perimeters are equal.

PROPOSITION XI.—PROBLEM.

27. *Given the radius and apothem of a regular polygon, to compute the radius and apothem of the isoperimetric polygon of double the number of sides.*

Let AB be a side of the given regular polygon, O the centre of this polygon, OA its radius, OD its apothem. Produce DO to meet the circumference of the circumscribed circle in O'; join $O'A$, $O'B$; let fall OA' perpendicular to $O'A$, and through A' draw $A'B'$ parallel to AB.

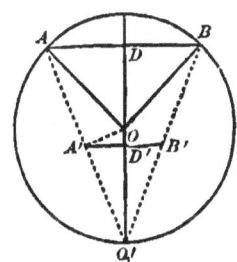

Since the new polygon is to have twice as many sides as the given polygon, the angle at its centre must be one-half the angle AOB;

therefore the angle $AO'B$, which is equal to one-half of AOB (II. 57), is equal to the angle at the centre of the new polygon.

Since the perimeter of the new polygon is to be equal to that of the given polygon, but is to be divided into twice as many sides, each of its sides must be equal to one-half of AB; therefore $A'B'$, which is equal to one-half of AB (I. 121), is a side of the new polygon; $O'A'$ is its radius, and $O'D'$ its apothem.

If, then, we denote the given radius OA by R, and the given apothem OD by r, the required radius $O'A'$ by R', and the apothem $O'D'$ by r', we have

$$O'D' = \frac{O'D}{2} = \frac{OO' + OD}{2},$$

or

$$r' = \frac{R + r}{2}. \qquad [1]$$

In the right triangle $OA'O'$, we have (III. 44),

$$\overline{O'A'}^2 = OO' \times O'D',$$

or

$$R' = \sqrt{R \times r'}; \qquad [2]$$

therefore, r' is an *arithmetic mean* between R and r, and R' is a *geometric mean* between R and r'.

MEASUREMENT OF THE CIRCLE.

The principle which we employed in the comparison of incommensurable ratios (II. 49) is fundamentally the same as that which we are about to apply to the measurement of the circle, but we shall now state it in a much more general form, better adapted for subsequent application.

28. *Definitions*. I. A *variable quantity*, or simply, a *variable*, is a quantity which has different successive values.

II. When the successive values of a variable, under the conditions imposed upon it, approach more and more nearly to the value of some fixed or constant quantity, so that the difference between the variable and the constant may become less than any assigned quantity, without becoming zero, the variable is said to *approach indefi-*

nitely to the constant; and the constant is called the *limit* of the variable.

Or, more briefly, the *limit* of a variable is a constant quantity to which the variable, under the conditions imposed upon it, approaches indefinitely.

As an example, illustrating these definitions, let a point be required to move from A to B under the following conditions: it shall first move over one-half of AB, that is to C; then over one-half of CB, to C'; then over one-half of $C'B$, to C''; and so on indefinitely; then the distance of the point from A is a *variable*, and this variable approaches indefinitely to the *constant AB*, as its *limit*, without ever reaching it.

As a second example, let A denote the angle of any regular polygon, and n the number of sides of the polygon; then, a right angle being taken as the unit, we have (8),

$$A = 2 - \frac{4}{n}.$$

The value of A is a variable depending upon n; and since n may be taken so great that $\frac{4}{n}$ shall be less than any assigned quantity however small, the value of A approaches to two right angles as its limit, but evidently never reaches that limit.

29. PRINCIPLE OF LIMITS. *Theorem. If two variable quantities are always equal to each other and each approaches to a limit, the two limits are necessarily equal.*

For, two variables always equal to each other present in fact but one value, and it is evidently impossible that one variable value shall at the same time approach indefinitely to two unequal limits.

30. *Theorem. The limit of the product of two variables is the product of their limits.* Thus, if x approaches indefinitely to the limit a, and y approaches indefinitely to the limit b, the product xy must approach indefinitely to the product ab; that is, the limit of the product xy is the product ab of the limits of x and y.

31. *Theorem. If two variables are in a constant ratio and each approaches to a limit, these limits are in the same constant ratio.*

Let x and y be two variables in the constant ratio m, that is, let

$x = my$; and let their limits be a and b respectively, Since y approaches indefinitely to b, my approaches indefinitely to mb; therefore we have x and my, two variables, always equal to each other, whose limits are a and mb, respectively, whence, by (29), $a = mb$; that is, a and b are in the constant ratio m.

PROPOSITION XII.—THEOREM.

32. *An arc of a circle is less than any line which envelops it and has the same extremities.*

Let AKB be an arc of a circle, AB its chord; and let ALB, AMB, etc., be any lines enveloping it and terminating at A and B.

Of all the lines AKB, ALB, AMB, etc. (each of which includes the segment, or *area*, AKB, between itself and the chord AB), there must be at least one *minimum*, or shortest line.* Now, no one of the lines ALB, AMB, etc., enveloping AKB, can be such a minimum; for, drawing a tangent CKD to the arc AKB, the line $ACKDB$ is less than $ACLDB$; therefore ALB is not the minimum; and in the same way it is shown that no other enveloping line can be the minimum. Therefore, the arc AKB is the minimum.

33. *Corollary.* The circumference of a circle is less than the perimeter of any polygon circumscribed about it.

34. *Scholium.* The demonstration is applicable when AKB is any convex curve whatever.

PROPOSITION XIII.—THEOREM.

35. *If the number of sides of a regular polygon inscribed in a circle be increased indefinitely, the apothem of the polygon will approach to the radius of the circle as its limit.*

---

* If we choose to admit the possibility of two or more equal shortest lines, still we say that of all the lines, AKB, ALB, etc., there must be one which is either *the* minimum line, or one of the minimum lines.

Let AB be a side of a regular polygon inscribed in the circle whose radius is OA; and let OD be its apothem.

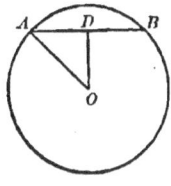

In the triangle OAD we have (I. 67),

$$OA - OD < AD.$$

Now, by increasing the number of sides of the polygon, the length of a side AB may evidently be made as small as we please, or less than any quantity that may be assigned. Hence AD, or $\frac{1}{2}AB$, and still more $OA - OD$, which is still less than AD, may become less than any assigned quantity; that is, the apothem OD approaches to the radius OA as its limit (28).

PROPOSITION XIV.—THEOREM.

36. *The circumference of a circle is the limit to which the perimeters of the inscribed and circumscribed polygons approach when the number of their sides is increased indefinitely; and the area of the circle is the limit of the areas of these polygons.*

Let AB and CD be sides of an inscribed and a similar circumscribed polygon; let r denote the apothem OE, R the radius OF, p the perimeter of the inscribed polygon, P the perimeter of the circumscribed polygon. Then, we have (10),

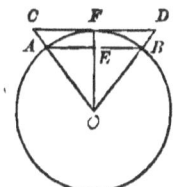

$$\frac{P}{p} = \frac{R}{r},$$

whence, by division (III. 10),

$$\frac{P-p}{P} = \frac{R-r}{R}, \text{ or } P - p = \frac{P}{R} \times (R - r).$$

Now we have seen, in the preceding proposition, that by increasing the number of sides of the polygons, the difference $R - r$ may be made less than any assigned quantity; consequently the quantity $\frac{P}{R} \times (R - r)$, or $P - p$, may also be made less than any assigned quantity. But P being always greater, and p always less, than the circumference of the circle, the difference between this circumference and either P or p is less than the difference $P - p$, and consequently may also be made less than any assigned quantity. Therefore, the circumference is the common limit of P and p.

Again, let s and S denote the areas of two similar inscribed and circumscribed polygons. The difference between the triangles COD and AOB is the trapezoid $CABD$, the measure of which is $\frac{1}{2}(CD + AB) \times EF$; therefore, the difference between the areas of the polygons is

$$S - s = \tfrac{1}{2}(P + p) \times (R - r);$$

consequently,

$$S - s < P \times (R - r).$$

Now by increasing the number of sides of the polygons, the quantity $P \times (R - r)$, and consequently also $S - s$, may be made less than any assigned quantity. But S being always greater, and s always less, than the area of the circle, the difference between the area of the circle and either S or s is less than the difference $S - s$, and consequently may also be made less than any assigned quantity. Therefore, the area of the circle is the common limit of S and s.

PROPOSITION XV.—THEOREM.

37. *The circumferences of two circles are to each other as their radii, and their areas are to each other as the squares of their radii.*

Let R and R' be the radii of the circles, C and C' their circumferences, S and S' their areas.

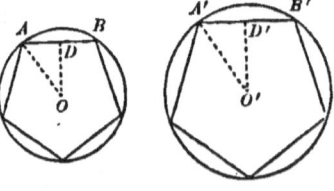

Inscribe in the two circles similar regular polygons; let P and P' denote the perimeters, A and A' the areas of these polygons; then, the polygons being similar, we have (10),

$$\frac{P}{P'} = \frac{R}{R'}, \quad \frac{A}{A'} = \frac{R^2}{R'^2}.$$

These relations remain the same whatever may be the number of sides in the polygons, provided there is the same number in each (9). When this number is indefinitely increased, P approaches C as its limit, and P' approaches C' as its limit (36); and since P and P' are in the constant ratio of R to R', their limits are in the same ratio (31); therefore

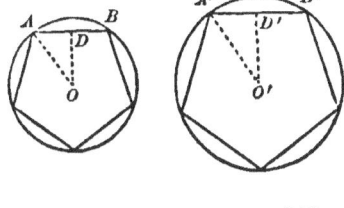

$$\frac{C}{C'} = \frac{R}{R'}. \qquad [1]$$

And since the limit of A is S, and the limit of A' is S', it follows in the same manner that

$$\frac{S}{S'} = \frac{R^2}{R'^2}. \qquad [2]$$

38. Corollary I. The proportion [1] is by (III. 9) the same as

$$\frac{C}{C'} = \frac{2R}{2R'}, \qquad [3]$$

and the proportion [2] is the same as

$$\frac{S}{S'} = \frac{4R^2}{4R'^2} = \frac{(2R)^2}{(2R')^2};$$

therefore, *the circumferences of circles are to each other as their diameters, and their areas are to each other as the squares of their diameters.*

39. Corollary II. Similar arcs, as AB, $A'B'$, are those which subtend equal angles at the centres of the circles to which they belong; they are therefore like parts of their respective circumferences,

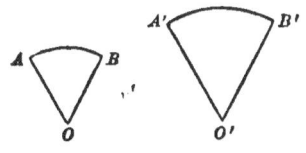

and are in the same ratio as the circumferences. Also the similar sectors AOB and $A'O'B'$ are like parts of the circles to which they belong. Therefore, *similar arcs are to each other as their radii, and similar sectors are to each other as the squares of their radii.*

40. Corollary III. *The ratio of the circumference of a circle to its diameter is constant;* that is, it is the same for all circles. For, from the proportion [3] we have

$$\frac{C}{2R} = \frac{C'}{2R'}.$$

This constant ratio is usually denoted by π, so that for any circle whose diameter is $2R$ and circumference C, we have

$$\frac{C}{2R} = \pi, \text{ or } C = 2\pi R.$$

41. *Scholium.* The ratio π is incommensurable (as can be proved by the higher mathematics), and can therefore be expressed in numbers only approximately. The letter π, however, is used to *symbolize* its *exact* value.

PROPOSITION XVI.—THEOREM.

42. *The area of a circle is equal to half the product of its circumference by its radius.*

Let the area of any regular polygon circumscribed about the circle be denoted by A, its perimeter by P, and its apothem which is equal to the radius of the circle by R; then (22),

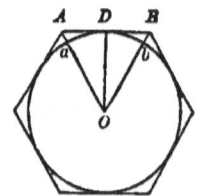

$$A = \tfrac{1}{2} P \times R, \text{ or } \frac{A}{P} = \tfrac{1}{2} R.$$

Let the number of the sides of the polygon be continually doubled, then A approaches the area S of the circle as its limit, and P approaches the circumference C as its limit; but A and P are in the constant ratio $\tfrac{1}{2} R$; therefore their limits are in the same ratio (31), and we have

$$\frac{S}{C} = \tfrac{1}{2} R, \text{ or } S = \tfrac{1}{2} C \times R. \qquad [1]$$

43. *Corollary* I. *The area of a circle is equal to the square of its radius multiplied by the constant number π.* For, substituting for C its value $2\pi R$ in [1], we have

$$S = \pi R^2.$$

44. *Corollary* II. *The area of a sector is equal to half the product of its arc by the radius.* For, denote the arc ab of the sector aOb by c, and the area of the sector by s; then, since c and s are like parts of C and S, we have (III. 9),

$$\frac{s}{S} = \frac{c}{C} = \frac{\tfrac{1}{2}c \times R}{\tfrac{1}{2}C \times R}.$$

But $S = \tfrac{1}{2} C \times R$; therefore $s = \tfrac{1}{2} c \times R$.

45. *Scholium.* A circle may be regarded as a regular polygon of an *infinite number of sides.* In proving that the circle is the *limit* towards which the inscribed regular polygon approaches when the number of its sides is increased indefinitely, it was tacitly assumed that the number of sides is always *finite.* It was shown that the difference between the polygon and the circle may be made less than any assigned quantity by making the number of sides sufficiently great; but an *assigned* difference being necessarily a finite quantity, there is also some finite number of sides sufficiently great to satisfy the imposed condition. Conversely, so long as the number of sides is finite, there is some finite difference between the polygon and the circle. But if we make the hypothesis that the number of sides of the inscribed regular polygon is *greater than any finite number,* that is, *infinite,* then it must follow that the difference between the polygon and the circle is *less than any finite quantity,* that is, *zero;* and consequently, the circle is identical with the inscribed polygon of an infinite number of sides.

This conclusion, it will be observed, is little else than an abridged statement of the theory of limits as applied to the circle; the abridgment being effected by the hypothetical introduction of *the infinite* into the statement.

PROPOSITION XVII.—PROBLEM.

46. *To compute the ratio of the circumference of a circle to its diameter, approximately.*

First Method, called the Method of Perimeters. In this method, we take the diameter of the circle as given and compute the perimeters of some inscribed and a similar circumscribed regular polygon. We then compute the perimeters of inscribed and circumscribed regular polygons of double the number of sides, by Proposition X. Taking the last-found perimeters as given, we compute the perimeters of polygons of double the number of sides by the same method; and so on. As the number of sides increases, the lengths of the perimeters approach to that of the circumference (36); hence, their successively computed values will be successive nearer and nearer approximations to the value of the circumference.

Taking, then, the diameter of the circle as given $= 1$, let us begin by inscribing and circumscribing a square. The perimeter of the inscribed square $= 4 \times \frac{1}{2} \times \sqrt{2} = 2\sqrt{2}$ (13); that of the circumscribed square $= 4$; therefore, putting

$$P = 4.$$
$$p = 2\sqrt{2} = 2.8284271,$$

we find, by Proposition X., for the perimeters of the circumscribed and inscribed regular octagons,

$$P' = \frac{2p \times P}{P + p} = 3.3137085,$$
$$p' = \sqrt{p \times P'} = 3.0614675.$$

Then taking these as given quantities, we put

$$P = 3.3137085, \ p = 3.0614675,$$

and find by the same formulæ for the polygons of 16 sides

$$P' = 3.1825979, \ p' = 3.1214452.$$

Continuing this process, the results will be found as in the following

TABLE.*

Number of sides.	Perimeter of circumscribed polygon.	Perimeter of inscribed polygon.
4	4.0000000	2.8284271
8	3.3137085	3.0614675
16	3.1825979	3.1214452
32	3.1517249	3.1365485
64	3.1441184	3.1403312
128	3.1422236	3.1412773
256	3.1417504	3.1415138
512	3.1416321	3.1415729
1024	3.1416025	3.1415877
2048	3.1415951	3.1415914
4096	3.1415933	3.1415923
8192	3.1415928	3.1415926

* The computations have been carried out with ten decimal places in order to ensure the accuracy of the seventh place as given in the table.

From the last two numbers of this table, we learn that the circumference of the circle whose diameter is unity is less than 3.1415928 and greater than 3.1415926; and since, when the diameter $= 1$, we have $C = \pi$, (40), it follows that

$$\pi = 3.1415927$$

within a unit of the seventh decimal place.

SECOND METHOD, called the METHOD OF ISOPERIMETERS. This method is based upon Proposition XI. Instead of taking the diameter as given and computing its circumference, we take the circumference as given and compute the diameter; or we take the semi-circumference as given and compute the radius.

Suppose we assume the semi-circumference $\frac{1}{2}C = 1$; then since $C = 2\pi R$, we have

$$\pi = \frac{\frac{1}{2}C}{R} = \frac{1}{R};$$

that is, the value of π is the reciprocal of the value of the radius of the circle whose semi-circumference is unity.

Let $ABCD$ be a square whose semi-perimeter $= 1$; then each of its sides $= \frac{1}{4}$. Denote its radius OA by R, and its apothem OE by r; then we have

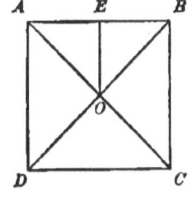

$$r = \tfrac{1}{4} \quad\quad = 0.2500000,$$
$$R = \tfrac{1}{4}\sqrt{2} = 0.3535534.$$

Now, by Proposition XI., we compute the apothem r' and the radius R' of the regular polygon of 8 sides having the same perimeter as this square; we find

$$r' = \frac{R + r}{2} = 0.3017767,$$
$$R' = \sqrt{R \times r'} = 0.3266407.$$

Again, taking these as given, we put

$$r = 0.3017767, \ R = 0.3266407,$$

and find by the same formulæ, for the apothem and radius of the isoperimetric regular polygon of 16 sides, the values

$$r' = 0.3142087, \ R' = 0.3203644.$$

Continuing this process, the results are found as in the following

TABLE.

Number of sides.	Apothem.	Radius.
4	0.2500000	0.3535534
8	0.3017767	0.3266407
16	0.3142087	0.3203644
32	0.3172866	0.3188218
64	0.3180541	0.3184376
128	0.3182460	0.3183418
256	0.3182939	0.3183179
512	0.3183059	0.3183119
1024	0.3183089	0.3183104
2048	0.3183096	0.3183100
4096	0.3183098	0.3183099
8192	0.3183099	0.3183099

Now, a circumference described with the radius r is inscribed in the polygon, and a circumference described with a radius R is circumscribed about the polygon; and the first circumference is less, while the second is greater, than the perimeter of the polygon. Therefore the circumference which is equal to the perimeter of the polygon has a radius greater than r and less than R; and this is true for each of the successive isoperimetric polygons. But the r and R of the polygon of 8192 sides do not differ by so much as .0000001; therefore, the radius of the circumference which is equal to the perimeter of the polygons, that is, to 2, is 0.3183099 within less than .0000001; and we have

$$\pi = \frac{1}{0.3183099} = 3.141593$$

within a unit of the sixth decimal place.

47. *Scholium* I. Observing that in this second method the value of $r = \frac{1}{4}$, for the square, is the arithmetic mean of 0 and $\frac{1}{2}$, and that $R = \frac{1}{4}\sqrt{2}$ is the geometric mean between $\frac{1}{2}$ and $\frac{1}{4}$, we arrive at the following proposition:

The value of $\frac{1}{\pi}$ is the limit approached by the successive numbers obtained by starting from the numbers 0 and $\frac{1}{2}$ and taking alternately the arithmetic mean and the geometric mean between the two which precede.

164 GEOMETRY.

48. *Scholium* II. ARCHIMEDES (born 287 B. C.) was the first to assign an approximate value of π. By a method similar to the above "first method," he proved that its value is between $3\frac{1}{7}$ and $3\frac{10}{71}$, or, in decimals, between 3.1428 and 3.1408; he therefore assigned its value correctly within a unit of the third decimal place. The number $3\frac{1}{7}$, or $\frac{22}{7}$, usually cited as Archimedes' value of π (although it is but one of the two limits assigned by him), is often used as a sufficient approximation in rough computations.

METIUS (A. D. 1640) found the much more accurate value $\frac{355}{113}$, which correctly represents even the sixth decimal place. It is easily remembered by observing that the denominator and numerator written consecutively, thus 113|355, present the first three odd numbers each written twice.

More recently, the value has been found to a very great number of decimals, by the aid of series demonstrated by the Differential Calculus. CLAUSEN and DASE of Germany (about A. D. 1846), computing independently of each other, carried out the value to 200 decimal places, and their results agreed to the last figure. The mutual verification thus obtained stamps their results as thus far the best established value to the 200th place. (See SCHUMACHER's *Astronomische Nachrichten*, No. 589.) Other computers have carried the value to over 500 places, but it does not appear that their results have been verified.

The value to fifteen decimal places is

$$\pi = 3.141592653589793.$$

For the greater number of practical applications, the value $\pi = 3.1416$ is sufficiently accurate.

MAXIMA AND MINIMA OF PLANE FIGURES.

49. *Definition.* Among quantities of the same kind, that which is greatest is called a *maximum*; that which is least, a *minimum*.

Thus, the diameter of a circle is a maximum among all straight lines joining two points of the circumference; the perpendicular is a minimum among all the lines drawn from a given point to a given straight line.

BOOK V. 165

An enclosed figure is said to be a maximum or a minimum, when its *area* is a maximum or a minimum.

50. *Definition.* Any two figures are called *isoperimetric* when their perimeters are equal.

PROPOSITION XVIII.—THEOREM.

51. *Of all triangles having the same base and equal areas, that which is isosceles has the minimum perimeter.*

Let ABC be an isosceles triangle, and $A'BC$ any other triangle having the same base and an equal area; then, $AB + AC < A'B + A'C$.

For, the altitudes of the triangles must be equal (IV. 15), and their vertices A and A' lie in the same straight line MN parallel to BC. Draw CND perpendicular to MN, meeting BA produced in D; join $A'D$. Since the angle $NAC = ACB = ABC = DAN$, the right triangles ACN, ADN, are equal; therefore, AN is perpendicular to CD at its middle point N, and we have $AD = AC$, $A'D = A'C$. But $BD < A'B + A'D$; that is, $AB + AC < A'B + A'C$.

52. *Corollary. Of all triangles having the same area, that which is equilateral has the minimum perimeter.* For, the triangle having the minimum perimeter enclosing a given area must be isosceles whichever side is taken as the base.

PROPOSITION XIX.—THEOREM.

53. *Of all triangles having the same base and equal perimeters, that which is isosceles is the maximum.*

Let ABC be an isosceles triangle, and let $A'BC$, standing on the same base BC, have an equal perimeter; that is, let $A'B + A'C = AB + AC$; then, the area of ABC is greater than the area of $A'BC$.

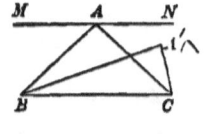

For, the vertex A' must fall between BC and the parallel MN drawn through A; since, if it fell upon MN, we should have, as in the preceding demonstration, $A'B + A'C > AB + AC$, and if it fell above MN, the sum $A'B + A'C$ would be still greater. Therefore the altitude of the triangle ABC is greater than that of $A'BC$, and hence also its area is the greater.

54. Corollary. *Of all isoperimetric triangles, that which is equilateral is the maximum.* For, the maximum triangle having a given perimeter must be isosceles whichever side is taken as the base.

PROPOSITION XX.—THEOREM.

55. *Of all triangles formed with the same two given sides, that in which these sides are perpendicular to each other is the maximum.*

Let ABC, $A'BC$, be two triangles having the sides AB, BC, respectively equal to $A'B$, BC; then, if the angle ABC is a right angle, the area of the triangle ABC is greater than that of the triangle $A'BC$.

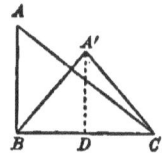

For, taking BC as the common base, the altitude AB of the triangle ABC is evidently greater than the altitude $A'D$ of the triangle $A'BC$.

PROPOSITION XXI.—THEOREM.

56. *Of all isoperimetric plane figures, the circle is the maximum.*

1st. With a given perimeter, there may be constructed an infinite number of figures of different forms and various areas. The area may be made as small as we please (IV. 35), but obviously cannot be increased indefinitely. Therefore, among all the figures of the same perimeter there must be one maximum figure, or several maximum figures of different forms and equal areas.

2d. Every closed figure of given perimeter containing a maximum area must necessarily be *convex*, that is, such that any straight line joining two points of the perimeter lies wholly within the figure.

Let $ACBNA$ be a non-convex figure, the straight line AB, joining two of the points in its perimeter, lying without the figure; then, if the re-entrant portion ACB be revolved about the line AB into the position $AC'B$, the figure $AC'BNA$ has the same perimeter as the first

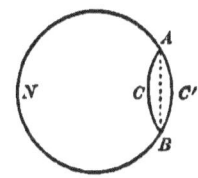

figure, but a greater area. Therefore, the non-convex figure cannot be a maximum among figures of equal perimeters.

3d. Now let $ACBFA$ be a maximum figure formed with a given perimeter; then we say that, taking any point A in its perimeter and drawing AB so as to divide the perimeter into two equal parts, this line also divides the area of the figure into two equal parts. For, if the area of one of the parts, as AFB, were greater than that of the other part, ACB, then, if the part AFB were revolved upon the line AB into the position $AF''B$, the area of the figure $AF''BFA$ would be greater than that of the figure $ACBFA$, and yet would have the same perimeter; thus the figure $ACBFA$ would not be a maximum.

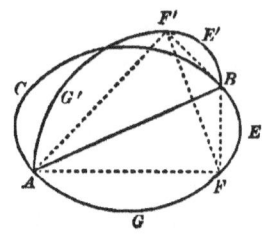

Hence also it appears that, $ACBFA$ being a maximum figure, $AF''BFA$ is also one of the maximum figures, for it has the same perimeter and area as the former figure. This latter figure is symmetrical with respect to the line AB, since by the nature of the revolution about AB, every line FF'' perpendicular to AB, and terminated by the perimeter, is bisected by AB (I. 140). Hence F and F'' being any two symmetrical points in the perimeter of this figure, the triangles AFB and $AF''B$ are equal.

Now the angles AFB and $AF''B$ must be right angles; for if they were not right angles the areas of the triangles AFB and $AF''B$ could be increased without varying the lengths of the chords AF, FB, AF'', $F''B$ (55), and then (the segments AGF, FEB, $AG'F'$, $F'E'B$, still standing on these chords), the whole figure would have its area increased without changing the length of its perimeter; consequently the figure $AF''BFA$ would not be a maximum. Therefore, the angles F and F' are right angles. But F is any point in the curve AFB; therefore, this curve is a semi-circumference (II. 59, 97).

Hence, if a figure $ACBFA$ of a given perimeter is a maximum, its half AFB, formed by drawing AB from any arbitrarily chosen point A in the perimeter, is a semicircle. Therefore the whole figure is a circle.*

---

* This demonstration is due to STEINER, *Crelle's Journal für die reine und angewandte Mathematik*, vol. 24. (Berlin, 1842.)

PROPOSITION XXII.—THEOREM.

57. *Of all plane figures containing the same area, the circle has the minimum perimeter.*

Let C be a circle, and A any other figure having the same area as C; then, the perimeter of C is less than that of A.

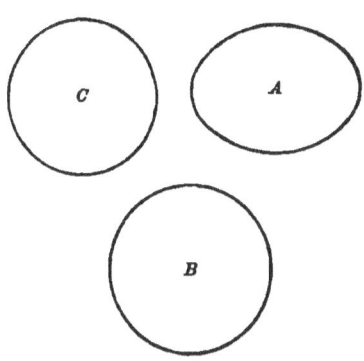

For, let B be a circle having the same perimeter as the figure A; then, by the preceding theorems $A < B$, or $C < B$. Now, of two circles, that which has the less area has the less perimeter; therefore, the perimeter of C is less than that of B, or less than that of A.

PROPOSITION XXIII.—THEOREM.

58. *Of all the polygons constructed with the same given sides, that is the maximum which can be inscribed in a circle.*

Let P be a polygon constructed with the sides a, b, c, d, e, and inscribed in a circle S, and let P' be any other polygon constructed with the same sides and not inscriptible in a circle; then, $P > P'$.

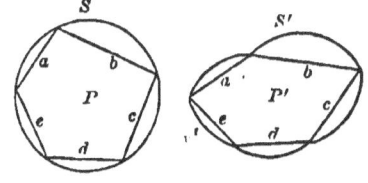

For, upon the sides a, b, c, etc., of the polygon P' construct circular segments equal to those standing on the corresponding sides of P. The whole figure S' thus formed has the same perimeter as the circle S; therefore, area of $S >$ area of S' (56); subtracting the circular segments from both, we have $P > P'$.

PROPOSITION XXIV.—PROBLEM.

59. *Of all isoperimetric polygons having the same number of sides, the regular polygon is the maximum.*

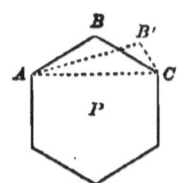

1st. The maximum polygon P, of all the isoperimetric polygons of the same number of sides must have its sides equal; for if two of its sides, as AB', $B'C$, were unequal, we could, by (53), substitute for the triangle $AB'C$ the isosceles triangle ABC having the same perimeter as $AB'C$ and a greater area, and thus the area of the whole polygon could be increased without changing the length of its perimeter or the number of its sides.

2d. The maximum polygon constructed with the same number of equal sides must, by (58), be inscriptible in a circle; therefore it must be a regular polygon.

PROPOSITION XXV.—THEOREM.

60. *Of all polygons having the same number of sides and the same area, the regular polygon has the minimum perimeter.*

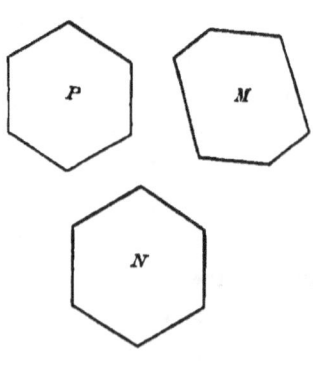

Let P be a regular polygon, and M any irregular polygon having the same number of sides and the same area as P; then, the perimeter of P is less than that of M.

For, let N be a regular polygon having the same perimeter and the same number of sides as M; then, by (59), $M < N$, or $P < N$. But of two regular polygons having the same number of sides, that which has the less area has the less perimeter; therefore the perimeter of P is less than that of N, or less than that of M.

PROPOSITION XXVI.—THEOREM.

61. *If a regular polygon be constructed with a given perimeter, its area will be the greater, the greater the number of its sides.*

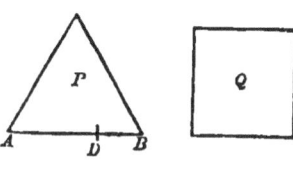

Let P be the regular polygon of three sides, and Q the regular polygon of four sides, constructed with the same given perimeter. In any side AB of P take any arbitrary point D; the polygon P may be regarded as an irregular polygon of four sides, in which the sides AD, DB, make an angle with each other equal to two right angles (I. 16); then, the irregular polygon P of four sides is less than the regular isoperimetric polygon Q of four sides (59). In the same manner it follows that Q is less than the regular isoperimetric polygon of five sides, and so on.

PROPOSITION XXVII.—THEOREM.

62. *If a regular polygon be constructed with a given area, its perimeter will be the less, the greater the number of its sides.*

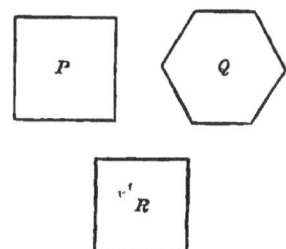

Let P and Q be regular polygons having the same area, and let Q have the greater number of sides; then, the perimeter of P will be greater than that of Q.

For, let R be a regular polygon having the same perimeter as Q and the same number of sides as P; then, by (61), $Q > R$, or $P > R$; therefore the perimeter of P is greater than that of R, or greater than that of Q.

GEOMETRY OF SPACE.

BOOK VI.

THE PLANE. POLYEDRAL ANGLES.

1. DEFINITION. A *plane* has already been defined as a surface such that the straight line joining any two points in it lies wholly in the surface.

Thus, the surface MN is a plane, if, A and B being *any two* points in it, the straight line AB lies wholly in the surface.

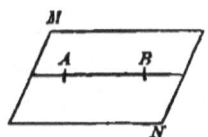

The plane is understood to be indefinite in extent, so that, however far the straight line is produced, all its points lie in the plane. But to represent a plane in a diagram, we are obliged to take a limited portion of it, and we usually represent it by a parallelogram supposed to lie in the plane.

DETERMINATION OF A PLANE.

PROPOSITION I.—THEOREM.

2. *Through any given straight line an infinite number of planes may be passed.*

Let AB be a given straight line. A straight line may be drawn in any plane, and the position of that plane may be changed until the line drawn in it is brought into coincidence with AB. We shall then have one plane

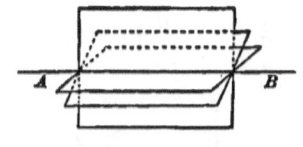

passed through AB; and this plane may be turned upon AB as an axis and made to occupy an infinite number of positions.

3. *Scholium.* Hence, a plane subjected to the single condition that it shall pass through a given straight line, is not fixed, or determinate, in position. But it will become determinate if it is required to pass through an additional point, or line, as shown in the next proposition.

A plane is said to be *determined* by given lines, or points, when it is the only plane which contains such lines or points.

·PROPOSITION II.—THEOREM.

4. *A plane is determined, 1st, by a straight line and a point without that line; 2d, by two intersecting straight lines; 3d, by three points not in the same straight line; 4th, by two parallel straight lines.*

1st. A plane MN being passed through a given straight line AB, and then turned upon this line as an axis until it contains a given point C not in the line AB, is evidently determined; for, if it is then turned in either direction about AB, it will cease to contain the point C. The plane is therefore determined by the given straight line and the point without it.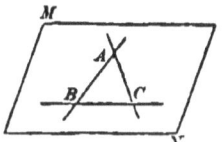

2d. If two intersecting straight lines AB, AC, are given, a plane passed through AB and any point C (other than the point A) of AC, contains the two straight lines, and is determined by these lines.

3d. If three points are given, A, B, C, not in the same straight line, any two of them may be joined by a straight line, and then the plane passed through this line and the third point, contains the three points, and is thus determined by them.

4th. Two parallel lines, AB, CD, are by definition (I. 42) necessarily in the same plane, and there is but one plane containing them, since a plane passed through one of them, AB, and any point E of the other, is determined in position.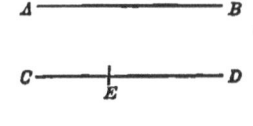

5. *Corollary.* The intersection of two planes is a straight line. For, the intersection cannot contain three points not in the same straight line, since only one plane can contain three such points.

BOOK VI.

PERPENDICULARS AND OBLIQUE LINES TO PLANES.

6. *Definition.* A straight line is *perpendicular to a plane* when it is perpendicular to every straight line drawn in the plane through its *foot*, that is, through the point in which it meets the plane.

In the same case, the plane is said to be perpendicular to the line.

PROPOSITION III.—THEOREM.

7. *From a given point without a plane, one perpendicular to the plane can be drawn, and but one.*

Let A be the given point, and MN the plane.

If any straight line, as AB, is drawn from A to a point B of the plane, and the point B is then supposed to move in the plane, the length of AB will vary. Thus, if B move along a straight line BB' in the plane, the distance AB will vary according to the distance of B from 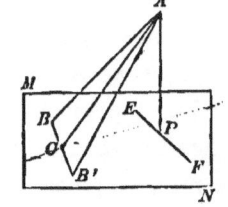 the foot C of the perpendicular AC let fall from A upon BB'. Now, of all the lines drawn from A to points in the plane, there must be one *minimum*, or shortest line. There cannot be two equal shortest lines; for if AB and AB' are two equal straight lines from A to the plane, each is greater than the perpendicular AC let fall from A upon BB'; hence they are not minimum lines. There is therefore one, and but one, minimum line from A to the plane. Let AP be that minimum line; then, AP is perpendicular to any straight line EF drawn in the plane through its foot P. For, in the plane of the lines AP and EF, AP is the shortest line that can be drawn from A to any point in EF, since it is the shortest line that can be drawn from A to any point in the plane MN; therefore, AP is perpendicular to EF (I. 28). Thus AP is perpendicular to *any*, that is, to *every*, straight line drawn in the plane through its foot, and is therefore perpendicular to the plane. Moreover, by the nature of the proof just given, AP is the only perpendicular that can be drawn from A to the plane MN.

8. *Corollary.* At a given point P in a plane MN, a perpendicular can be erected to the plane, and but one.

For, let $M'N'$ be any other plane, A' any point without it, and $A'P'$ the perpendicular from A' to this plane. Suppose the plane $M'N'$ to be applied to the plane MN with the point P' upon P, and let AP be the position then occupied by the perpendicular $A'P'$. We then have one perpendicular, AP, to the plane

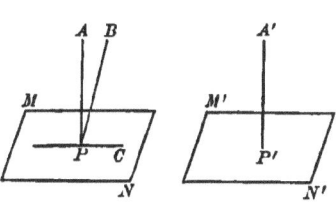

MN, erected at P. There can be no other: for let PB be any other straight line drawn through P; let the plane determined by the two lines PA, PB, intersect the plane MN in the line PC; then, since APC is a right angle, BPC is not a right angle, and therefore BP is not perpendicular to the plane.

9. *Scholium.* By the *distance of a point from a plane* is meant the *shortest distance;* hence it is the perpendicular distance from the point to the plane.

PROPOSITION IV.—THEOREM.

10. *Oblique lines drawn from a point to a plane, at equal distances from the perpendicular, are equal; and of two oblique lines unequally distant from the perpendicular the more remote is the greater.*

1st. Let AB, AC be oblique lines from the point A to the plane MN, meeting the plane at the equal distances PB, PC, from the foot of the perpendicular AP; then, $AB = AC$. For, the right triangles APB, APC, are equal (I. 76).

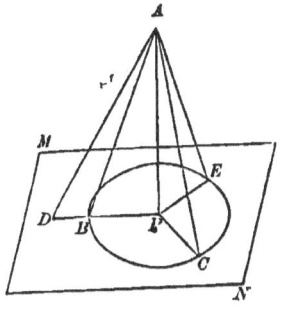

2d. Let AD meet the plane at a distance, PD, from P, greater than PC; then, $AD > AC$. For, upon PD take $PB = PC$, and join AB: then $AD > AB$ (I. 35); but $AB = AC$; therefore, $AD > AC$.

11. *Corollary* I. Conversely, equal oblique lines from a point to a plane meet the plane at equal distances from the perpendicular; and of two unequal oblique lines, the greater meets the plane at the greater distance from the perpendicular.

BOOK VI. 175

12. *Corollary* II. Equal straight lines from a point to a plane meet the plane in the circumference of a circle whose centre is the foot of the perpendicular from the point to the plane. Hence we derive a method of drawing a perpendicular from a given point A to a given plane MN: find any three points, B, C, E, in the plane, equidistant from A, and find the centre P of the circle passing through these points; the straight line AP will be the required perpendicular.

PROPOSITION V.—THEOREM.

13. *If a straight line is perpendicular to each of two straight lines at their point of intersection, it is perpendicular to the plane of those lines.*

Let AP be perpendicular to PB and PC, at their intersection P; then, AP is perpendicular to the plane MN which contains those lines.

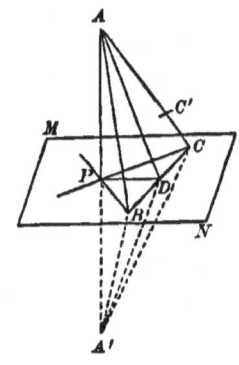

For, let PD be any other straight line drawn through P in the plane MN. Draw any straight line BDC intersecting PB, PC, PD, in B, C, D; produce AP to A' making $PA' = PA$, and join A and A' to each of the points B, C, D.

Since BP is perpendicular to AA', at its middle point, we have $BA = BA'$, and for a like reason $CA = CA'$; therefore, the triangles ABC, $A'BC$, are equal (I. 80). If, then, the triangle ABC is turned about its base BC until its plane coincides with that of the triangle $A'BC$, the vertex A will fall upon A'; and as the point D remains fixed, the line AD will coincide with $A'D$; therefore, D and P are each equally distant from the extremities of AA', and DP is perpendicular to AA' or AP (I. 41). Hence AP is perpendicular to *any* line PD, that is, to *every* line, passing through its foot in the plane MN, and is consequently perpendicular to the plane.

14. *Corollary* I. At a given point P of a straight line AP, a plane can be passed perpendicular to that line, and but one. For, two perpendiculars, PB, PC, being drawn to AP in any two different planes APB, APC, passed through AP, the plane of the lines PB, PC, will

176   GEOMETRY.

be perpendicular to the line AP. Moreover, no other plane passed through P can be perpendicular to AP; for, any other plane not containing the point C would cut the oblique line AC in a point C' different from C, and we should have the angle APC' different from APC, and therefore not a right angle.

15. *Corollary* II. All the perpendiculars PB, PC, PD, etc., drawn to a line AP at the same point, lie in one plane perpendicular to AP. Hence, if an indefinite straight line PQ, perpendicular to AP, be made to revolve, always remaining perpendicular to AP, it is said to *generate* the plane MN perpendicular to AP; for the line PQ passes successively, during its revolution, through every point of this plane.

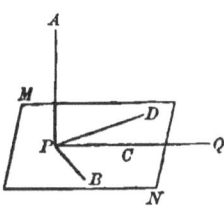

16. *Corollary* III. Through any point C without a given straight line AP, a plane can be passed perpendicular to AP, and but one. For, in the plane determined by the line AP and the point C, the perpendicular CP can be drawn to AP, and then the plane generated by the revolution of PC about AP as an axis will, by the preceding corollary, be perpendicular to AP; and it is evident that there can be but one such perpendicular plane.

PROPOSITION VI.—THEOREM.

17. *If from the foot of a perpendicular to a plane a straight line is drawn at right angles to any line of the plane, and its intersection with that line is joined to any point of the perpendicular, this last line will be perpendicular to the line of the plane.*

Let AP be perpendicular to the plane MN; from its foot P let PD be drawn at right angles to any line BC of the plane; then, A being any point in AP, the straight line AD is perpendicular to BC.

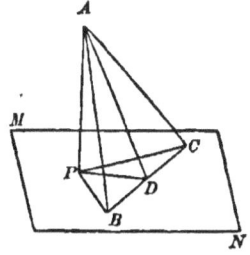

For, lay off $DB = DC$, and join PB, PC, AB, AC. Since $DB = DC$, we have $PB = PC$ (I. 30), and hence $AB = AC$ (10). Therefore, A and D being each equally distant from B and C, the line AD is perpendicular to BC (I. 41).

PARALLEL STRAIGHT LINES AND PLANES.

18. *Definitions.* A straight line is *parallel to a plane* when it cannot meet the plane though both be indefinitely produced.

In the same case, the plane is said to be parallel to the line.

Two planes are parallel when they do not meet, both being indefinite in extent.

PROPOSITION VII.—THEOREM.

19. *If two straight lines are parallel, every plane passed through one of them is parallel to the other.*

Let AB and CD be parallel lines, and MN any plane passed through CD; then, the line AB and the plane MN are parallel.

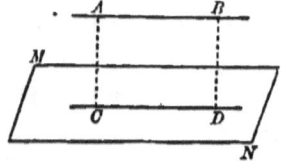

For, the parallels AB, CD, are in the same plane, $ACDB$, which intersects the plane MN in the line CD; and if AB could meet the plane MN, it could meet it only in some point of CD; but AB cannot meet CD, since it is parallel to it; therefore AB cannot meet the plane MN.

20. *Corollary* I. *Through any given straight line HK, a plane can be passed parallel to any other given straight line AB.*

For, in the plane determined by AB and any point H of HK, let HL be drawn parallel to AB; then, the plane MN, determined by HK and HL, is parallel to AB.

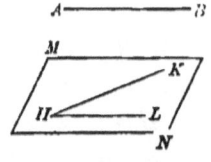

21. *Corollary* II. *Through any given point O, a plane can be passed parallel to any two given straight lines AB, CD, in space.*

For, in the plane determined by the given point O and the line AB let aOb be drawn through O parallel to AB; and in the plane determined by the point O and the line CD, let cOd be drawn through O parallel to CD; then, the plane determined by the lines ab and cd is parallel to each of the lines AB and CD.

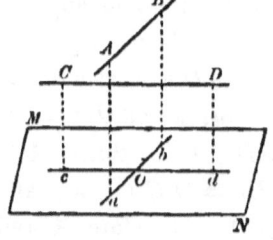

PROPOSITION VIII.—THEOREM.

22. *If a straight line and a plane are parallel, the intersections of the plane with planes passed through the line are parallel to that line and to each other.*

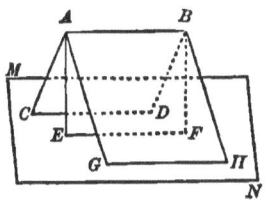

Let the line AB be parallel to the plane MN, and let CD, EF, etc., be the intersections of MN with planes passed through AB; then, these intersections are parallel to AB and to each other.

For, the line AB cannot meet CD, since it cannot meet the plane in which CD lies; and since these lines are in the same plane, AD, and cannot meet, they are parallel. For the same reason, EF, GH, are parallel to AB.

Moreover, no two of these intersections, as CD, EF, can meet; for if they met, their point of meeting and the line AB would be at once in two different planes, AD and AF, which is impossible (4).

23. *Corollary. If a straight line AB is parallel to a plane MN, a parallel CD to the line AB, drawn through any point C of the plane, lies in the plane.*

For, the plane passed through the line AB and the point C intersects the plane MN in a parallel to AB, which must coincide with CD, since there cannot be two parallels to AB drawn through the same point C.

PROPOSITION IX.—THEOREM.

24. *Planes perpendicular to the same straight line are parallel to each other.*

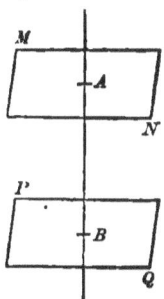

The planes MN, PQ, perpendicular to the same straight line AB, cannot meet; for, if they met, we should have through a point of their intersection two planes perpendicular to the same straight line, which is impossible (16); therefore these planes are parallel.

PROPOSITION X.—THEOREM.

25. *The intersections of two parallel planes with any third plane are parallel.*

Let MN and PQ be parallel planes, and AD any plane intersecting them in the lines AB and CD; then, AB and CD are parallel.

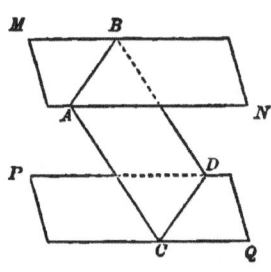

For, the lines AB and CD cannot meet, since the planes in which they are situated cannot meet, and they are lines in the same plane AD; therefore they are parallel.

26. *Corollary.* Parallel lines AC, BD, intercepted between parallel planes MN, PQ, are equal. For, the plane of the parallels AC, BD, intersects the parallel planes MN, PQ, in the parallel lines AB, CD; therefore, the figure $ABDC$ is a parallelogram, and $AC = BD$.

PROPOSITION XI.—THEOREM.

27. *A straight line perpendicular to one of two parallel planes is perpendicular to the other.*

Let MN and PQ be parallel planes, and let the straight line AB be perpendicular to PQ; then, it will also be perpendicular to MN.

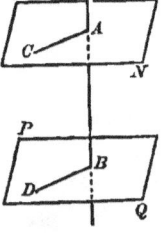

For, through A draw any straight line AC in the plane MN, pass a plane through AB and AC, and let BD be the intersection of this plane with PQ. Then AC and BD are parallel (25); but AB is perpendicular to BD (6), and consequently also to AC; therefore AB, being perpendicular to any line AC of the plane MN, is perpendicular to the plane MN.

28. *Corollary. Through any given point A, one plane can be passed parallel to a given plane PQ, and but one.* For, from A a perpendicular AB can be drawn to the plane PQ (7), and then through A

a plane MN can be passed perpendicular to AB (14); the plane MN is parallel to PQ (24).

No other plane can be passed through A parallel to PQ; for every plane parallel to PQ must be perpendicular to the line AB (27), and there can be but one plane perpendicular to AB passed through the same point A (14).

PROPOSITION XII.—THEOREM.

29. *The locus of all the straight lines drawn through a given point parallel to a given plane, is a plane passed through the point parallel to the given plane.*

Let A be the given point, and PQ the given plane; then, every straight line AB, drawn through A parallel to the plane PQ, lies in the plane MN passed through A parallel to PQ.

For, pass any plane through AB, intersecting the plane PQ in a straight line CD; then AB is parallel to CD (22). But CD is parallel to the plane MN, since it is in the parallel plane PQ and cannot meet MN; therefore, the line AB drawn through the point A parallel to CD lies in the plane MN (23).

30. *Scholium.* In the geometry of space, the term *locus* has the same general signification as in plane geometry (I. 40); only it is not limited to *lines*, but may, as in this proposition, be extended to a surface. In the present case, the locus is the *assemblage of all the points of all the lines* which satisfy the two conditions of passing through a given point and being parallel to a given plane.

31. *Corollary.* Since two straight lines are sufficient to determine a plane (4), if two intersecting straight lines are each parallel to a given plane, the plane of these lines is parallel to the given plane.

PROPOSITION XIII.—THEOREM.

32. *If two angles, not in the same plane, have their sides respectively parallel and lying in the same direction, they are equal and their planes are parallel.*

Let BAC, $B'A'C'$, be two angles lying in the planes MN, $M'N'$; and let AB, AC, be parallel respectively to $A'B'$, AC', and in the same directions.*

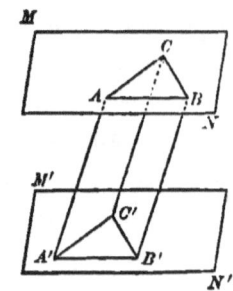

1st. The angles BAC and $B'A'C'$ are equal. For, through the parallels AB, $A'B'$, pass a plane AB', and through the parallels AC $A'C'$, pass a plane AC', intersecting the first in the line AA'. Let BC' be any plane parallel to AA', intersecting the planes AB', AC', in the lines BB', CC', and the planes MN, $M'N'$, in the lines BC, $B'C'$, respectively. Since AA' is parallel to the plane BC', the intersections BB', CC', are parallel to AA' and to each other (22); hence, the quadrilaterals AB' and AC' are parallelograms, and we have $AB = A'B'$, $AC = A'C'$, and $BB' = AA' = CC'$. Therefore, BB' and CC' are equal and parallel, and the quadrilateral BC' is a parallelogram, and we have $BC = B'C'$. The triangles ABC, $A'B'C'$, therefore, have their three sides equal each to each, and consequently the angles BAC and $B'A'C'$ are equal.

2d. The planes of these angles are parallel. For, each of the lines AB, AC, being parallel to a line of the plane $M'N'$, is parallel to that plane, therefore the plane MN of these lines is parallel to the plane $M'N'$ (31).

PROPOSITION XIV.—THEOREM.

33. *If one of two parallel lines is perpendicular to a plane, the other is also perpendicular to that plane.*

Let AB, $A'B'$, be parallel lines, and let AB be perpendicular to the plane MN; then, $A'B'$ is also perpendicular to MN.

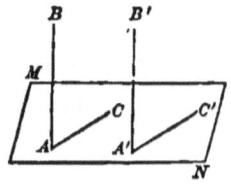

For, let A and A' be the intersections of these lines with the plane; through A' draw any line $A'C'$ in the plane MN, and through A draw AC parallel to $A'C'$ and in the same direction. The angles

---

* Two parallels AB, $A'B'$, lie in the same direction when they lie on the same side of the line AA' joining their *origins* A and A'. Compare note (I. 60).

BAC, $B'A'C'$, are equal (32); but BAC is a right angle, since BA is perpendicular to the plane; hence, $B'A'C'$ is a right angle; that is, $B'A'$ is perpendicular to any line $A'C'$ drawn through its foot in the plane MN, and is consequently perpendicular to the plane.

34. Corollary I. *Two straight lines AB, $A'B'$, perpendicular to the same plane MN, are parallel to each other.* For, if through any point of $A'B'$ a parallel to AB is drawn, it will be perpendicular to the plane MN, since AB is perpendicular to that plane; but through the same point there cannot be two perpendiculars to the plane; therefore, the parallel drawn to AB coincides with $A'B'$.

35. Corollary II. *If two straight lines A and B are parallel to a third C, they are parallel to each other.* For, let MN be a plane perpendicular to C; then (33), A and B are each perpendicular to this plane and are parallel to each other (34).

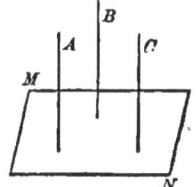

36. Corollary III. *Two parallel planes are everywhere equally distant.* All perpendiculars to one of two parallel planes are also perpendicular to the other (27); and since they are parallels (34) intercepted between parallel planes, they are equal (26).

PROPOSITION XV.—THEOREM.

37. *If two straight lines are intersected by three parallel planes, their corresponding segments are proportional.*

Let AB, CD, be intersected by the parallel planes MN, PQ, RS, in the points A, E, B, and C, F, D; then,

$$\frac{AE}{EB} = \frac{CF}{FD}.$$

For, draw AD cutting the plane PQ in G, and join EG and FG. The plane of the lines AB, AD, cuts the parallel planes PQ and RS in the lines EG and BD; therefore, EG and BD are parallel (25), and we have (III. 15),

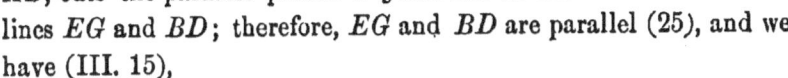

$$\frac{AE}{EB} = \frac{AG}{GD}.$$

The plane of the lines DA and DC cuts the parallel planes MN and PQ in the lines AC and GF; therefore, AC and GF are parallel, and we have

$$\frac{AG}{GD} = \frac{CF}{FD}.$$

Comparing these two proportions, we obtain

$$\frac{AE}{EB} = \frac{CF}{FD}.$$

DIEDRAL ANGLES.—ANGLE OF A LINE AND PLANE, ETC.

38. *Definition.* When two planes meet and are terminated by their common intersection, they form a *diedral angle.*

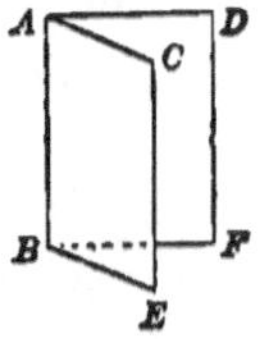

Thus, the planes AE, AF, meeting in AB, and terminated by AB, form a diedral angle.

The planes AE, AF, are called the *faces*, and the line AB the *edge*, of the diedral angle.

A diedral angle may be named by four letters, one in each face and two on its edge, the two on the edge being written between the other two; thus, the angle in the figure may be named $DABC$.

When there is but one diedral angle formed at the same edge, it may be named by two letters on its edge; thus, in the preceding figure, the diedral angle $DABC$ may be named the diedral angle AB.

39. *Definition.* The angle CAD formed by two straight lines AC, AD, drawn, one in each face of the diedral angle, perpendicular to its edge AB at the same point, is called the *plane angle* of the diedral angle.

The plane angle thus formed is the same at whatever point of the edge of the diedral angle it is constructed. Thus, if at B, we draw BE and BF in the two faces respectively, and perpendicular to AB, the angle EBF is equal to the angle CAD, since the sides of these angles are parallel each to each (32).

It is to be observed that the plane of the plane angle CAD is perpendicular to the edge AB (13); and conversely, a plane perpendicular to the edge of a diedral angle cuts its faces in lines which are perpendicular to the edge and therefore form the plane angle of the diedral angle.

40. A diedral angle $DABC$ may be conceived to be *generated* by a plane, at first coincident with a fixed plane AE, revolving upon the line AB as an axis until it comes into the position AF. In this revolution, a straight line CA, perpendicular to AB, generates the plane angle CAD.

41. *Definition.* Two diedral angles are equal when they can be placed so that their faces shall coincide.

Thus, the diedral angles $CABD$, $C'A'B'D'$, are equal, if, when the edge $A'B'$ is applied to the edge AB and the face $A'F'$ to the face AF, the face $A'E'$ also coincides with the face AE.

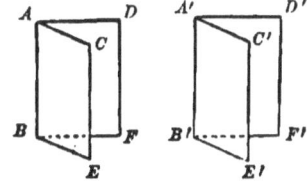

Since the faces continue to coincide when produced indefinitely, it is apparent that the *magnitude* of the diedral angle does not depend upon the extent of its faces, but only upon their relative position.

Two diedral angles are evidently equal when their plane angles are equal.

42. *Definition.* Two diedral angles $CABD$, $DABE$, which have a common edge AB and a common plane BD between them, are called *adjacent*.

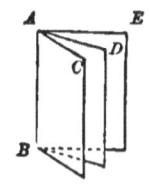

Two diedral angles are added together by placing them adjacent to each other. Thus, the diedral angle $CABE$ is the sum of the two diedral angles $CABD$ and $DABE$.

43. *Definition.* When a plane CAB meets another MN, forming two equal adjacent diedral angles, $CABM$ and $CABN$, each of these angles is called a *right diedral angle*, and the plane CAB is *perpendicular* to the plane MN.

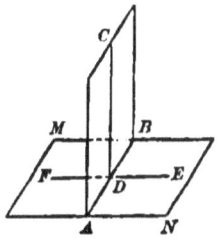

It is evident that in this case the plane angles CDE, CDF, of the two equal diedral angles, are right angles.

Through any straight line AB in a plane MN, a plane CAB can be passed perpendicular to the plane MN. The proof is similar to that of the corresponding proposition in plane geometry (I. 9).

PROPOSITION XVI.—THEOREM.

44. *Two diedral angles are in the same ratio as their plane angles.*

Let $CABD$ and $GEFH$ be two diedral angles; and let CAD and GEH be their plane angles.

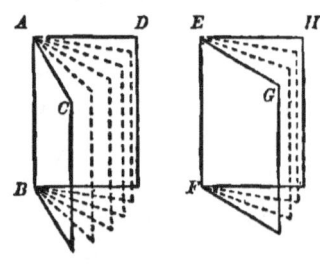

Suppose the plane angles have a common measure, contained, for example, 5 times in CAD and 3 times in GEH; the ratio of these angles is then $5:3$. Let straight lines be drawn from the vertices of these angles, dividing the angle DAC into 5 equal parts, and the angle HEG into 3 equal parts, each equal to the common measure; let planes be passed through the edge AB and the several lines of division of the plane angle CAD, and also planes through the edge EF and the several lines of division of the plane angle GEH. The given diedral angles are thus divided into partial diedral angles which are all equal to each other since their plane angles are equal. The diedral angle $CABD$ contains 5 of these partial angles, and the diedral angle $GEFH$ contains 3 of them; therefore, the given diedral angles are also in the ratio $5:3$; that is, they are in the same ratio as their plane angles.

The proof is extended to the case in which the given plane angles are incommensurable, by the method exemplified in (II. 51).

45. Corollary I. Since the diedral angle is proportional to its plane angle (that is, varies proportionally with it), the plane angle is taken as the *measure* of the diedral angle, just as an arc is taken as the measure of a plane angle. Thus, a diedral angle will be expressed by $45°$ if its plane angle is expressed by $45°$, etc.

46. Corollary II. *The sum of two adjacent diedral angles, formed by one plane meeting another, is equal to two right diedral angles.* For, the sum of the plane angles which measure them is equal to two right angles.

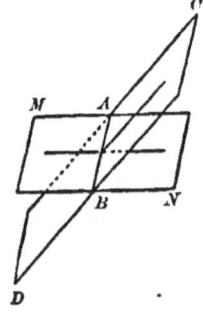

In a similar manner, a number of properties of diedral angles can be proved, which are analogous to propositions relating to plane angles. The student can establish the following:

Opposite or vertical diedral angles are equal; as *CABN* and *DABM*, in the preceding figure.

When a plane intersects two parallel planes, the alternate diedral angles are equal, and the corresponding diedral angles are equal; (the terms *alternate* and *corresponding* having significations similar to those given in plane geometry.)

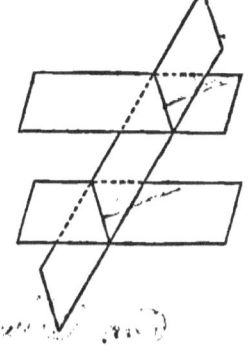

Two diedral angles which have their faces respectively parallel, or respectively perpendicular to each other, are either equal or supplementary.

PROPOSITION XVII.—THEOREM.

47. *If a straight line is perpendicular to a plane, every plane passed through the line is also perpendicular to that plane.*

Let *AB* be perpendicular to the plane *MN*; then, any plane *PQ*, passed through *AB*, is also perpendicular to *MN*.

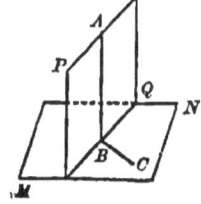

For, at *B* draw *BC*, in the plane *MN*, perpendicular to the intersection *BQ*. Since *AB* is perpendicular to the plane *MN*, it is perpendicular to *BQ* and *BC*; therefore, the angle *ABC* is the plane angle of the diedral angle formed by the planes *PQ* and *MN*; and since the angle *ABC* is a right angle, the planes are perpendicular to each other.

48. *Corollary.* If *AO*, *BO* and *CO*, are three straight lines perpendicular to each other at a common point *O*, each is perpendicular to the plane of the other two, and the three planes are perpendicular to each other.

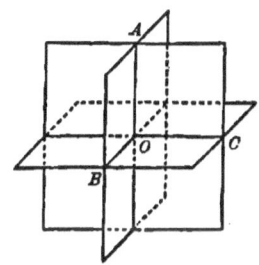

PROPOSITION XVIII.—THEOREM.

49. *If two planes are perpendicular to each other, a straight line drawn in one of them, perpendicular to their intersection, is perpendicular to the other.*

Let the planes PQ and MN be perpendicular to each other; and at any point B of their intersection BQ, let BA be drawn, in the plane PQ, perpendicular to BQ; then, BA is perpendicular to the plane MN.

For, drawing BC, in the plane MN, perpendicular to BQ, the angle ABC is a right angle, since it is the plane angle of the right diedral angle formed by the two planes; therefore, AB, perpendicular to the two straight lines BQ, BC, is perpendicular to their plane MN (13).

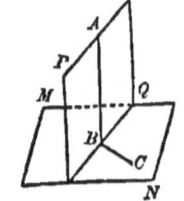

50. *Corollary* I. If two planes, PQ and MN, are perpendicular to each other, a straight line BA drawn through any point B of their intersection perpendicular to one of the planes MN, will lie in the other plane PQ (8).

51. *Corollary* II. If two planes, PQ and MN, are perpendicular to each other, a straight line drawn from any point A of PQ, perpendicular to MN, lies in the plane PQ (7).

PROPOSITION XIX.—THEOREM.

52. *Through any given straight line, a plane can be passed perpendicular to any given plane.*

Let AB be the given straight line and MN the given plane. From any point A of AB let AC be drawn perpendicular to MN, and through AB and AC pass a plane AD. This plane is perpendicular to MN (47).

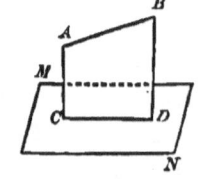

Moreover, since, by (51), any plane passed through AB perpendicular to MN must contain the perpendicular AC, the plane AD is the only plane perpendicular to MN that can be passed through AB, unless AB is itself perpendicular to MN, in which case an infinite number of planes can be passed through it perpendicular to MN (47).

PROPOSITION XX.—THEOREM.

53. *If two intersecting planes are each perpendicular to a third plane, their intersection is also perpendicular to that plane.*

Let the planes PQ, RS, intersecting in the line AB, be perpendicular to the plane MN; then, AB is perpendicular to the plane MN.

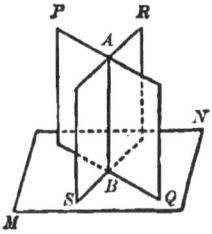

For, if from any point A of AB, a perpendicular be drawn to MN, this perpendicular will lie in each of the planes PQ and RS (50), and must therefore be their intersection AB.

54. *Scholium.* This proposition may be otherwise stated as follows: *If a plane* (MN) *is perpendicular to each of two intersecting planes* (PQ *and* RS), *it is perpendicular to the intersection* (AB) *of those planes.*

PROPOSITION XXI.—THEOREM.

55. *Every point in the plane which bisects a diedral angle is equally distant from the faces of that angle.*

Let the plane AM bisect the diedral angle $CABD$; let P be any point in this plane; PE and PF the perpendiculars from P upon the planes ABC and ABD; then, $PE = PF$.

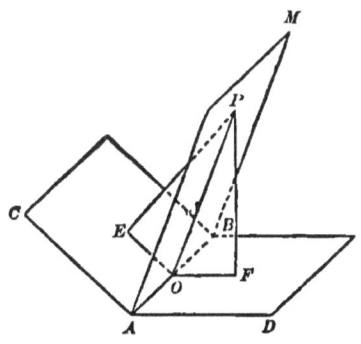

For, through PE and PF pass a plane, intersecting the planes ABC and ABD in OE and OF; join PO. The plane PEF is perpendicular to each of the planes ABC, ABD (47), and consequently perpendicular to their intersection AB (54). Therefore the angles POE and POF measure the diedral angles $MABC$ and $MABD$, which by hypothesis are equal. Hence the right triangles POE and POF are equal (I. 83), and $PE = PF$.

56. Definitions. The *projection of a point A upon a plane MN*, is the foot a of the perpendicular let fall from A upon the plane.

The *projection of a line ABCDE...*, *upon a plane MN*, is the line $abcde...$ formed by the projections of all the points of the line $ABCDE...$ upon the plane.

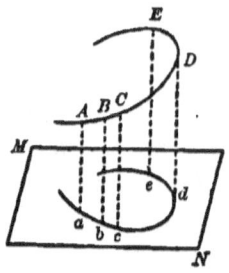

PROPOSITION XXII.—THEOREM.

57. *The projection of a straight line upon a plane is a straight line.*

Let AB be the given straight line, and MN the given plane. The plane Ab, passed through AB perpendicular to the plane MN, contains all the perpendiculars let fall from points of AB upon MN (50); therefore, these perpendiculars all meet the plane MN in the intersection ab of the perpendicular plane with MN. The projection of AB upon the plane MN is, consequently, the straight line ab.

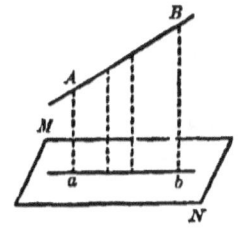

58. Scholium. The plane Ab is called the *projecting plane* of the straight line AB upon the plane MN.

PROPOSITION XXIII.—THEOREM.

59. *The acute angle which a straight line makes with its own projection upon a plane, is the least angle which it makes with any line of that plane.*

Let Ba be the projection of the straight line BA upon the plane MN, the point B being the point of intersection of the line BA with the plane; let BC be any other straight line drawn through B in the plane; then, the angle ABa is less than the angle ABC.

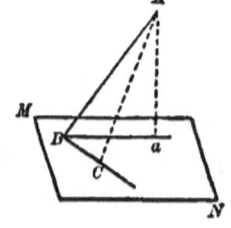

For, take $BC = Ba$, and join AC. In the triangles ABa, ABC, we have AB common, and $Ba = BC$, but $Aa < AC$, since the perpendicular is less than any oblique line; therefore, the angle ABa is less than the angle ABC (I. 85).

60. *Definition.* The acute angle which a straight line makes with its own projection upon a plane is called the *inclination of the line to the plane*, or the *angle of the line and plane*.

61. *Definition.* Two straight lines *AB*, *CD*, not in the same plane, are regarded as making an angle with each other which is equal to the angle between two straight lines *Ob*, *Od*, drawn through any point *O* in space, parallel respectively to the two lines and in the same directions.

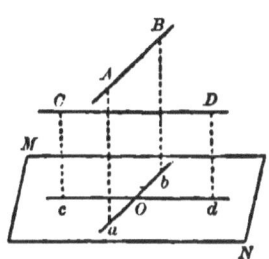

Since every straight line has two opposite directions (I. 4), the angle which one line makes with another is either acute or obtuse, according to the directions considered. Thus, if *Ob* is drawn in the direction expressed by *AB* (that is, on the same side of a straight line joining *A* and *O*), and if *Od* is drawn in the direction expressed by *CD*, then *dOb* is equal to the angle which *CD* makes with *AB*; but if *Oa* is drawn in the direction expressed by *BA* (which is the opposite of *AB*), while *Od* is still in the direction of *CD*, then *dOa* is equal to the angle which *CD* makes with *BA*.

If *MN* is any plane parallel to the two lines *AB*, *CD* (21), then the angle of these lines is the same as the angle of their projections *ab*, *cd*, upon this plane.

62. From the preceding definition, it follows that *when a straight line is perpendicular to a plane, it is perpendicular to all the lines of the plane, whether the lines pass through its foot or not.* For, let *AB* be perpendicular to the plane *MN*, and *CD* any line of the plane. At any point *B'* in *CD*, let *A'B'* be drawn perpendicular to the plane; then, *A'B'* being parallel to *AB*, the right angle *A'B'C* is equal to the angle of the lines *AB* and *CD*, that is, *AB* is perpendicular to *CD*.

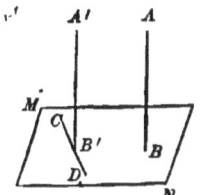

PROPOSITION XXIV.—THEOREM.

63. *Two straight lines not in the same plane being given:* 1st, *a common perpendicular to the two lines can be drawn;* 2d, *but one such common perpendicular can be drawn;* 3d, *the common perpendicular is the shortest distance between the two lines.*

Let AB and CD be the given straight lines.

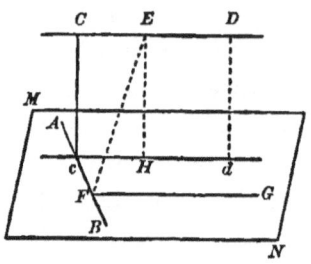

1st. Through one of the given lines, say AB, pass a plane MN, parallel to the other (20); let cd be the projection of CD upon this plane. Then, cd will be parallel to CD (22), and therefore not parallel to AB; hence it will meet AB in some point c. At c draw cC perpendicular to cd in the projecting plane Cd; then Cc is a common perpendicular to AB and CD.

For, CD and cd being parallel, Cc drawn perpendicular to cd is perpendicular to CD. Also, since Cc is the line which projects the point C upon the plane MN, it is perpendicular to that plane, and therefore perpendicular to AB.

2d. The line Cc is the only common perpendicular. For, if another line EF, drawn between AB and CD, could be perpendicular to AB and CD, it would be perpendicular also to a line FG drawn parallel to CD in the plane MN, and consequently perpendicular to the plane MN; but EH, drawn in the plane Cd, parallel to Cc, is perpendicular to the plane MN; hence we should have two perpendiculars from the point E to the plane MN, which is impossible.

3d. The common perpendicular Cc is the shortest distance between AB and CD. For, any other distance EF is greater than the perpendicular EH, or than its equal Cc.

64. *Scholium.* The preceding construction furnishes also the angle between AB and CD, namely, the angle Bcd.

POLYEDRAL ANGLES.

65. *Definition.* When three or more planes meet in a common point, they form a *polyedral angle*.

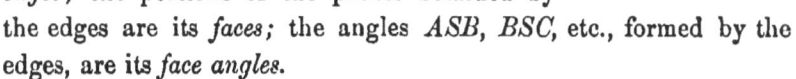

Thus, the figure *S–ABCD*, formed by the planes *ASB, BSC, CSD, DSA*, meeting in the common point *S*, is a polyedral angle.

The point *S* is the vertex of the angle; the intersections of the planes, *SA, SB*, etc., are its *edges;* the portions of the planes bounded by the edges are its *faces;* the angles *ASB, BSC*, etc., formed by the edges, are its *face angles.*

A *triedral* angle is a polyedral angle having but three faces, which is the least number of faces that can form a polyedral angle.

66. *Definition.* Two polyedral angles are *equal* when they can be applied to each other so as to coincide in all their parts.

Since two equal polyedral angles coincide however far their edges and faces are produced, the *magnitude* of a polyedral angle does not depend upon the extent of its faces; but in order to represent the angle clearly in a diagram we usually pass a plane, as *ABCD*, cutting all its faces in straight lines *AB, BC*, etc.; and by the face *ASB* is not meant the triangle *ASB*, but the indefinite surface included between the lines *SA* and *SB* indefinitely produced.

67. *Definition.* A polyedral angle *S–ABCD* is *convex*, when any section, *ABCD*, made by a plane cutting all its faces, is a convex polygon (I. 95).

68. *Symmetrical polyedral angles.* If we produce the edges *AS*, *BS*, etc., through the vertex *S*, we obtain another polyedral angle *S–A'B'C'D'*, which is symmetrical with the first, the vertex *S* being the centre of symmetry.

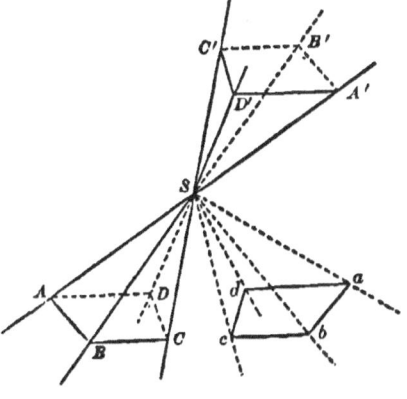

If we pass a plane *A'B'C'D'*, parallel to *ABCD*, so as to make *SA' = SA*, we shall also have *SB' = SB, SC' = SC*, etc.; for we may suppose a third parallel plane passing through

S, and then AA', BB', etc., being divided proportionally by three parallel planes (37), if any one of them is bisected at S, the others are also bisected at that point. The points A', B', etc., are, then, symmetrical with A, B, etc., the definition of symmetry in a plane (I. 138), being extended to symmetry in space.

The two symmetrical polyedral angles are equal in all their parts; for their face angles, ASB and $A'SB'$, BSC and $B'SC'$, are equal, each to each, being vertical plane angles; and the diedral angles at the edges SA and SA', SB and SB', etc., are equal, being vertical diedral angles formed by the same planes. But the equal parts are arranged in inverse order in the two figures, as will appear more plainly, if we turn the polyedral angle S-$A'B'C'D'$ about, until the polygon $A'B'C'D'$ is brought into the same plane with $ABCD$, the vertex S remaining fixed; the polygon $A'B'C'D'$ is then in the position $abcd$, and it is apparent that while in the polyedral angle S-$ABCD$ the parts ASB, BSC, etc., succeed each other in the order *from right to left*, their corresponding equal parts aSb, bSc, etc., in the polyedral angle S-$abcd$ succeed each other in the order *from left to right*. The two figures, therefore, cannot be made to coincide by superposition, and are not regarded as equal in the strict sense of the definition (I. 75), but are said to be *equal by symmetry*.

PROPOSITION XXV.—THEOREM.

69. *The sum of any two face angles of a triedral angle is greater than the third.*

The theorem requires proof only when the third angle considered is greater than each of the others.

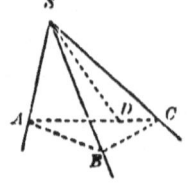

Let S-ABC be a triedral angle in which the face angle ASC is greater than either ASB or BSC; then, $ASB + BSC > ASC$.

For, in the face ASC draw SD making the angle ASD equal to ASB, and through any point D of SD draw any straight line ADC cutting SA and SC; take $SB = SD$, and join AB, BC.

The triangles ASD and ASB are equal, by the construction (I. 76), whence $AD = AB$. Now, in the triangle ABC, we have

$$AB + BC > AC,$$

and subtracting the equals AB and AD,

$$BC > DC;$$

therefore, in the triangles BSC and DSC, we have the angle $BSC > DSC$ (I. 85), and adding the equal angles ASB and ASD, we have $ASB + BSC > ASC$.

PROPOSITION XXVI.—THEOREM.

70. *The sum of the face angles of any convex polyedral angle is less than four right angles.*

Let the polyedral angle S be cut by a plane, making the section $ABCDE$, by hypothesis, a convex polygon. From any point O within this polygon draw OA, OB, OC, OD, OE.

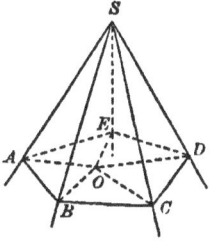

The sum of the angles of the triangles ASB, BSC, etc., which have the common vertex S, is equal to the sum of the angles of the same number of triangles AOB, BOC, etc., which have the common vertex O. But in the triedral angles formed at A, B, C, etc., by the faces of the polyedral angle and the plane of the polygon, we have (69).

$$SAE + SAB > EAB,$$
$$SBA + SBC > ABC, \text{ etc.};$$

hence, taking the sum of all these inequalities, it follows that the sum of the angles at the bases of the triangles whose vertex is S is greater than the sum of the angles at the bases of the triangles whose vertex is O; therefore, the sum of the angles at S is less than the sum of the angles at O, that is, less than four right angles.

PROPOSITION XXVII.—THEOREM.

71. *Two triedral angles are either equal or symmetrical, when the three face angles of one are respectively equal to the three face angles of the other.*

In the triedral angles S and s, let $ASB = asb$, $ASC = asc$, and

$BSC = bsc$; then, the diedral angle SA is equal to the diedral angle sa.

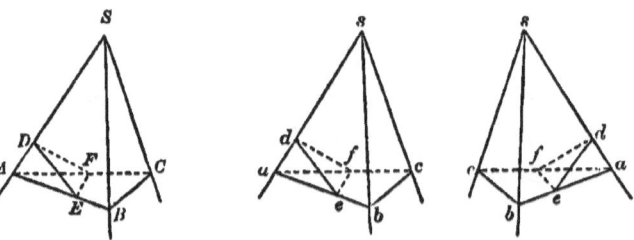

On the edges of these angles take the six equal distances SA, SB, SC, sa, sb, sc, and draw AB, BC, AC, ab, bc, ac. The isosceles triangles SAB and sab are equal, having an equal angle included by equal sides, hence $AB = ab$; and for the same reason, $BC = bc$, $AC = ac$; therefore, the triangles ABC and abc are equal.

At any point D in SA, draw DE in the face ASB and DF in the face ASC, perpendicular to SA; these lines meet AB and AC, respectively, for, the triangles ASB and ASC being isosceles, the angles SAB and SAC are acute; let E and F be the points of meeting, and join EF. Now on sa take $sd = SD$, and repeat the same construction in the triedral angle s.

The triangles ADE and ade are equal, since $AD = ad$, and the angles at A and D are equal to the angles at a and d; hence, $AE = ae$ and $DE = de$. In the same manner, we have $AF = af$ and $DF = df$. Therefore, the triangles AEF and aef are equal (I. 76), and we have $EF = ef$. Finally, the triangles EDF and edf, being mutually equilateral, are equal; therefore, the angle EDF, which measures the diedral angle SA, is equal to the angle edf, which measures the diedral angle sa, and the diedral angles SA and sa are equal (41). In the same manner, it may be proved that the diedral angles SB and SC are equal to the diedral angles sb and sc, respectively.

So far the demonstration applies to either of the two figures denoted in the diagram by s–abc, which are symmetrical with each other. If the first of these figures is given, it follows that S and s are equal, since they can evidently be applied to each other so as to coincide in all their parts (66); if the second is given, it follows that S and s are symmetrical (68).

BOOK VII.

POLYEDRONS.

1. DEFINITION. A *polyedron* is a geometrical solid bounded by planes.

The bounding planes, by their mutual intersections, limit each other, and determine the *faces* (which are polygons), the *edges*, and the *vertices*, of the polyedron. A *diagonal* of a polyedron is a straight line joining any two of its vertices not in the same face.

The least number of planes that can form a polyedral angle is three; but the space within the angle is indefinite in extent, and it requires a fourth plane to enclose a finite portion of space, or to form a solid; hence, the least number of planes that can form a polyedron is four.

2. *Definition.* A polyedron of four faces is called a *tetraedron;* one of six faces, a *hexaedron;* one of eight faces, an *octaedron;* one of twelve faces, a *dodecaedron;* one of twenty faces, an *icosaedron*.

3. *Definition.* A polyedron is *convex* when the section, formed by any plane intersecting it, is a convex polygon.

All the polyedrons treated of in this work will be understood to be convex.

4. *Definition.* The *volume* of any polyedron is the numerical measure of its magnitude, referred to some other polyedron as the *unit*. The polyedron adopted as the unit is called the *unit of volume.*

To *measure* the volume of a polyedron is, then, to find its ratio to the unit of volume.

5. *Definition. Equivalent* solids are those which have equal volumes.

PRISMS AND PARALLELOPIPEDS.

6. *Definitions.* A *prism* is a polyedron two of whose faces are equal polygons lying in parallel planes and having their homologous sides parallel, the other faces being parallelograms formed by the intersections of planes passed through the homologous sides of the equal polygons.

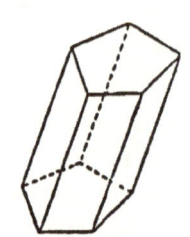

The parallel faces are called the *bases* of the prism; the parallelograms taken together constitute its *lateral or convex surface;* the intersections of the lateral faces are its *lateral edges.*

The *altitude* of a prism is the perpendicular distance between the planes of its bases.

A *triangular* prism is one whose base is a triangle; a *quadrangular* prism, one whose base is a quadrilateral; etc.

7. *Definitions.* A *right prism* is one whose lateral edges are perpendicular to the planes of its bases.

In a right prism, any lateral edge is equal to the altitude.

An *oblique prism* is one whose lateral edges are oblique to the planes of its bases.

In an oblique prism, a lateral edge is greater than the altitude.

8. *Definition.* A *regular prism* is a right prism whose bases are regular polygons.

9. *Definition.* If a prism, *ABCDE–F*, is intersected by a plane *GK*, not parallel to its base, the portion of the prism included between the base and this plane, namely *ABCDE–GHIKL*, is called a *truncated prism.*

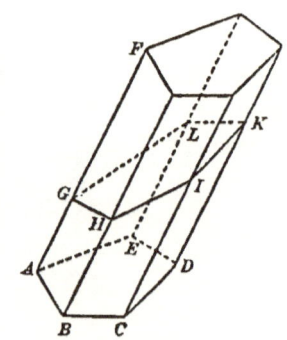

10. *Definition.* If a plane intersects a prism at right angles to its lateral edges, the section is called a *right section* of the prism.

198 GEOMETRY.

11. *Definition.* A *parallelopiped* is a prism whose bases are parallelograms. It is therefore a polyhedron all of whose faces are parallelograms.

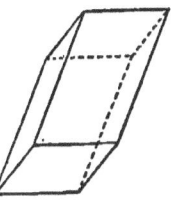

From this definition and (VI. 32) it is evident that any two opposite faces of a parallelopiped are equal parallelograms.

12. *Definition.* A *right parallelopiped* is a parallelopiped whose lateral edges are perpendicular to the planes of its bases. Hence, by (VI. 6), its lateral faces are rectangles; but its bases may be either rhomboids or rectangles.

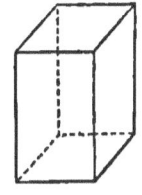

A *rectangular parallelopiped* is a right parallelopiped whose bases are rectangles. Hence it is a parallelopiped all of whose faces are rectangles.

Since the perspective of figures in space distorts the angles, the diagram may represent either a right, or a rectangular, parallelopiped.

13. *Definition.* A *cube* is a rectangular parallelopiped whose six faces are all squares.

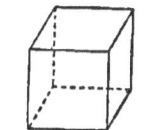

PROPOSITION I.—THEOREM.

14. *The sections of a prism made by parallel planes are equal polygons.*

Let the prism AD' be intersected by the parallel planes GK, $G'K'$; then, the sections, $GHIKL$, $G'H'I'K'L'$, are equal polygons.

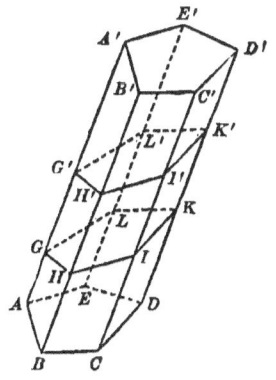

For, the sides of these polygons are parallel, each to each, as for example, GH and $G'H'$, being the intersections of parallel planes with a third plane (VI. 25), and they are equal, being parallels included between parallels (I. 104); hence, also, the angles of the polygons are equal, each to each (VI. 32). Therefore, the two sections, being both mutually equilateral and mutually equiangular, are equal.

15. *Corollary.* Any section of a prism, made by a plane parallel to the base, is equal to the base.

PROPOSITION II.—THEOREM.

16. *The lateral area of a prism is equal to the product of the perimeter of a right section of the prism by a lateral edge.*

Let AD' be a prism, and $GHIKL$ a right section of it; then, the area of the convex surface of the prism is equal to the perimeter $GHIKL$ multiplied by a lateral edge AA'.

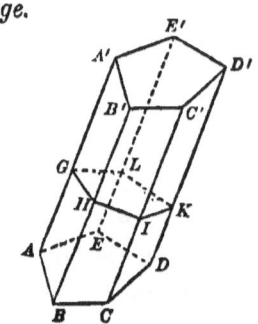

For, the sides of the section $GHIKL$ being perpendicular to the lateral edges AA', BB', etc., are the altitudes of the parallelograms which form the convex surface of the prism, if we take as the bases of these parallelograms the lateral edges, AA' BB', etc., which are all equal. Hence, the area of the sum of these parallelograms is (IV. 10),

$$GH \times AA' + HI \times BB' + \text{etc.}$$
$$= (GH + HI + \text{etc.}) \times AA'.$$

17. *Corollary.* The lateral area of a right prism is equal to the product of the perimeter of its base by its altitude.

PROPOSITION III.—THEOREM.

18. *The four diagonals of a parallelopiped bisect each other.*

Let $ABCD\text{-}G$ be a parallelopiped; its four diagonals, AG, EC, BH, DF, bisect each other.

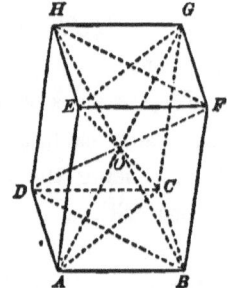

Through the opposite and parallel edges AE, CG, pass a plane which intersects the parallel faces $ABCD$, $EFGH$, in the parallel lines AC and EG. The figure $ACGE$ is a parallelogram, and its diagonals AG and EC bisect each other in the point O. In the same manner it is shown that AG and BH, AG and DF, bisect each other; therefore, the four diagonals bisect each other in the point O.

19. *Scholium.* The point O, in which the four diagonals intersect, is called the *centre* of the parallelopiped; and it is easily proved that any straight line drawn through O and terminated by two opposite faces of the parallelopiped is bisected in that point.

PROPOSITION IV.—THEOREM.

20. *The sum of the squares of the four diagonals of a parallelopiped is equal to the sum of the squares of its twelve edges.*

In the parallelogram $ACGE$ we have (III. 64),

$$\overline{AG}^2 + \overline{CE}^2 = 2\overline{AE}^2 + 2\overline{AC}^2,$$

and in the parallelogram $DBFH$,

$$\overline{BH}^2 + \overline{DF}^2 = 2\overline{BF}^2 + 2\overline{BD}^2.$$

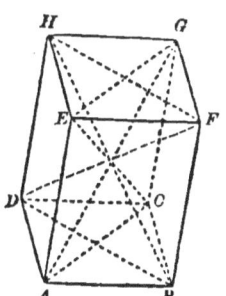

Adding, and observing that $BF = AE$, and also that in the parallelogram $ABCD$, $2\overline{AC}^2 + 2\overline{BD}^2 = 4\overline{AB}^2 + 4\overline{AD}^2$, we have

$$\overline{AG}^2 + \overline{CE}^2 + \overline{BH}^2 + \overline{DF}^2 = 4\overline{AE}^2 + 4\overline{AB}^2 + 4\overline{AD}^2,$$

which proves the theorem.

21. *Corollary.* In a rectangular parallelopiped, the four diagonals are equal to each other; and the square of a diagonal is equal to the sum of the squares of the three edges which meet at a common vertex. Thus, if AG is a rectangular parallelopiped, we have, by dividing the preceding equation by 4,

$$\overline{AG}^2 = \overline{AE}^2 + \overline{AB}^2 + \overline{AD}^2.$$

22. *Scholium.* If any three straight lines AB, AE, AD, not in the same plane, are given, meeting in a common point, a parallelopiped can be constructed upon them. For, pass a plane through the extremity of each line parallel to the plane of the other two; these planes, together with the planes of the given lines, determine the parallelopiped.

In a rectangular parallelopiped, if the plane of two of the three edges which meet at a common vertex is taken as a base, the third edge is the altitude. These three edges, or the three perpendicular distances between the opposite faces of a rectangular parallelopiped, are called its three *dimensions*.

PROPOSITION V.—THEOREM.

23. *Two prisms are equal, if three faces including a triedral angle of the one are respectively equal to three faces similarly placed including a triedral angle of the other.*

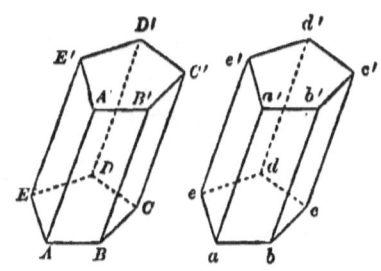

Let the triedral angles A and a of the prisms $ABCDE\text{-}A'$, $abcde\text{-}a'$, be contained by equal faces similarly placed, namely, $ABCDE$ equal to $abcde$, AB' equal to ab', and AE' equal to ae'; then, the prisms are equal.

For, the triedral angles A and a are equal (VI. 71), and can be applied, the one to the other, so as to coincide; and then the bases $ABCDE$, $abcde$, coinciding, the face AB' will coincide with ab', and the face AE' with ae'; therefore the sides $A'B'$, $A'E'$, of the upper base of one prism, will coincide with the sides $a'b'$, $a'e'$, of the upper base of the other prism, and since these bases are equal they will coincide throughout; consequently also the lateral faces of the two prisms will coincide, each to each, and the prisms will coincide throughout; therefore, the prisms are equal.

24. *Corollary* I. *Two truncated prisms are equal, if three faces including a triedral angle of the one are respectively equal to three faces similarly placed including a triedral angle of the other.* For, the preceding demonstration applies whether the planes $A'B'C'D'E'$ and $a'b'c'd'e'$ are parallel or inclined to the lower bases.

25. *Corollary* II. *Two right prisms are equal, if they have equal bases and equal altitudes.*

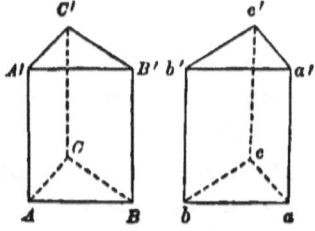

In the case of right prisms, it is not necessary to add the condition that the faces shall be similarly placed; for, if the two right prisms $ABC\text{-}A'$ $abc\text{-}a'$, cannot be made to coincide by placing the base ABC upon the equal base abc; yet, by inverting one of the prisms and applying the base ABC to the base $a'b'c'$, they will coincide.

PROPOSITION VI.—THEOREM.

26. *Any oblique prism is equivalent to a right prism whose base is a right section of the oblique prism, and whose altitude is equal to a lateral edge of the oblique prism.*

Let $ABCDE-A'$ be the oblique prism. At any point F in the edge AA', pass a plane perpendicular to AA' and forming the right section $FGHIK$. Produce AA' to F'', making $FF''=AA'$, and through F' pass a second plane perpendicular to the edge AA', intersecting all the faces of the prism produced, and forming another right section $F'G'H'I'K'$ parallel and equal to the first. The prism $FGHIK-F'$ is a right prism whose base is the right section and whose altitude FF' is equal to the lateral edge of the oblique prism.

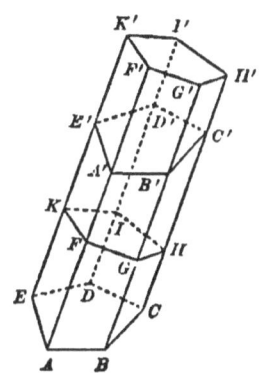

The solid $ABCDE-F$ is a truncated prism which is equal to the truncated prism $A'B'C'D'E'-F''$ (24). Taking the first away from the whole solid $ABCDE-F''$, there remains the right prism; taking the second away from the same solid, there remains the oblique prism; therefore, the right prism and the oblique prism have the same volume, that is, they are equivalent.

PROPOSITION VII.—THEOREM.

27. *The plane passed through two diagonally opposite edges of a parallelopiped divides it into two equivalent triangular prisms.*

Let $ABCD-A'$ be any parallelopiped; the plane $ACC'A'$, passed through its opposite edges AA' and CC', divides it into two equivalent triangular prisms $ABC-A'$ and $ACD-A'$.

Let $FGHI$ be any right section of the parallelopiped, made by a plane perpendicular to the edge AA'. The intersection, FH, of this plane with the plane AC', is the di-

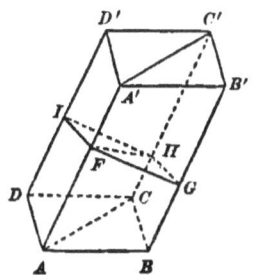

BOOK VII.   203

agonal of the parallelogram *FGHI*, and divides that parallelogram into two equal triangles, *FGH* and *FIH*. The oblique prism *ABC-A'* is equivalent to a right prism whose base is the triangle *FGH* and whose altitude is *AA'* (26); and the oblique prism *ADC-A'* is equivalent to a right prism whose base is the triangle *FIH* and whose altitude is *AA'*. The two right prisms are equal (25); therefore, the oblique prisms, which are respectively equivalent to them, are equivalent to each other.

PROPOSITION VIII.—THEOREM.

28. *Two rectangular parallelopipeds having equal bases are to each other as their altitudes.*

Let *P* and *Q* be two rectangular parallelopipeds having equal bases, and let *AB* and *CD* be their altitudes.

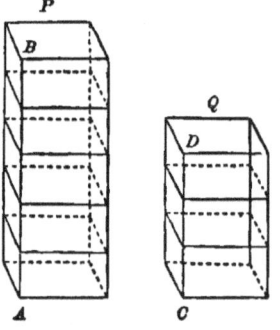

1st. Suppose the altitudes have a common measure, which is contained, for example, 5 times in *AB* and 3 times in *CD*, so that if *AB* is divided in 5 equal parts, *CD* will contain 3 of these parts; then we have

$$\frac{AB}{CD} = \frac{5}{3}.$$

If now we pass planes through the several points of division of *AB* and *CD*, perpendicular to these lines, the parallelopiped *P* will be divided into 5 equal parallelopipeds, and *Q* into 3 parallelopipeds, each equal to those in *P*; hence,

$$\frac{P}{Q} = \frac{5}{3},$$

and, therefore,

$$\frac{P}{Q} = \frac{AB}{CD}.$$

2d. If the altitudes are incommensurable, the proof may be given by the method exemplified in (II. 51) and (III. 15), or, according to the method of limits, as follows.

Let *CD* be divided into any number of equal parts, and let one of these parts be applied to *AB* as often as *AB* will contain it.

Since AB and CD are incommensurable, a certain number of these parts will extend from A to a point B', leaving a remainder BB' less than one of the parts. Through B' pass a plane perpendicular to AB, and denote the parallelopiped whose base is the same as that of P or Q, and whose altitude is AB', by P'; then, since AB' and CD are commensurable,

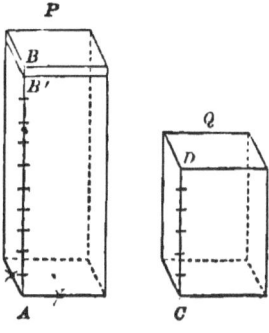

$$\frac{P'}{Q} = \frac{AB'}{CD}.$$

Now, suppose the number of parts into which CD is divided to be continually increased; the length of each part will become less and less, and the point B' will approach nearer and nearer to B. The limit of AB' will be AB, and the limit of P' will be P (V. 28). The limit of $\frac{P'}{Q}$ will therefore be $\frac{P}{Q}$, and that of $\frac{AB'}{CD}$ will be $\frac{AB}{CD}$. Since, then, the variables $\frac{P'}{Q}$ and $\frac{AB'}{CD}$ are constantly equal and approach two limits, these limits are equal (V. 29), and we have

$$\frac{P}{Q} = \frac{AB}{CD}.$$

29. *Scholium.* The three edges of a rectangular parallelopiped which meet at a common vertex being called its *dimensions*, the preceding theorem may also be expressed as follows:

Two rectangular parallelopipeds which have two dimensions in common are to each other as their third dimensions.

PROPOSITION IX.—THEOREM.

30. *Two rectangular parallelopipeds having equal altitudes are to each other as their bases.*

Let a, b and c be the three dimensions of the rectangular parallelopiped P; m, n and c those of the rectangular parallelopiped Q; the dimension c, or the altitude, being common.

BOOK VII.   205

Let R be a third rectangular parallelopiped whose dimensions are m, b and c; then, R has the two dimensions b and c in common with P, and the two dimensions m and c in common with Q; hence (29),

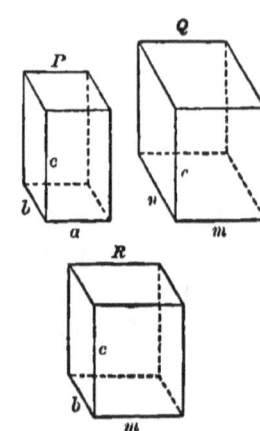

$$\frac{P}{R} = \frac{a}{m}, \qquad \frac{R}{Q} = \frac{b}{n},$$

and multiplying these ratios together,

$$\frac{P}{Q} = \frac{a \times b}{m \times n}.$$

But $a \times b$ is the area of the base of P, and $m \times n$ is the area of the base of Q; therefore, P and Q are in the ratio of their bases.

31. *Scholium.* This proposition may also be expressed as follows:

Two rectangular parallelopipeds which have one dimension in common, are to each other in the products of the other two dimensions.

PROPOSITION X.—THEOREM.

32. *Any two rectangular parallelopipeds are to each other as the products of their three dimensions.*

Let a, b and c be the three dimensions of the rectangular parallelopiped P; m, n and p those of the rectangular parallelopiped Q.

Let R be a third rectangular parallelopiped whose dimensions are a, b and p; then R has two dimensions in common with P and one dimension in common with Q; hence, by (29) and (31),

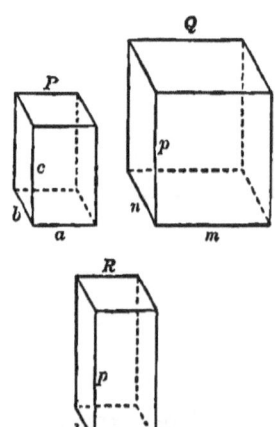

$$\frac{P}{R} = \frac{c}{p}, \qquad \frac{R}{Q} = \frac{a \times b}{m \times n},$$

and multiplying these ratios together,

$$\frac{P}{Q} = \frac{a \times b \times c}{m \times n \times p}.$$

18

PROPOSITION XI.—THEOREM.

33. *The volume of a rectangular parallelopiped is equal to the product of its three dimensions, the unit of volume being the cube whose edge is the linear unit.*

Let a, b, c, be the three dimensions of the rectangular parallelopiped P; and let Q be the cube whose edge is the linear unit. The three dimensions of Q are each equal to unity, and we have, by the preceding proposition.

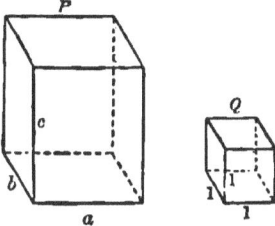

$$\frac{P}{Q} = \frac{a \times b \times c}{1 \times 1 \times 1} = a \times b \times c.$$

Now, Q being taken as the unit of volume, $\frac{P}{Q}$ is the numerical measure, or volume of P, in terms of this unit (4); therefore the volume of P is equal to the product $a \times b \times c$.

34. *Scholium* I. Since the product $a \times b$ represents the base, when c is called the altitude, of the parallelopiped, this proposition may also be expressed as follows:

The volume of a rectangular parallelopiped is equal to the product of its base by its altitude.

35. *Scholium* II. When the three dimensions of the parallelopiped are each exactly divisible by the linear unit, the truth of the proposition is rendered evident by dividing the solid into cubes, each of which is equal to the unit of volume. Thus, if the three edges which meet at a common vertex A are, respectively, equal to 3, 4 and 5, times the linear unit, these edges may be divided respectively into 3, 4 and 5 equal parts, and then planes passed through the several points of division at right angles to these edges will divide the solid into cubes, each equal to the unit cube, the number of which is evidently $3 \times 4 \times 5$.

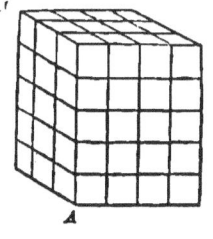

But the more general demonstration, above given, includes also the cases in which one of the dimensions, or two of them, or all three, are incommensurable with the linear unit.

36. *Scholium* III. If the three dimensions of a rectangular parallelopiped are each equal to a, the solid is a cube whose edge is a, and its volume is $a \times a \times a = a^3$; or, *the volume of a cube is the third power of its edge.* Hence it is that in arithmetic and algebra, the expression "cube of a number" has been adopted to signify the "third power of a number."

PROPOSITION XII.—THEOREM.

37. *The volume of any parallelopiped is equal to the product of its base by its altitude.*

Let $ABCD-A'$ be any oblique parallelopiped, whose base is $ABCD$, and altitude $B'O$.

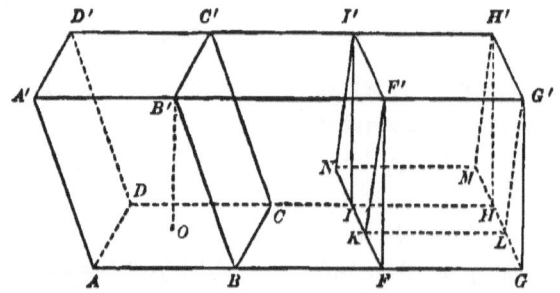

Produce the edges AB, $A'B'$, DC, $D'C'$; in AB produced take $FG = AB$, and through F and G pass planes, $FF''I'I$, $GG'H'H$, perpendicular to the produced edges, forming the right parallelopiped $FGHI-F'$, with the base $FF''I'I$ and altitude FG, equivalent to the given oblique parallelopiped $ABCD-A'$ (26).

From F', draw FK perpendicular to FI or $F'I'$. Since AF is perpendicular to the plane FI', the plane of the base and the plane FI' are perpendicular to each other (VI. 47); therefore, $F'K$ is perpendicular to the plane of the base (VI. 49) and is equal to $B'O$.

Now the three lines $F'G'$, $F'I'$ and $F'K$ are perpendicular to each other; consequently the parallelopiped $KLMN-F''$, constructed upon them, is rectangular. The parallelopiped $FGHI-F''$, regarded as an oblique prism whose base is $FGG'F'$ and lateral edge $F''I'$, is equivalent to the right prism, or rectangular parallelopiped, $KLMN-F''$, whose base is the right section $F'L$ and whose altitude is $F'I'$ (26). Therefore, the given parallelopiped $ABCD-A'$ is also equivalent to the rectangular parallelopiped $KLMN-F'$. The volume of this rectangular parallelopiped is equal to the product of its base KM by its altitude $F'K$; its base KM is equal to $F'H'$, or FH, which is equivalent to AC, and its altitude $F'K$ is equal to $B'O$; therefore the volume of the parallelopiped $ABCD-A'$ is equal to the product of its base AC by its altitude $B'O$.

PROPOSITION XIII.—THEOREM.

38. *The volume of any prism is equal to the product of its base by its altitude.*

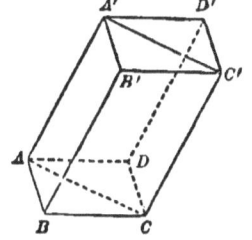

1st. Let $ABC-A'$ be a triangular prism. This prism is equivalent to one-half the parallelopiped $ABCD-A'$ constructed upon the edges AB, BC and BB' (27), and it has the same altitude. The volume of the parallelopiped is equal to its base BD multiplied by its altitude; therefore, the volume of the triangular prism is equal to its base ABC, the half of BD, multiplied by its altitude.

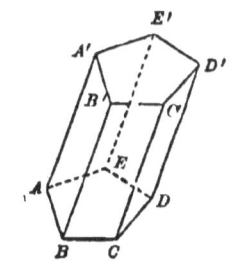

2d. Let $ABCDE-A'$ be any prism. It may be divided into triangular prisms by planes passed through a lateral edge AA' and the several diagonals of its base. The volume of the given prism is the sum of the volumes of the triangular prisms, or the sum of their bases multiplied by their common altitude, which is the base $ABCDE$ of the given prism multiplied by its altitude.

39. *Corollary.* Prisms having equivalent bases are to each other as their altitudes; prisms having equal altitudes are to each other as their bases; and any two prisms are to each other as the products of their bases and altitudes.

PYRAMIDS.

40. Definitions. A *pyramid* is a polyedron bounded by a polygon and triangular faces formed by the intersections of planes passed through the sides of the polygon and a common point out of its plane; as *S–ABCDE*.

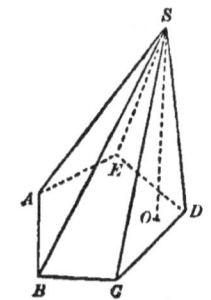

The polygon, *ABCDE*, is the *base* of the pyramid; the point, *S*, in which the triangular faces meet, is its *vertex;* the triangular faces taken together constitute its *lateral, or convex, surface;* the area of this surface is the *lateral area;* the lines *SA*, *SB*, etc., in which the lateral faces intersect, are its *lateral edges*. The *altitude* of the pyramid is the perpendicular distance *SO* from the vertex to the base.

A *triangular* pyramid is one whose base is a triangle; a *quadrangular* pyramid, one whose base is a quadrilateral; etc.

A triangular pyramid, having but four faces (all of which are triangles), is a *tetraedron*; and any one of its faces may be taken as its base.

41. Definitions. A *regular pyramid* is one whose base is a regular polygon, and whose vertex is in the perpendicular to the base erected at the centre of the polygon. This perpendicular is called the *axis* of the regular pyramid.

From this definition and (VI. 10) it follows that all the lateral faces of a regular pyramid are equal isosceles triangles.

The *slant height* of a regular pyramid is the perpendicular from the vertex to the base of any one of its lateral faces.

42. Definitions. A *truncated pyramid* is the portion of a pyramid included between its base and a plane cutting all its lateral edges.

When the cutting plane is parallel to the base, the truncated pyramid is called a *frustum of a pyramid*. The *altitude* of a frustum is the perpendicular distance between its bases.

210   GEOMETRY.

In a frustum of a regular pyramid, the lateral faces are equal trapezoids; and the perpendicular distance between the parallel sides of any one of these trapezoids is the slant height of the frustum.

PROPOSITION XIV.—THEOREM.

43. *If a pyramid is cut by a plane parallel to its base:* 1st, *the edges and the altitude are divided proportionally;* 2d, *the section is a polygon similar to the base.*

Let the pyramid $S{-}ABCDE$, whose altitude is SO, be cut by the plane $abcde$ parallel to the base, intersecting the lateral edges in the points a, b, c, d, e, and the altitude in o; then,

1st. The edges and the altitude are divided proportionally.

For, suppose a plane passed through the vertex S parallel to the base; then, the edges and altitude, being intersected by three parallel planes, are divided proportionally (VI. 37), and we have

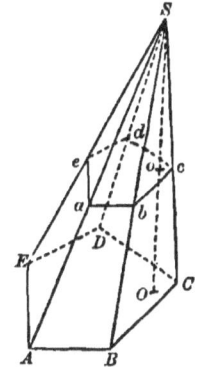

$$\frac{Sa}{SA} = \frac{Sb}{SB} = \frac{Sc}{SC} = \ldots = \frac{So}{SO}.$$

2d. The section $abcde$ is similar to the base $ABCDE$.

For, the sides ab, bc, etc., are parallel respectively to AB, BC, etc. (VI. 25), and in the same directions: therefore the angles of the two polygons are equal, each to each (VI. 32).

Also, since ab is parallel to AB, and bc parallel to BC, the triangles Sab and SAB are similar, and the triangles Sbc and SBC are similar; therefore,

$$\frac{ab}{AB} = \frac{Sb}{SB}, \text{ and } \frac{bc}{BC} = \frac{Sb}{SB},$$

whence

$$\frac{ab}{AB} = \frac{bc}{BC};$$

and in the same manner we should find

$$\frac{bc}{BC} = \frac{cd}{CD} = \frac{de}{DE} = \frac{ea}{EA}.$$

Therefore, the polygons $abcde$ and $ABCDE$ are similar (III. 24).

44. *Corollary* I. The polygons $abcde$ and $ABCDE$ being similar, their surfaces are proportional to the squares of their homologous sides; hence

$$\frac{abcde}{ABCDE} = \frac{\overline{ab}^2}{\overline{AB}^2} = \frac{\overline{Sa}^2}{\overline{SA}^2} = \frac{\overline{So}^2}{\overline{SO}^2};$$

that is, *the surface of any section of a pyramid parallel to its base is proportional to the square of its distance from the vertex.*

45. *Corollary* II. *If two pyramids, S–$ABCDE$ and S'–$A'B'C'D'$, having equal altitudes SO and $S'O'$, are cut by planes parallel to their bases and at equal distances, So and $S'o'$, from their vertices, the sections $abcde$ and $a'b'c'd'$ will be proportional to the bases.*

For, by the preceding corollary,

$$\frac{abdce}{ABCDE} = \frac{\overline{So}^2}{\overline{SO}^2};$$

and

$$\frac{a'b'c'd'}{A'B'C'D'} = \frac{\overline{S'o'}^2}{\overline{S'O'}^2};$$

whence, since $So = S'o'$ and $SO = S'O'$,

$$\frac{abcde}{ABCDE} = \frac{a'b'c'd'}{A'B'C'D'}.$$

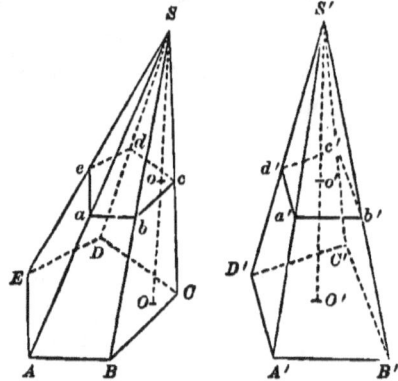

46. *Corollary* III. *If two pyramids have equal altitudes and equivalent bases, sections made by planes parallel to their bases and at equal distances from their vertices are equivalent.*

PROPOSITION XV.—THEOREM.

47. *The lateral area of a regular pyramid is equal to the product of the perimeter of its base by one-half its slant height.*

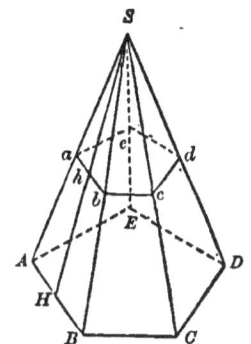

For, let $S-ABCDE$ be a regular pyramid; the lateral faces SAB, SAC, etc., being equal isosceles triangles, whose bases are the sides of the regular polygon $ABCDE$ and whose common altitude is the slant height SH, the sum of their areas, or the lateral area of the pyramid, is equal to the sum of AB, BC, etc., multiplied by $\frac{1}{2}SH$ (IV. 13).

48. *Corollary.* The lateral area of the frustum of a regular pyramid is equal to the half sum of the perimeters of its bases multiplied by the slant height of the frustum. For, this product is the measure of the sum of the areas of the trapezoids $ABba$, $BCcb$, etc., whose common altitude is the slant height hH (IV. 17).

PROPOSITION XVI.—LEMMA.

49. *A series of prisms may be inscribed in any given triangular pyramid whose total volume shall differ from the volume of the pyramid by less than any assigned volume.*

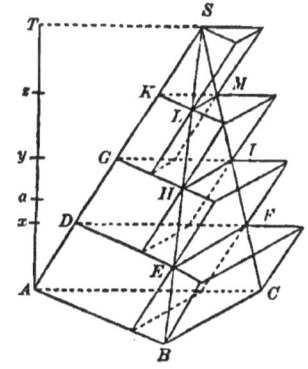

Let $S-ABC$ be the given triangular pyramid, whose altitude is AT. Divide the altitude AT into any number of equal parts Ax, xy, etc., and denote one of these parts by h. Through the points of division x, y, etc., pass planes parallel to the base, cutting from the pyramid the sections DEF, GHI, etc. Upon the triangles DEF, GHI, etc., as *upper* bases, construct prisms whose lateral edges are parallel to SA, and whose altitudes are each equal to h. This is effected by passing planes through EF, HI, etc., parallel to SA. There will thus be formed a series of prisms DEF-A, GHI-D, etc., *inscribed* in the pyramid.

Again, upon the triangles ABC, DEF, GHI, etc., as *lower* bases, construct prisms whose lateral edges are parallel to SA, and whose altitudes are each equal to h. This also is effected by passing planes through BC, EF, HI, etc., parallel to SA. There will thus be formed a series of prisms ABC-D, DEF-G, etc., which may be said to be *circumscribed* about the pyramid.

Now, the first inscribed prism DEF-A is equivalent to the second circumscribed prism DEF-G, since they have the same base DEF and equal altitudes (39); the second inscribed prism GHI-D is equivalent to the third circumscribed prism GHI-K; and so on. Therefore, the sum of all the inscribed prisms differs from the sum of all the circumscribed prisms only by the first circumscribed prism ABC-D. But the pyramid is greater than the sum of the inscribed prisms and less than the sum of the circumscribed prisms; therefore, the difference between the total volume of the inscribed prisms and the volume of the pyramid is less than the volume of the prism ABC-D.

The volume of the prism ABC-D may be made as small as we please, or less than any assigned volume, by dividing the altitude AT into a sufficiently great number of equal parts; for, if the assigned volume is represented by a prism whose base is ABC and altitude Aa, we have only to divide AT into a number of equal parts each less than Aa.

Therefore, the difference between the total volume of the inscribed prisms and the volume of the pyramid may be made less than any assigned volume.

50. *Corollary.* If the number of parts into which the altitude is divided is increased indefinitely, the difference between the volume of the inscribed prisms and that of the pyramid approaches indefinitely to zero; and therefore the pyramid is the *limit* of the sum of the inscribed prisms, as their number is indefinitely increased (V. 28).

PROPOSITION XVII.—THEOREM.

51. *Two triangular pyramids having equivalent bases and equal altitudes are equivalent.*

Let S–ABC and S'–$A'B'C'$ be two triangular pyramids having

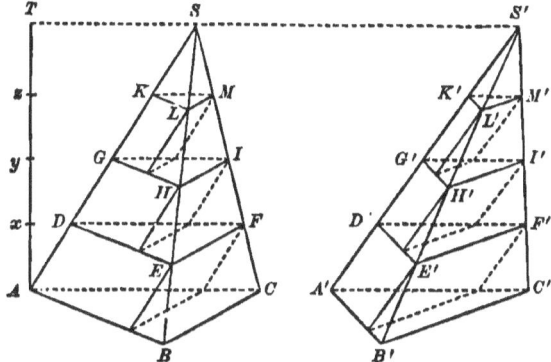

equivalent bases, ABC, $A'B'C'$, in the same plane, and a common altitude AT.

Divide the altitude AT into a number of equal parts Ax, xy, yz, etc., and through the points of division pass planes parallel to the plane of the bases, intersecting the two pyramids. In the pyramid S–ABC inscribe a series of prisms whose upper bases are the sections DEF, GHI, etc., and in the pyramid S'–$A'B'C'$ inscribe a series of prisms whose upper bases are the sections $D'E'F'$, $G'H'I'$, etc. Since the corresponding sections are equivalent (46), the corresponding prisms, having equivalent bases and equal altitudes, are equivalent (39); therefore, the sum of the prisms inscribed in the pyramid S–ABC is equivalent to the sum of the prisms inscribed in the pyramid S'–$A'B'C'$; that is, if we denote the total volumes of the two series of prisms by V and V', we have

$$V = V'.$$

Now let the number of equal parts into which the altitude is divided be supposed to be indefinitely increased; the volume V approaches to the volume of the pyramid S–ABC as its limit, and the volume V' approaches to the volume of the pyramid S'–$A'B'C'$ as its limit (50). Since, then, the variables V and V' are always equal to each other and approach two limits, these limits are equal (V. 29); that is, the volumes of the pyramids are equal.

PROPOSITION XVIII.—THEOREM.

52. *A triangular pyramid is one-third of a triangular prism of the same base and altitude.*

Let $S\text{-}ABC$ be a triangular pyramid. Through one edge of the base, as AC, pass a plane $ACDE$ parallel to the opposite lateral edge SB, and through S pass a plane SED parallel to the base; the prism $ABC\text{-}E$ has the same base and altitude as the given pyramid, and we are to prove that the pyramid is one-third of the prism.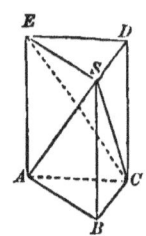

Taking away the pyramid $S\text{-}ABC$ from the prism, there remains a quadrangular pyramid whose base is the parallelogram $ACDE$ and vertex S. The plane SEC, passed through SE and SC, divides this pyramid into two triangular pyramids, $S\text{-}AEC$ and $S\text{-}ECD$, which are equivalent to each other, since their triangular bases AEC and ECD are the halves of the parallelogram $ACDE$, and their common altitude is the perpendicular from S upon the plane $ACDE$ (51). The pyramid $S\text{-}ECD$ may be regarded as having ESD as its base and its vertex at C; therefore, it is equivalent to the pyramid $S\text{-}ABC$ which has an equivalent base and the same altitude. Therefore, the three pyramids into which the prism is divided are equivalent to each other, and the given pyramid is one-third of the prism.

53. *Corollary.* The volume of a triangular pyramid is equal to one-third of the product of its base by its altitude.

PROPOSITION XIX.—THEOREM.

54. *The volume of any pyramid is equal to one-third of the product of its base by its altitude.*

For, any pyramid, $S\text{-}ABCDE$, may be divided into triangular pyramids by passing planes through an edge SA and the diagonals AD, AC, etc., of its base. The bases of these pyramids are the triangles which compose the base of the given pyramid, and their common altitude is the altitude SO of the given pyramid. The volume of the given pyramid is equal to the sum of the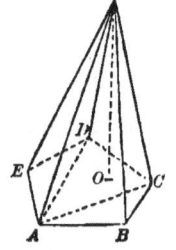

volumes of the triangular pyramids, which is one-third of the sum of their bases multiplied by their common altitude, or one-third the product of the base $ABCDE$ by the altitude SO.

55. Corollary. *Pyramids having equivalent bases are to each other as their altitudes. Pyramids having equal altitudes are to each other as their bases. Any two pyramids are to each other as the products of their bases and altitudes.*

56. Scholium. The volume of any polyedron may be found by dividing it into pyramids, and computing the volumes of these pyramids separately. The division may be effected by drawing all the diagonals that can be drawn from a common vertex; the bases of the pyramids will be all the faces of the polyedron except those which meet at the common vertex. Or, a point may be taken within the polyedron and joined to all the vertices; the polyedron will then be decomposed into pyramids whose bases will be the faces of the polyedron, and whose common vertex will be the point taken within it.

PROPOSITION XX.—THEOREM.

57. *Two tetraedrons which have a triedral angle of the one equal to a triedral angle of the other, are to each other as the products of the three edges of the equal triedral angles.*

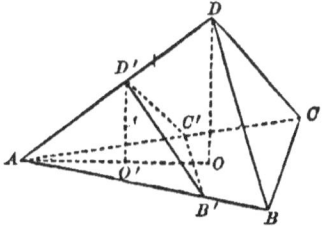

Let $ABCD$, $AB'C'D'$, be the given tetraedrons, placed with their equal triedral angles in coincidence at A. From D and D', let fall DO and $D'O'$ perpendicular to the face ABC. Then, taking the faces ABC, $AB'C'$, as the bases of the triangular pyramids $D-ABC$, $D'-AB'C'$, and denoting the volumes by V and V', we have (55),

$$\frac{V}{V'} = \frac{ABC \times DO}{AB'C' \times D'O'} = \frac{ABC}{AB'C'} \times \frac{DO}{D'O'}.$$

By (IV. 22) and (III. 25), we have

$$\frac{ABC}{AB'C'} = \frac{AB \times AC}{AB' \times AC'} \text{ and } \frac{DO}{D'O'} = \frac{AD}{AD'},$$

therefore,
$$\frac{V}{V'} = \frac{AB \times AC \times AD}{AB' \times AC' \times AD'}.$$

PROPOSITION XXI.—THEOREM.

58. *A frustum of a triangular pyramid is equivalent to the sum of three pyramids whose common altitude is the altitude of the frustum, and whose bases are the lower base, the upper base, and a mean proportional between the bases, of the frustum.*

Let $ABC\text{-}D$ be a frustum of a triangular pyramid, formed by a plane DEF parallel to the base ABC.

Through the vertices A, E and C, pass a plane AEC; and through the vertices E, D and C, pass a plane EDC, dividing the frustum into three pyramids. For brevity, denote the pyramid

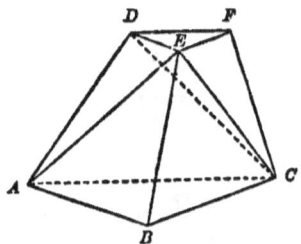

$E\text{-}ABC$ by P, the pyramid $E\text{-}DFC$ by p, and the pyramid $E\text{-}ADC$ by Q.

The pyramids P and Q, regarded as having the common vertex C and their bases in the same plane BD, have a common altitude and are to each other as their bases AEB and AED (55). But the triangles AEB and AED, having a common altitude, namely, the altitude of the trapezoid $ABED$, are to each other as their bases AB and DE; hence we have
$$\frac{P}{Q} = \frac{AB}{DE}.$$

The pyramids Q and p, regarded as having the common vertex E and their bases in the same plane AF, have a common altitude, and are to each other as their bases ADC and DCF. But the triangles ADC and DCF, having a common altitude, namely, the altitude of the trapezoid $ACFD$, are to each other as their bases AC and DF; hence we have
$$\frac{Q}{p} = \frac{AC}{DF}.$$

Moreover, the section DEF being similar to ABC (43), we have

$$\frac{AB}{DE} = \frac{AC}{DF},$$

and therefore

$$\frac{P}{Q} = \frac{Q}{p},$$

whence

$$Q^2 = P \times p, \quad Q = \sqrt{P \times p};$$

that is (III. 5), the *pyramid Q is a mean proportional between the pyramids P and p*.

Now, denote the lower base ABC of the frustum by B, its upper base by b, and its altitude by h. The pyramid P, regarded as having its vertex at E, has the altitude h and the base B; the pyramid p, regarded as having its vertex at C, has the altitude h and the base b; hence (54),

$$P = \tfrac{1}{3} h \times B, \quad p = \tfrac{1}{3} h \times b,$$

and

$$Q = \sqrt{\tfrac{1}{3} h \times B \times \tfrac{1}{3} h \times b} = \tfrac{1}{3} h \times \sqrt{B \times b};$$

consequently, Q is equivalent to a pyramid whose altitude is h and whose base is a mean proportional between the bases B and b; and since the given frustum is the sum of P, p and Q, the proposition is established.

If V denotes the volume of the frustum, the proposition is expressed by the formula

$$V = \tfrac{1}{3} h \times B + \tfrac{1}{3} h \times b + \tfrac{1}{3} h \times \sqrt{B \times b},$$

or

$$V = \tfrac{1}{3} h \, (B + b + \sqrt{B \times b}).$$

59. Corollary. *A frustum of any pyramid is equivalent to the sum of three pyramids whose common altitude is the altitude of the frustum, and whose bases are the lower base, the upper base, and a mean proportional between the bases, of the frustum.*

For, let $ABCDE\text{-}F$ be a frustum of any pyramid $S\text{-}ABCDE$. Let $S'\text{-}A'B'C'$ be a triangular pyramid, having the same altitude as the pyramid $S\text{-}ABCDE$, and a base $A'B'C'$ equivalent to the base $ABCDE$, and in the same plane with it. The volumes of the two pyramids are equivalent (55). Let the plane of the upper base of the given frustum be produced to cut the triangular pyramid.

The section $F'G'I'$ being equivalent to the section $FGHIK$ (46), the pyramid S'-$F'G'I'$ is equivalent to the pyramid S-$FGHIK$;

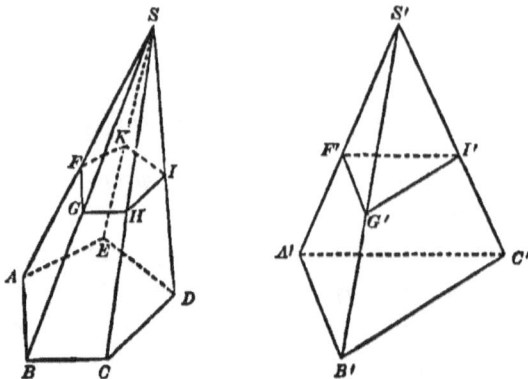

and taking away these pyramids from the whole pyramids, the frustums that remain are equivalent; therefore, denoting by B the area of $ABCDE$ or of $A'B'C'$, by b that of $FGHIK$ or of $F'G'I'$, and by h the common altitude of the two frustums, we have for the volume of the given frustum the same expression as for that of the triangular frustum; namely,

$$V = \tfrac{1}{3}h\,(B + b + \sqrt{B \times b}).$$

TRUNCATED TRIANGULAR PRISM.

PROPOSITION XXII.—THEOREM.

60. *A truncated triangular prism is equivalent to the sum of three pyramids whose common base is the base of the prism, and whose vertices are the three vertices of the inclined section.*

Let ABC-DEF be a truncated triangular prism whose base is ABC and inclined section DEF.

Pass the planes AEC and DEC, dividing the truncated prism into the three pyramids, E-ABC, E-ACD and E-CDF.

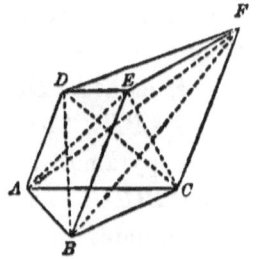

The first of these pyramids, E-ABC, has the base ABC and the vertex E.

220   GEOMETRY.

The second pyramid, E-ACD, is equivalent to the pyramid B-ACD; for they have the same base ACD, and the same altitude, since their vertices E and B are in the line EB parallel to this base. But the pyramid B-ACD is the same as D-ABC; that is, it has the base ABC and the vertex D.

The third pyramid, E-CDF, is equivalent to the pyramid B-ACF; for they have equivalent bases CDF and ACF in the same plane, and also the same altitude, since their vertices E and B are in the line EB parallel to that plane. But the pyramid B-ACF is the same as F-ABC; that is, it has the base ABC and the vertex F.

Therefore the truncated prism is equivalent to three pyramids whose common base is ABC and whose vertices are E, D and F.

61. Corollary I. *The volume of a truncated right triangular prism is equal to the product of its base by one-third the sum of its lateral edges.* For, the lateral edges AD, BE, CF, being perpendicular to the base, are the altitudes of the three pyramids to which the truncated prism has been proved to be equivalent; therefore, the volume is

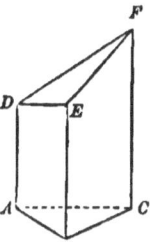

$$ABC \times \tfrac{1}{3}AD + ABC \times \tfrac{1}{3}BE + ABC \times \tfrac{1}{3}CF,$$

or

$$ABC \times \frac{AD + BE + CF}{3}.$$

62. Corollary II. *The volume of any truncated triangular prism is equal to the product of its right section by one-third the sum of its lateral edges.* For, let ABC-$A'B'C'$ be any truncated triangular prism; the right section DEF divides it into two truncated right prisms whose volumes are, by the preceding corollary,

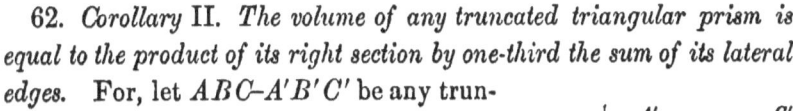

$$DEF \times \frac{AD + BE + CF}{3}$$

and

$$DEF \times \frac{A'D + B'E + C'F}{3},$$

the sum of which is

$$DEF \times \frac{AA' + BB' + CC'}{3}.$$

SIMILAR POLYEDRONS.

63. *Definition.* *Similar polyedrons* are those which are bounded by the same number of faces similar each to each and similarly placed, and which have their homologous polyedral angles equal.

Parts similarly placed in two similar polyedrons, whether faces, lines, or angles, are *homologous.*

64. *Corollary* I. Since homologous edges are in the ratio of similitude of the polygons of which they are homologous sides (III. 24), and every edge belongs to two faces, in each polyedron, it follows that the ratio of similitude of any two homologous faces is the same as that of any other two homologous faces, and this ratio may be called the *ratio of similitude of the two polyedrons.*

Therefore, *any two homologous edges of two similar polyedrons are in the ratio of similitude of the polyedrons; or, homologous edges are proportional to each other.*

65. *Corollary* II. *The ratio of the surfaces of any two homologous faces is the square of the ratio of similitude of the polyedrons* (IV. 24); *or, any two homologous faces are to each other as the squares of any two homologous edges.*

Hence, by the theory of proportions (III. 12), *the entire surfaces of two similar polyedrons are to each other as the squares of any two homologous edges.*

PROPOSITION XXIII.—THEOREM.

66. *If a tetraedron is cut by a plane parallel to one of its faces, the tetraedron cut off is similar to the first.*

Let the tetraedron $ABCD$ be cut by the plane $B'C'D'$ parallel to BCD; then, the tetraedrons $AB'C'D'$ and $ABCD$ are similar.

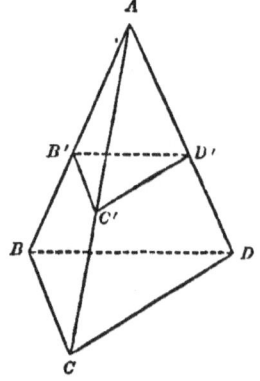

For, since the edges AB, AC, AD, are divided proportionally at B', C', D', the face $AB'C'$ is similar to the face ABC, $AC'D'$ to ACD, and $AB'D'$ to ABD; also, $B'C'D'$ is similar to BCD (43). Moreover, the homologous triedral angles, being contained by equal face angles similarly placed, are equal, each to each (VI. 71). Therefore, by the definition (63), the tetraedrons are similar.

PROPOSITION XXIV.—THEOREM.

67. *Two tetraedrons are similar, when a diedral angle of the one is equal to a diedral angle of the other, and the faces including these angles are similar each to each, and similarly placed.*

Let $ABCD$, $A'B'C'D'$, be two tetraedrons in which the diedral angle AB is equal to the diedral angle $A'B'$, and the faces ABC and ABD are respectively similar to the faces $A'B'C'$ and $A'B'D'$; then, the tetraedrons are similar.

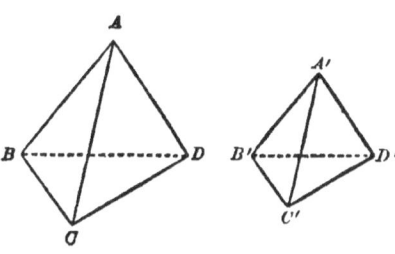

The triedral angles A and A' are equal, since they may evidently be placed with their vertices in coincidence so as to coincide in all their parts. Therefore, the angles CAD and $C'A'D'$ are equal. The given similar faces furnish the proportions

$$\frac{AC}{A'C'} = \frac{AB}{A'B'}, \qquad \frac{AD}{A'D'} = \frac{AB}{A'B'},$$

whence

$$\frac{AC}{A'C'} = \frac{AD}{A'D'};$$

therefore, the faces ACD and $A'C'D'$ are similar (III. 32).

In like manner it is shown that the triedral angles B and B' are equal, and the faces BCD and $B'C'D'$ are similar.

Finally, the triedral angles C and C' are equal, since their face angles are equal each to each and are similarly placed (VI. 71); and the triedral angles D and D' are equal for the same reason. Therefore, the two tetraedrons are similar (63).

PROPOSITION XXV.—THEOREM.

68. *Two similar polyedrons may be decomposed into the same number of tetraedrons similar each to each and similarly placed.*

Let $ABCDEFGH$ and $abcdefgh$ be similar polyedrons, of which A and a are homologous vertices.

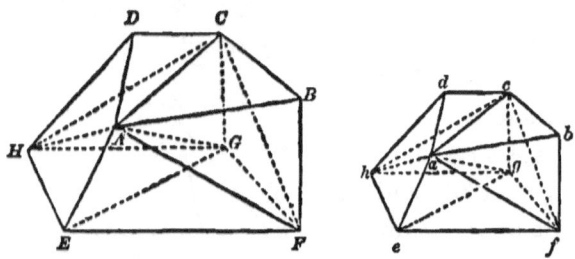

Let all the faces not adjacent to A, in the first polyedron, be decomposed into triangles, and let straight lines be drawn from A to the vertices of these triangles; the polyedron is then divided into tetraedrons having these triangles as bases and the common vertex A.

Also decompose the faces not adjacent to a, in the second polyedron, into triangles similar to those in the first polyedron and similarly placed (III. 39), and let straight lines be drawn from a to the vertices of these triangles; the second polyedron is then divided into the same number of tetraedrons as the first, and it is readily proved that two tetraedrons similarly placed in the two polyedrons are similar.

We leave the details of the proof to the student. See (III. 39).

69. *Corollary.* Homologous diagonals, and in general any two homologous lines, in two similar polyedrons, are in the same ratio as any two homologous edges, that is, in the ratio of similitude of the polyedrons.

PROPOSITION XXVI.—THEOREM.

70. *Two polyedrons composed of the same number of tetraedrons, similar each to each and similarly placed, are similar.*

The proof is left to the student. See (III. 38).

PROPOSITION XXVII.—THEOREM.

71. *Similar polyedrons are to each other as the cubes of their homologous edges.*

1st. Let $ABCD$, $abcd$, be two similar tetraedrons; let the similar faces BCD, bcd, be taken as bases, and let AO, ao be their altitudes.

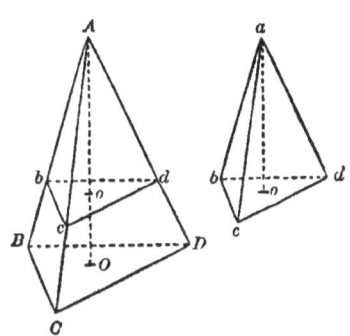

Since the tetraedrons are similar, they may be placed with their equal homologous polyedral angles A and a in coincidence, and the base bcd will then be parallel to the base BCD, since their planes make equal angles with the plane of the face ABC. The perpendicular AO, to BCD, will also be perpendicular to bcd, and Ao will be the altitude of the tetraedron $Abcd$ or $abcd$. Denoting the volumes of the tetraedrons by V and v, we have (55),

$$\frac{V}{v} = \frac{BCD \times AO}{bcd \times Ao} = \frac{BCD}{bcd} \times \frac{AO}{Ao}.$$

The bases being similar, we have

$$\frac{BCD}{bcd} = \frac{\overline{BC}^2}{\overline{bc}^2},$$

and by (69), we have

$$\frac{AO}{Ao} = \frac{AC}{ac} = \frac{BC}{bc};$$

hence

$$\frac{V}{v} = \frac{\overline{BC}^2}{\overline{bc}^2} \times \frac{BC}{bc} = \frac{\overline{BC}^3}{\overline{bc}^3},$$

or, since any two homologous edges are in the same ratio as any other two, the two similar tetraedrons are to each other as the cubes of any two homologous edges.

2d. Two similar polyedrons may be decomposed into the same number of tetraedrons, similar each to each; and any two homologous tetraedrons are to each other as the cubes of their homologous edges; but the ratio of the homologous edges of the two similar tetraedrons is equal to ratio of any two homologous edges of the polyedron (69); therefore, any two homologous tetraedrons are to each other as the cubes of two homologous edges of the polyedron, and by the theory of proportion, their sums, or the polyedrons themselves, are in the same ratio, or as the cubes of their homologous edges.

72. *Corollary* I. Similar prisms, or pyramids are to each other as the cubes of their altitudes.

73. *Corollary* II. Two similar polyedrons are to each other as the cubes of any two homologous lines.

SYMMETRICAL POLYEDRONS.

a. *Symmetry with respect to a plane.*

74. *Definitions.* Two points, A and A', are *symmetrical* with respect to a plane, MN, when this plane bisects at right angles the straight line AA' joining the points.

Two figures are symmetrical with respect to a plane, when every point of one figure has its symmetrical point in the other.

We leave the proof of the following simple theorems to the student.

75. *Theorem.* The symmetrical figure of a finite straight line, AB, is an equal straight line, $A'B'$.

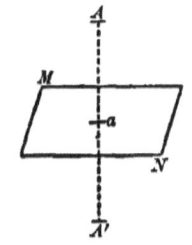

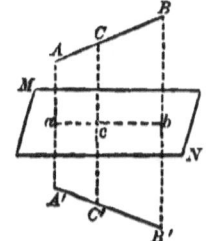

76. *Theorem.* The symmetrical figure of an indefinite straight line, AB, is another indefinite straight line, $A'B'$, which intersects the first in the plane of symmetry, and makes the same angle with the plane.

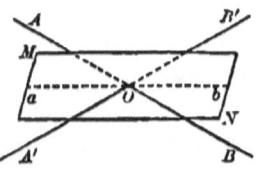

77. Theorem. *The symmetrical figure of a plane angle, BAC, is an equal plane angle, B'A'C'* (Fig. 1).

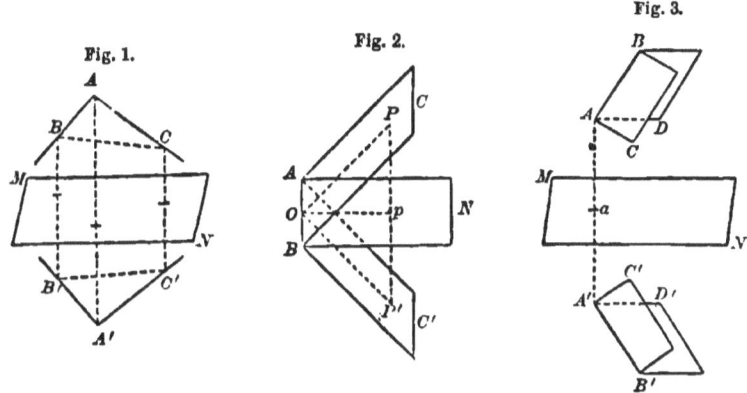

78. Theorem. *The symmetrical figure of a plane ABC, is a plane ABC'; and the two planes intersect in the plane of symmetry ABN, and make equal angles with it* (Fig. 2).

Corollary. If a plane is parallel to the plane of symmetry, its symmetrical plane is also parallel to the plane of symmetry, and at the same distance from it.

79. Theorem. *The symmetrical figure of a diedral angle, CABD, is an equal diedral angle, C'A'B'D'* (Fig. 3).

PROPOSITION XXVIII.—THEOREM.

80. *If two polyedrons are symmetrical with respect to a plane,* 1st, *their homologous faces are equal;* 2d, *their homologous polyedral angles are symmetrical.*

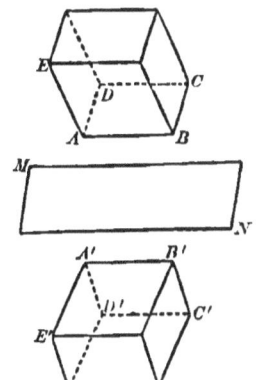

1st. Let A, B, C, D, be the vertices of a face of one of the polyedrons; their symmetrical points, A', B', C', D', are in the same plane (78); the homologous sides of the polygons $ABCD$, $A'B'C'D'$, are equal (75), and their homologous angles are equal (77); therefore the homologous faces are equal.

2d. The homologous face angles of two polyedral angles, A and A', are equal (77), and their homologous diedral angles are equal (79); but if one of the face angles as

BAD be applied to its equal B'A'D', so as to bring the other edges of the polyedral angles A and A' on the same side of the common plane B'A'D', it will be apparent that the face angles succeed each other in inverse orders in the two figures; therefore, the homologous polyedral angles of the two polyedrons are symmetrical (VI. 68).

81. *Corollary.* Two symmetrical polyedrons may be decomposed into the same number of tetraedrons symmetrical each to each. For one of the polyedrons being divided into tetraedrons by drawing diagonals from a common vertex, and the homologous diagonals being drawn in the other polyedron, any two corresponding tetraedrons thus formed will have their vertices symmetrical each to each, and will consequently be symmetrical tetraedrons.

82. *Scholium.* Two polyedrons whose faces are equal each to each and whose polyedral angles are symmetrical each to each, are called symmetrical polyedrons, whatever may be their position with respect to each other, since they admit of being placed on opposite sides of a plane so as to make their homologous vertices symmetrical with respect to that plane.

PROPOSITION XXIX.—THEOREM.

83. *Two symmetrical polyedrons are equivalent.*

Since two symmetrical polyedrons may be decomposed into the same number of tetraedrons symmetrical each to each, it is only necessary to prove that two symmetrical tetraedrons are equivalent.

Let $SABC$ be a tetraedron; let the plane of one of its faces, ABC, be taken as a plane of symmetry, and construct its symmetrical tetraedron $S'ABC$. The tetraedrons, having the same base ABC and equal altitudes SO, $S'O$, are equivalent (55).

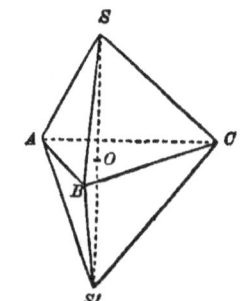

b. *Symmetry with respect to a centre.*

84. *Definitions.* Two points A and A', are symmetrical with respect to a fixed point, O, called the centre of symmetry, when this point bisects the straight line, AA', joining the two points.

228 GEOMETRY.

Any two figures are symmetrical with respect to a centre, when every point of one figure has its symmetrical point on the other.

These definitions are identical with those given in (I. 138), but are here extended to figures in space.

The student can readily establish the following theorems on figures symmetrical with respect to a centre.

85. Theorem. *The symmetrical figure of a finite straight line, AB, is an equal straight line, $A'B'$, parallel to the first* (Fig. 1).

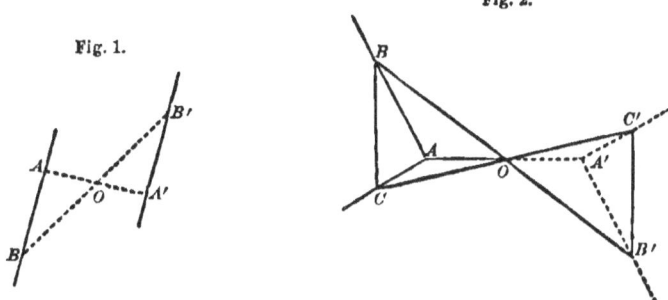

86. Theorem. *The symmetrical figure of a plane angle, BAC, is an equal plane angle, $B'A'C'$* (Fig. 2).

87. Theorem. *The symmetrical figure of a plane, BAC, is a parallel plane, $B'A'C'$* (Fig. 2).

88. Theorem. *The symmetrical figure of a diedral angle, DABC, is an equal diedral angle, $D'A'B'C'$.*

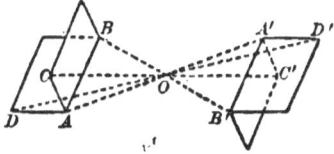

89. Theorem. *If two polyedrons are symmetrical with respect to a centre, 1st, their homologous faces are equal; 2d, their homologous angles are symmetrical.*

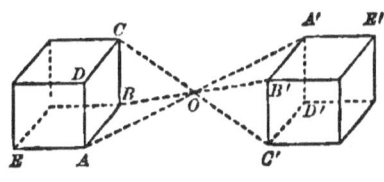

Corollary I. *The symmetrical figure of a polyedron is the same, whether the symmetry be with respect to a plane or with respect to a centre.*

Corollary II. *Two polyedrons, symmetrical with respect to a centre, are equivalent.*

c. Symmetry of a single figure.

90. Definition. Any figure in space is called *a symmetrical figure*, 1st, if it can be divided by a plane into two figures which are symmetrical with respect to that plane; 2d, if it has a centre which bisects all straight lines drawn through it, and terminated by the surface of the figure; 3d, if it has an *axis* which contains the centres of all the sections perpendicular to that axis.

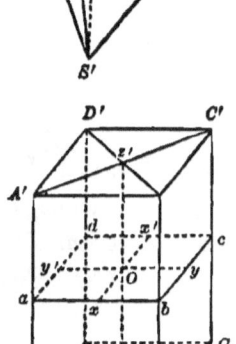

For example, 1st, the hexaedron $SABCS'$ is symmetrical with respect to the plane ABC, which divides the solid into the two symmetrical tetraedrons $SABC, S'ABC$.

2d. The intersection of the four diagonals of a parallelopiped is the centre of symmetry of the parallelopiped (18).

3d. The straight line zz', joining the centres of the bases of a right parallelopiped AC', is an axis of symmetry of the figure, since it evidently contains the centre O of any section $abcd$ perpendicular to it, or parallel to the bases. If the parallelopiped is rectangular, it has three axes xx', yy', zz', perpendicular to each other which intersect in its centre.

We leave the demonstration of the following theorems to the student.

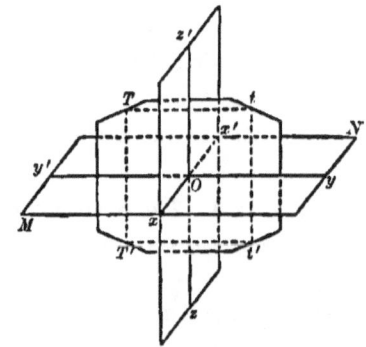

91. Theorem. *If a figure has two planes of symmetry, MN and PQ, the intersection, xx', of these planes, is an axis of symmetry of the figure.*
See (I. 141).

230 GEOMETRY.

92. Theorem. *If a figure has three planes of symmetry perpendicular to each other* (VI. 48), *the intersections of these planes are three axes of symmetry, and the common intersection of these axes is the centre of symmetry of the figure.*

THE REGULAR POLYEDRONS.

93. Definition. A *regular polyedron* is one whose faces are all equal regular polygons and whose polyedral angles are all equal to each other.

PROPOSITION XXX.—PROBLEM.

94. *To construct a regular polyedron, having given one of its edges.*

There are five regular polyedrons, which we shall consider in their order.

Construction of the regular tetraedron.

Let AB be the given edge. Upon AB construct the equilateral triangle ABC. At the centre O of this triangle erect a perpendicular, OD, to its plane, and take the point D so that $AD = AB$; join DA, DB, DC. The faces of the tetraedron $ABCD$ are each equal to the face ABC (VI. 10), and its polyedral angles are all equal (VI. 71); therefore, $ABCD$ is a regular tetraedron.

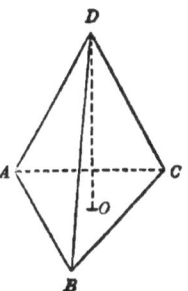

Construction of the regular hexaedron.

Upon the given edge AB, construct the square $ABCD$. The cube $ABCDE$, whose faces are each equal to this square, is a regular hexaedron, and the method of constructing it is obvious.

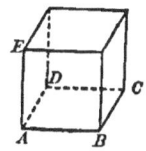

Construction of the regular octaedron.

Let AB be the given edge. Upon AB construct the square $ABCD$, and at the centre O of the square erect the perpendicular FG to its plane. In this perpendicular, take the points F and G so that $OF = OA$ and $OG = OA$, and join FA, FB, FC, FD, GA, GB, GC, GD. These edges are equal to each other (VI. 10), and also to the edge AB, since AOF and AOB are equal triangles; therefore, the faces of the figure are eight equal equilateral triangles.

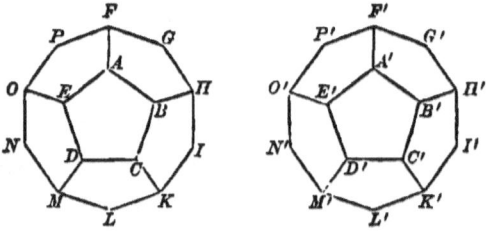

Since the triangles DFB and DAB are equal, $DFBG$ is a square, and it is evident that the pyramid $A-DFBG$ is equal in all its parts to the pyramid $F-ABCD$; therefore, the polyedral angles A and F are equal; whence, also, all the polyedral angles of the figure are equal to each other, and the figure is a regular octaedron.

Construction of the regular dodecaedron.

Upon the given edge AB, construct a regular pentagon $ABCDE$; to each of the sides of this pentagon apply the side of an equal

pentagon, and let the planes of these pentagons be so inclined to that of $ABCDE$ as to form triedral angles at A, B, C, D, E. There is thus formed a convex surface, $FGHI$, etc., composed of six regular pentagons.

Construct a second convex surface, $F'G'H'I'$, etc., equal to the first. The two surfaces may be combined so as to form a single convex surface. For, suppose the diagram to represent the exterior of the first surface and the interior of the second; let the point P of the first be placed on F' of the second; then the three equal angles OPF, $P'F'A'$, $A'F'G'$, can be united so as to form a triedral angle at F' equal to that at A', since the diedral angle $F''A'$ is already that which belongs to such a triedral angle. But when PF coincides with $F'G'$, there will be brought together at G' three angles PFA, AFG, $F'G'H'$, which will form a triedral angle equal to A',

232   GEOMETRY.

since the diedral angles at the edges FA and $F'G'$ are already those which belong to such an angle. Thus, it can be shown, successively, that all the edges PF, FG, etc., of the first figure, will coincide with the edges $F'G'$, $G'H'$, etc., of the second, and that all the polyedral angles of the whole convex surface thus formed are equal. This surface is therefore a regular dodecaedron.

Construction of the regular icosaedron.

Upon the given edge AB, construct a regular pentagon $ABCDE$, and at its centre O erect OS perpendicular to its plane, taking S so that $SA = AB$; then, joining SA, SB, etc., the pyramid S-$ABCDE$ is regular, and each of its faces is an equilateral triangle. Now let

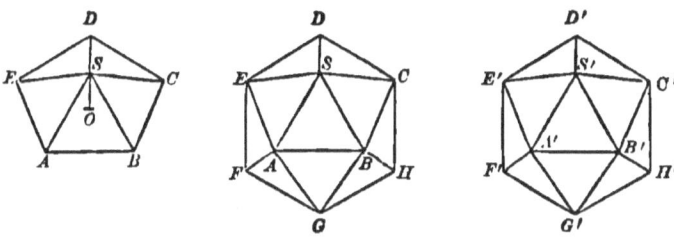

the vertices A and B be taken (as in the second figure) as the vertices of two other pyramids, A-$BSEFG$ and B-$ASCHG$, each equal to the first and having in common with it the faces ASB and ASE, ASB and BSC, respectively, and in common with each other the faces ASB and ABG. There is thus formed a convex surface $CDEFGH$, composed of ten equal equilateral triangles.

Construct a second convex surface $C'D'E'F'G'H'$, equal in all respects to the first; and let the figures represent the exterior of the first surface, and the interior of the second. Let the first surface be applied to the second by bringing the point D, where two faces meet, upon the point C', where three faces meet. The edges DE and DC can then be brought into coincidence with the edges $C'D'$ and $C'H'$, respectively, to form a polyedral angle of five faces equal to S, without in any way changing the form of either surface, since the diedral angles at the edges SD, $S'C'$, $B'C'$, are those which belong to such a polyedral angle. But when DC has been brought into coincidence with $C'H'$, there have been brought together, at the point H', five equal faces having the necessary diedral inclinations to form another polyedral angle equal to S; and thus, in succession, it can be shown that all the outer edges of the first surface coincide with those of the second, and that all the polyedral angles of the entire convex surface thus formed are equal. This surface is therefore a regular icosaedron.

PROPOSITION XXXI.—THEOREM.

95. *Only five regular (convex) polyedrons are possible.*

The faces of a regular polyedron must be regular polygons, and at least three faces are necessary to form a polyedral angle.

1st. The simplest regular polygon is the equilateral triangle. Three angles of an equilateral triangle can be combined to form a convex polyedral angle, and this combination, as shown in the preceding proposition, gives the regular tetraedron.

The combination of four such angles gives the regular octaedron; and that of five gives the regular icosaedron. The combination of six or more (each being $\frac{2}{3}$ of a right angle) gives a sum equal to, or greater than, four right angles, and therefore cannot form a convex polyedral angle (VI. 70). Therefore, only three regular convex polyedrons are possible whose surfaces are composed of triangles.

2d. Three right angles can be combined to form a polyedral angle, and this combination gives the regular hexaedron, or cube. Four or more right angles cannot form a convex polyedral angle (VI. 70); therefore, but one regular convex polyedron is possible whose surface is composed of squares.

3d. Three angles of a regular pentagon, being less than four right angles (each being $\frac{6}{5}$ of a right angle), may form a polyedral angle, as in the case of the dodecaedron; but four or more would exceed four right angles. Therefore, but one regular convex polyedron is possible with pentagonal faces.

4th. Three or more angles of a regular hexagon (each being $\frac{4}{3}$ of a right angle) cannot form a convex polyedral angle; nor can angles of any regular polygon of a greater number of sides form such a polyedral angle.

Therefore, the five regular convex polyedrons constructed in the preceding proposition are the only ones possible.

96. *Scholium.* The student may derive some aid in comprehending the preceding discussion of the regular polyedrons by constructing models of them, which he can do in a very simple manner, and at the same time with great accuracy, as follows.

Draw on card-board the following diagrams; cut them out entire, and at the lines separating adjacent polygons cut the card-board half through; the figures will then readily bend into the form of the respective surfaces, and can be retained in that form by glueing the edges.

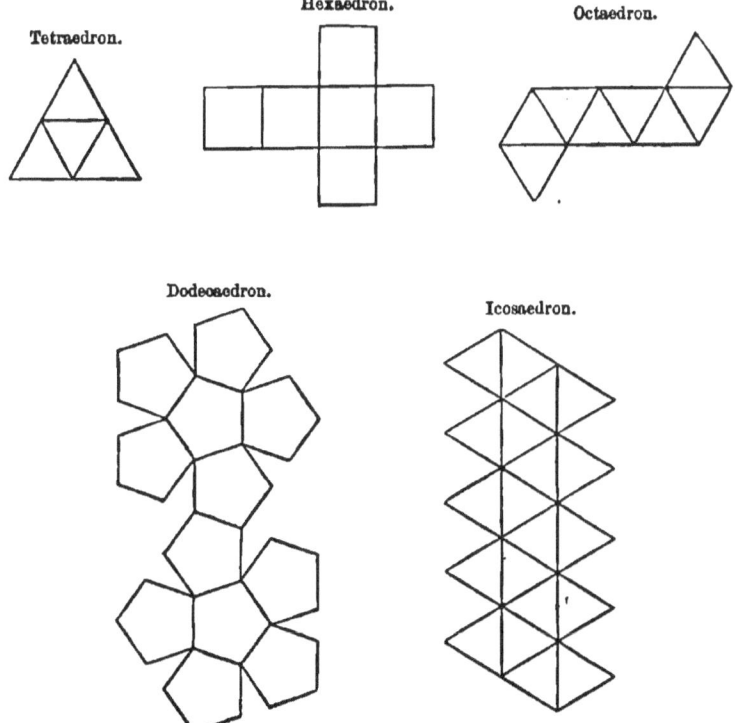

GENERAL THEOREMS ON POLYEDRONS.

PROPOSITION XXXII.—THEOREM.

97. *In any polyedron, the number of its edges increased by two is equal to the number of its vertices increased by the number of its faces.*

Let E denote the number of edges of any polyedron, V the number of its vertices, and F the number of its faces; then we are to prove that

$$E + 2 = V + F.$$

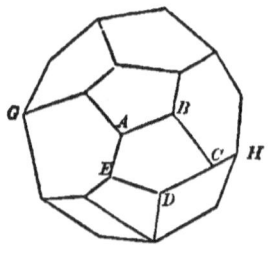

In the first place, we observe that if we remove a face, as $ABCDE$, from any convex polyedron GH, we leave an *open* surface, terminated by a broken line which was the contour of the face removed; and in this open surface the number of edges and the number of vertices remain the same as in the original surface.

Now let us form this open surface by putting together its faces successively, and let us examine the law of connection between the number of edges E, the number of vertices V, and the number of faces, at each successive step. Beginning with one face we have $E = V$. Annexing a second face, by applying one of its edges to an edge of the first, we form a surface having one edge and two vertices in common with the first; therefore, whatever the number of sides of the new face, the whole number of edges is now one more than the whole number of vertices; that is,

For 2 faces, $\qquad E = V + 1.$

Annexing a third face, adjacent to each of the former, the new surface will have two edges and three vertices in common with the preceding surface; therefore the increase in the number of edges is again one more than the increase in the number of vertices; and we have

For 3 faces, $\qquad E = V + 2.$

At different stages of this process the number of common edges to two successive open surfaces may vary, but in all cases it is apparent that the addition of a new face increases E by one more unit than it increases V; and hence we have the following series of results:

In an open surface of
$$\begin{cases} 1 \text{ face,} & E = V, \\ 2 \text{ faces,} & E = V + 1, \\ 3 \text{ "} & E = V + 2, \\ 4 \text{ "} & E = V + 3, \\ \text{etc.} & \text{etc.} \\ F - 1 \text{ faces,} & E = V + F - 2; \end{cases}$$

where the *law* is, that, in the successive values of E, the number to be added to V is a unit less than the number of faces. The last line expresses the relation for the open surface of $F - 1$ faces, that is, for the open surface which wants but one face to make the closed surface of F faces. But the number of edges and the number of vertices of this open surface are the same as in the closed surface. Therefore, in a closed surface of F faces, we have

$$E = V + F - 2,$$

or

$$E + 2 = V + F,$$

as was to be proved.

This theorem was discovered by Euler, and is called *Euler's Theorem on Polyedrons*.

PROPOSITION XXXIII.—THEOREM.

98. *The sum of all the angles of the faces of any polyedron is equal to four right angles taken as many times as the polyedron has vertices less two.*

Let E denote the number of edges, V the number of vertices, F the number of faces, and S the sum of all the angles of the faces, of any polyedron.

If we consider both the interior angles of a polygon and the exterior ones formed by producing its sides as in (I. 101), the sum of all the angles both interior and exterior is $2R \times n$, where R denotes a right angle, and n is the number of sides of the polygon. If, then, E denotes the number of edges of the polyedron, $2E$ denotes the whole number of sides of all its faces considered as independent polygons, and the sum S of the interior angles of all the F faces *plus* the sum of their exterior angles is $2R \times 2E$. But the sum of the exterior angles of one polygon is $4R$, and the sum of the exterior angles of the F polygons is $4R \times F$; that is,

$$S + 4R \times F = 2R \times 2E,$$

or, reducing,

$$S = 4R \times (E - F).$$

But by *Euler's Theorem*, $E - F = V - 2$; hence,

$$S = 4R \times (V - 2).$$

BOOK VIII.

THE THREE ROUND BODIES.

OF the various solids bounded by curved surfaces, but three are treated of in Elementary Geometry—namely, the *cylinder*, the *cone*, and the *sphere*, which are called the THREE ROUND BODIES.

THE CYLINDER.

2. *Definition.* A *cylindrical surface* is a curved surface generated by a moving straight line which continually touches a given curve, and in all of its positions is parallel to a given fixed straight line not in the plane of the curve.

Thus, if the straight line Aa moves so as continually to touch the given curve $ABCD$, and so that in any of its positions, as Bb, Cc, Dd, etc., it is parallel to a given fixed straight line Mm, the surface $ABCDdcba$ is a cylindrical surface. If the moving line is of indefinite length, a surface of indefinite extent is generated.

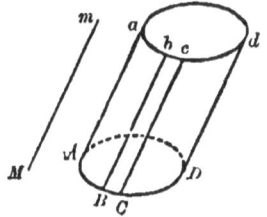

The moving line is called the *generatrix*; the curve which it touches is called the *directrix*. Any straight line in the surface, as Bb, which represents one of the positions of the generatrix, is called an *element* of the surface.

In this general definition of a cylindrical surface, the directrix may be any curve whatever. Hereafter we shall assume it to be a *closed* curve, and usually a circle, as this is the only curve whose properties are treated of in elementary geometry.

3. *Definition.* The solid *Ad* bounded by a cylindrical surface and two parallel planes, *ABD* and *abd*, is called a *cylinder;* its plane surfaces, *ABD*, *abd*, are called its *bases;* the curved surface is sometimes called its *lateral surface;* and the perpendicular distance between its bases is its *altitude.*

A cylinder whose base is a circle is called a *circular cylinder.*

4. *Definition.* A *right cylinder* is one whose elements are perpendicular to its base.

5. *Definition.* A *right cylinder with a circular base*, as *ABCa*, is called a *cylinder of revolution*, because it may be generated by the revolution of a rectangle *AOoa* about one of its sides, *Oo*, as an axis; the side *Aa* generating the curved surface, and the sides *OA* and *oa* generating the bases. The fixed side *Oo* is the *axis* of the cylinder. The radius of the base is called the *radius of the cylinder.*

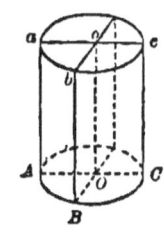

PROPOSITION I.—THEOREM.

6. *Every section of a cylinder made by a plane passing through an element is a parallelogram.*

Let *Bb* be an element of the cylinder *Ac*; then, the section *BbdD*, made by a plane passed through *Bb*, is a parallelogram.

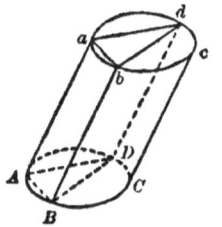

1st. The line *Dd* in which the cutting plane intersects the curved surface a second time is an element. For, if through any point *D* of this intersection a straight line is drawn parallel to *Bb*, this line by the definition of a cylindrical surface, is an element of the surface, and it must also lie in the plane *Bd*; therefore, this element, being common to both surfaces, is their intersection.

2d. The lines *BD* and *bd* are parallel (VI. 25), and the elements *Bb* and *Dd* are parallel; therefore, *Bd* is a parallelogram.

7. *Corollary.* Every section of a right cylinder made by a plane perpendicular to its base is a rectangle.

PROPOSITION II.—THEOREM.

8. *The bases of a cylinder are equal.*

Let BD be the straight line joining any two points of the perimeter of the lower base, and let a plane passing through BD and the element Bb cut the upper base in the line bd; then, $BD = bd$ (6).

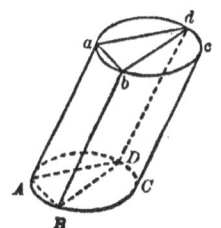

Let A be any third point in the perimeter of the lower base, and Aa the corresponding element. Join AB, AD, ab, ad. Then $AB = ab$ and $AD = ad$ (6); and the triangles ABD, abd, are equal. Therefore, if the upper base be applied to the lower base with the line bd in coincidence with its equal BD, the triangles will coincide and the point a will fall upon A; that is, *any* point a of the upper base will fall on the perimeter of the lower base, and consequently the perimeters will coincide throughout. Therefore, the bases are equal.

9. *Corollary.* I. Any two parallel sections MPN, mpn, of a cylindrical surface Ab, are equal.

For, these sections are the bases of the cylinder Mn.

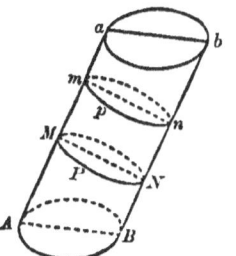

10. *Corollary* II. All the sections of a circular cylinder parallel to its bases are equal circles; and the straight line joining the centres of the bases passes through the centres of all the parallel sections. This line is called the *axis* of the cylinder.

11. *Definition.* A *tangent plane* to a cylinder is a plane which passes through an element of the curved surface without cutting this surface. The element through which it passes is called the *element of contact.*

BOOK VIII. 241

PROPOSITION III.—PROBLEM.

12. *Through a given point, to pass a plane tangent to a given circular cylinder.*

1st. When the given point is in the curved surface of the cylinder, in which case the element of contact is given, since it must be the element passing through the given point.

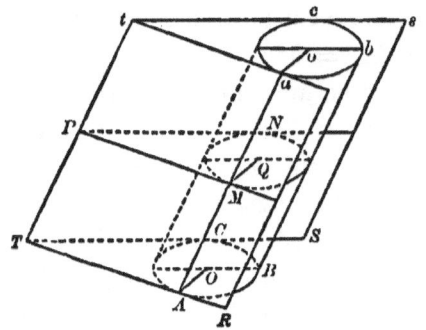

Let the given point be a point in the element Aa. At A, in the plane of the base, draw AT tangent to the base, and pass a plane Rt through Aa and AT; this plane is tangent to the cylinder. For, let P be any point in this plane not in the element Aa, and through P pass a plane parallel to the base, intersecting the cylinder in the circle MN and the plane Rt in the line MP. Let Q be the centre of the circle MN, and join QM. Since MP and MQ are parallel respectively to AT and AO (VI. 25), the angle PMQ is equal to the angle TAO, and PM is tangent to the circle MN at M; therefore, P lies without the circle MN and consequently without the cylinder. Hence the plane Rt does not cut the cylinder and is a tangent plane.

2d. When the given point is without the cylinder. Let P be the given point. Through P draw the straight line PT, parallel to the elements of the cylinder, meeting the plane of the base in T. From T draw TA and TC tangents to the base (II. 90); through PT and the tangent TA pass a plane Rt, and through PT and TC pass a plane Ts. The plane Rt, passing through PT and the point A, must contain the element Aa, since Aa is parallel to PT; and it is a tangent plane since it also contains the tangent AT. For a like reason the plane Ts is a tangent plane.

13. *Corollary.* The intersection of two tangent planes to a cylinder is parallel to the elements of the cylinder.

14. *Scholium.* Any straight line, drawn in a tangent plane and cutting the element of contact, is tangent to the cylinder.

THE CONE.

15. *Definition*. A *conical surface* is a curved surface generated by a moving straight line which continually touches a given curve, and passes through a given fixed point not in the plane of the curve.

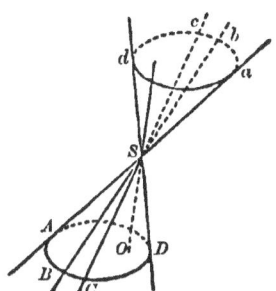

Thus, if the straight line *SA* moves so as continually to touch the given curve *ABCD*, and in all its positions, *SB*, *SC*, *SD*, etc., passes through the given fixed point *S*, the surface *S-ABCD* is a conical surface.

The moving line is called the *generatrix;* the curve which it touches is called the *directrix*. Any straight line in the surface, as *SB*, which represents one of the positions of the generatrix, is called an *element* of the surface. The point *S* is called the *vertex*.

If the generatrix is of indefinite length, as *ASa*, the whole surface generated consists of two symmetrical portions, each of indefinite extent, lying on opposite sides of the vertex, as *S-ABCD* and *S-abcd*, which are called *nappes;* one the *upper*, the other the *lower nappe*.

16. *Definition*. The solid *S-ABCD*, bounded by a conical surface and a plane *ABD* cutting the surface, is called a *cone;* its plane surface *ABD* is its *base*, the point *S* is its *vertex*, and the perpendicular distance *SO* from the vertex to the base is its *altitude*.

A cone whose base is a circle is called a *circular cone*. The straight line drawn from the vertex of a circular cone to the centre of its base is the *axis* of the cone.

17. *Definition*. A *right circular cone* is a circular cone whose axis is perpendicular to its base, as *S-ABCD*.

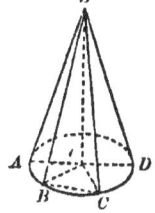

The right circular cone is also called a *cone of revolution*, because it may be generated by the revolution of a triangle, *SAO*, about one of its perpendicular sides, *SO*, as an axis; the hypotenuse *SA* generating the curved surface, and the remaining perpendicular side *OA* generating the base.

PROPOSITION IV.—THEOREM.

18. *Every section of a cone made by a plane passing through its vertex is a triangle.*

Let the cone S–$ABCD$ be cut by a plane SBC which passes through the vertex S and cuts the base in the straight line BC; then, the section SBC is a triangle, that is, the intersections SB and SC with the curved surface are straight lines.

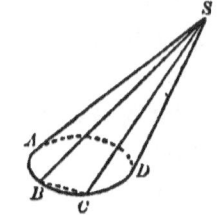

For, the straight lines joining S with B and C are elements of the surface, by the definition of a cone, and they also lie in the cutting plane; therefore they coincide with the intersections of that plane with the curved surface.

PROPOSITION V.—THEOREM.

19. *If the base of a cone is a circle, every section made by a plane parallel to the base is a circle.*

Let the section abc, of the circular cone S–ABC, be parallel to the base.

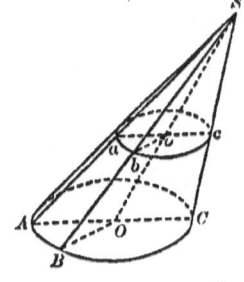

Let O be the centre of the base, and let o be the point in which the axis SO cuts the plane of the parallel section. Through SO and any number of elements SA, SB, etc., pass planes cutting the base in the radii OA, OB, etc., and the parallel section in the straight lines oa, ob, etc. Since oa is parallel to OA, and ob to OB, we have

$$\frac{oa}{OA} = \frac{So}{SO} \text{ and } \frac{ob}{OB} = \frac{So}{SO}, \text{ whence } \frac{oa}{OA} = \frac{ob}{OB}.$$

But $OA = OB$, therefore $oa = ob$; hence, all the straight lines drawn from o to the perimeter of the section are equal, and the section is a circle.

20. *Corollary.* The axis of a circular cone passes through the centres of all the sections parallel to the base.

21. *Definition.* A *tangent plane* to a cone is a plane which passes through an element of the curved surface without cutting this surface. The element through which it passes is called the *element of contact.*

PROPOSITION VI.—PROBLEM.

22. *Through a given point, to pass a plane tangent to a given circular cone.*

1st. **When the given point is in the curved surface of the cone.**

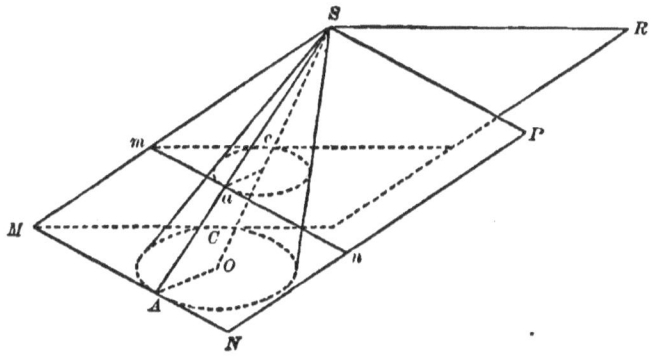

Let the given point be a point in the element SA. At A, in the plane of the base, draw AM tangent to the base, and pass a plane MP through SA and AM; this plane is tangent to the cone. The proof is the same as for the tangent plane to the cylinder.

2d. When the given point is a point m without the cone. Join the vertex S and the point m, and produce Sm to meet the plane of the base in M. From M draw MA and MC, tangents to the base, and through SM and these tangents pass the planes MP and MR. The plane MP, containing the element SA and the tangent MA, is a tangent plane to the cone, and it also passes through the given point m; and for a like reason, the plane MR also satisfies the conditions of the problem.

23. *Scholium* I. Any straight line, drawn in a tangent plane and cutting the element of contact, is tangent to the cone.

24. *Scholium* II. When the given point is without the cone, the problem may be stated in the following form:

Through any given straight line passing through the vertex of a cone, to pass a plane tangent to the cone.

THE SPHERE.

25. Definition. A *sphere* is a solid bounded by a surface all the points of which are equally distant from a point within called the *centre*.

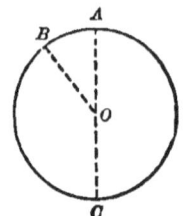

A sphere may be generated by the revolution of a semicircle ABC about its diameter AC as an axis; for the surface generated by the curve ABC will have all its points equally distant from the centre O.

A *radius* of the sphere is any straight line drawn from the centre to the surface. A *diameter* is any straight line drawn through the centre and terminated both ways by the surface.

Since all the radii are equal and every diameter is double the radius, all the diameters are equal.

26. Definition. It will be shown that every section of a sphere made by a plane is a circle; and as the greatest possible section is one made by a plane passing through the centre, such a section is called a *great circle*. Any section made by a plane which does not pass through the centre is called a *small circle*.

27. Definition. The *poles* of a circle of the sphere are the extremities of the diameter of the sphere which is perpendicular to the plane of the circle; and this diameter is called the *axis* of the circle.

PROPOSITION VII.—THEOREM.

28. *Every section of a sphere made by a plane is a circle.*

Let abc be a plane section of the sphere whose centre is O.

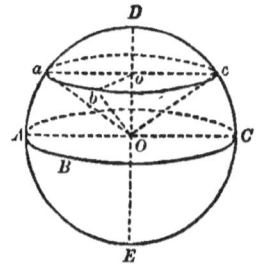

All the straight lines Oa, Ob, etc., drawn from O to points in the curve of intersection abc, are equal, being radii of the sphere; therefore, the curve abc is the circumference of a circle (VI. 12), and its centre is the foot o of the perpendicular Oo let fall from O upon the plane of the section.

29. Corollary I. All great circles, as ABC, $ADCE$, are equal;

for, since their planes pass through the centre of the sphere, their radii OA, Oa, are radii of the sphere.

30. *Corollary* II. A small circle abc is the less, the greater its distance Oo from the centre of the sphere.

31. *Corollary* III. Every great circle divides the sphere into two equal parts; for, if the parts be separated and then placed with their bases in coincidence and their convexities turned the same way, their surfaces will coincide; otherwise there would be points in the spherical surface unequally distant from its centre.

32. *Corollary* IV. Any two great circles $ACBD$, $AEBF$, bisect each other; for, the common intersection AB of their planes passes through the centre of the sphere and is a diameter of each circle.

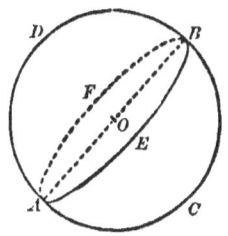

33. *Corollary* V. An arc of a great circle may be drawn through any two given points, A, E, of the surface of the sphere; for the two points, A and E, together with the centre O, determine the plane of a great circle whose circumference passes through A and E (VI. 4).

If, however, the two given points are the extremities A and B of a diameter of the sphere, the position of the circle is not determined, for the points A, O and B, being in the same straight line, an infinite number of planes can be passed through them (VI. 2).

34. *Corollary* VI. An arc of a circle may be drawn through any three given points on the surface of the sphere; for, the three points determine a plane which cuts the sphere in a circle.

PROPOSITION VIII.—THEOREM.

35. *All the points in the circumference of a circle of the sphere are equally distant from each of its poles.*

Let $abcd$ be any circle of the sphere and PP' the diameter of the sphere perpendicular to its plane; then, by the definition (27), P and P' are the poles of the circle $abcd$.

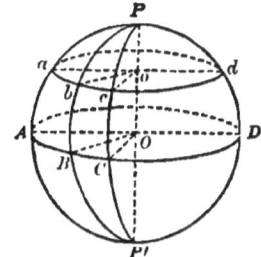

Since PP' passes through the centre o of the circle, the distances Pa, Pb, Pc, are oblique lines from P to points a, b, c, equally distant from the foot of the perpendicular, and are therefore equal (VI. 10). Hence, all the points of the circumference *abcd* are equally distant from the pole P. For the same reason, they are equally distant from the pole P'.

36. *Corollary* I. All the arcs of great circles drawn from a pole of a circle to points in its circumference, as the arcs Pa, Pb, Pc, are equal, since their chords are equal chords in equal circles.

By the distance of two points on the surface of a sphere is usually understood the arc of a great circle joining the two points. The arc of a great circle drawn from any point of a given circle *abc*, to one of its poles, as the arc Pa, is called the *polar distance* of the given circle, and the distance from the *nearest* pole is usually understood.

37. *Corollary* II. The polar distance of a great circle is a quadrant of a great circle; thus PA, PB, etc., $P'A$, $P'B$, etc., polar distances of the great circle $ABCD$, are quadrants; for, they are the measures of the right angles AOP, BOP, AOP', BOP', etc., whose vertices are at the centre of the great circles PAP', PBP', etc.

In connection with the sphere, by a *quadrant* is usually to be understood a *quadrant of a great circle*.

38. *Corollary* III. If a point P on the surface of the sphere is at the distance of a quadrant from two points, B and C, of an arc of a great circle, it is the pole of that arc. For, the arcs PB and PC being quadrants, the angles POB and POC are right angles; therefore, the radius OP is perpendicular to each of the lines OB, OC, and is consequently perpendicular to the plane of the arc BC (VI. 13); hence, P is the pole of the arc BC.

39. *Scholium*. By means of poles, arcs of circles may be drawn upon the surface of a sphere with the same ease as upon a plane surface. Thus, by revolving the arc Pa about the pole P, its extremity a will describe the small circle *abd*; and by revolving the quadrant PA about the pole P, the extremity A will describe the great circle ABD.

If two points, B and C, are given on the surface, and it is required to draw the arc BC, of a great circle, between them, it will be necessary first to find the pole P of this circle; for which purpose, take B and C as poles, and at a quadrant's distance describe two arcs on the surface intersecting in P. The arc BC can then be described with a pair of compasses, placing one foot of the compasses on P and tracing the arc with the other foot. The opening of the compasses (distance between their feet) must in this case be equal to the chord of a quadrant; and to obtain this it is necessary to know the radius of the sphere.

PROPOSITION IX.—PROBLEM.

40. *To find the radius of a given sphere.*

We here suppose that a material sphere is given, and that only measurements on the surface are possible.

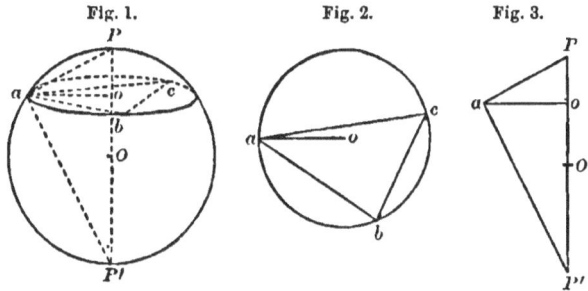

1st. With any point P (Fig. 1) of the given surface as a pole, and with any arbitrary opening of the compasses, describe a circumference abc on the surface. The rectilinear distance Pa, being the arbitrary opening of the compasses, is a known line.

Take any three points, a, b, c, in this circumference, and with the compasses measure the rectilinear distances ab, bc, ca.

2d. On a plane surface construct a triangle abc (Fig. 2), with the three distances ab, bc, ca, and find the centre o of the circle circumscribed about the triangle (II. 87). The radius ao of this circle is the radius of the circle abc of Fig. 1.

3d. With the radius ao as a side, and the known distance Pa as the hypotenuse, construct a right triangle aoP (Fig. 3). Draw aP' perpendicular to aP, meeting Po produced in P'. Then it is evident that PP', thus determined, is equal to the diameter of the given sphere, and its half PO is the required radius.

41. *Definition.* A plane is *tangent* to a sphere when it has but one point in common with the surface of the sphere.

42. *Definition.* Two spheres are tangent to each other when their surfaces have but one point in common.

PROPOSITION X.—THEOREM.

43. *A plane perpendicular to a radius of a sphere at its extremity is tangent to the sphere.*

Let O be the centre of a sphere, and let the plane MN be perpendicular to a radius OA at its extremity A; then, the plane MN is tangent to the sphere at the point A.

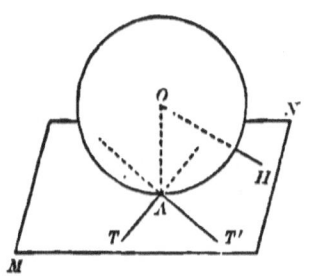

For, taking any other point, as H, in the plane, and joining OH, the oblique line OH is greater than the perpendicular OA; therefore the point H is without the sphere. Hence the plane MN has but the point A in common with the sphere, and is consequently tangent to the sphere.

44. *Corollary.* Conversely, a plane tangent to a sphere is perpendicular to the radius drawn to the point of contact. For, since every point of the plane except the point of contact is without the sphere, the radius drawn to the point of contact is the shortest line from the centre of the sphere to the plane, therefore it is perpendicular to the plane (VI. 9).

45. *Scholium.* Any straight line AT, drawn in the tangent plane through the point of contact, is tangent to the sphere.

Any two straight lines, AT, AT', tangent to the sphere at the same point A, determine the tangent plane at that point.

PROPOSITION XI.—PROBLEM.

46. *Through a given straight line without a given sphere, to pass a plane tangent to the sphere.*

Through the given straight line and the centre of the sphere, a plane can be passed which will cut the sphere in a great circle. Let the plane of the paper represent this plane; let MN be the given line, O the centre of the sphere, and $aPcP'$ the great circle in which the plane passed through MN and the centre O cuts the sphere.

From any point M in the given line draw a tangent MaT to the great circle aPc; draw MO cutting the circumference of the circle in P and P'; let fall ao perpendicular to MO, and join Oa.

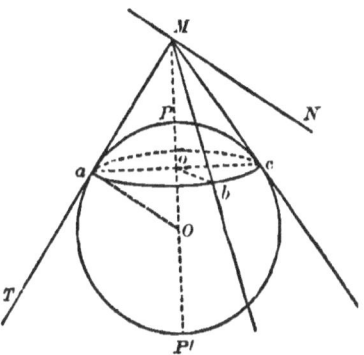

Conceive the sphere to be generated by the revolution of the semicircle PaP' about its diameter, and let the tangent Ma revolve with it. The line ao, perpendicular to the axis, will generate a small circle abc whose poles are P and P'; the tangent MaT will generate a conical surface; and the portion of this surface between the point M and the circumference abc is the surface of the cone whose vertex is M and whose base is the circle abc. Every element of this cone as Mb is a tangent to the sphere, since it has the point b, and that point only, in common with the sphere.

Now, every plane which is tangent to this cone is also tangent to the sphere; for any plane touching the cone in an element Mb, has the point b, and only the point b, in common with the sphere.

Therefore the solution of the present problem is reduced to passing a plane through the given line MN, tangent to the cone M-abc; which is done by Proposition VI. of this Book, observing the Scholium (24).

Since there are two tangent planes to the cone, there are also two tangent planes to the sphere, passing through the given line MN.

47. *Scholium.* The indefinite conical surface generated by the revolution of the tangent MT is *circumscribed about* the sphere; and the sphere is *inscribed in* this surface. The circle abc is called the *circle of contact* of the cone and sphere.

PROPOSITION XII.—THEOREM.

48. *The intersection of two spheres is a circle whose plane is perpendicular to the straight line joining the centres of the spheres, and whose centre is in that line.*

Through the centres O and O' of the two spheres, let any plane be passed, cutting the spheres in great circles which intersect each other in the points A and B; the chord AB is bisected at C by the line OO' at right angles (II. 34). If we now revolve the plane of these two circles about the line OO', the circles will generate the two spheres, and the point A will describe the line of intersection of their surfaces. Moreover, since the line AC

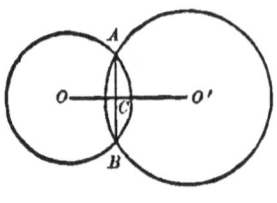

will, during this revolution, remain perpendicular to OO', it will generate a circle whose plane is perpendicular to OO' (VI. 15), and whose centre is C.

49. *Scholium.* Two spheres being given in any position whatever, if any plane is passed through their centres cutting them in two great circles, the spheres will intersect if these circles intersect, will be tangent to each other if these circles are tangent to each other, etc. For each of these positions, therefore, we shall have the same relations between the distance of the centres and the radii of the spheres, as have been established for the corresponding positions of two circles in Book II.

PROPOSITION XIII.—THEOREM.

50. *Through any four points not in the same plane, a spherical surface can be made to pass, and but one.*

Let A, B, C, D, be four given points not in the same plane. These four points may be taken as the vertices of a tetraedron $ABCD$.

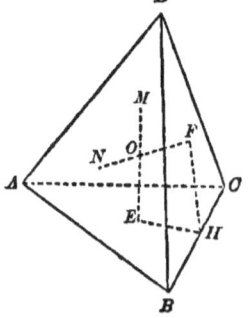

Let E be the centre of the circle circumscribed about the face ABC, and draw EM perpendicular to this face; every point in EM is equally distant from the points A, B and C (VI. 10).

Let F be the centre of the circle circumscribed about the face BCD, and draw FN perpendicular to this face; every point in FN is equally distant from the points B, C and D.

The two perpendiculars, EM and FN, intersect each other. For, let H be the middle point of BC, and draw EH, FH. The lines EH and FH are each perpendicular to BC (II. 16); therefore, the plane passed through EH and FH is perpendicular to BC (VI. 13) and consequently also to each of the faces ABC, BCD (VI. 47). Hence, the perpendiculars EM and FN lie in the same plane EHF (VI. 50), and must meet unless they are parallel; but they cannot be parallel unless the planes BCD and ABC are one and the same plane, which is contrary to the hypothesis that the four given points are not in the same plane.

The intersection O of the perpendiculars EM and FN, being equally distant from A, B and C, and also equally distant from B, C and D, is equally distant from the four points A, B, C and D; therefore, a spherical surface whose centre is O and whose radius is the distance of O from any one of these points, will pass through them all.

Moreover, since the centre of any spherical surface passing through the four points A, B, C and D is necessarily in each of the perpendiculars EM, FN, the intersection O is the centre of the only spherical surface that can be made to pass through the four given points.

51. *Corollary* I. The four perpendiculars to the planes of the faces of a tetraedron, erected at the centres of the faces, meet in the same point.

52. *Corollary* II. The six planes, perpendicular to the six edges of a tetraedron at their middle points, intersect in the same point.

PROPOSITION XIV.—THEOREM.

53. *A sphere may be inscribed in any given tetraedron.*

Let $ABCD$ be the given tetraedron.

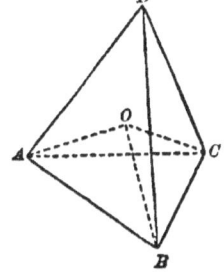

Let the planes OAB, OBC, OAC, bisect the diedral angles at the edges AB, BC, AC, respectively. Every point in the plane OAB is equally distant from the faces ABC and ABD (VI. 55); every point in the plane OBC is equally distant from the faces ABC and DBC; and every point in the plane OAC is equally distant from the faces ABC and ADC; therefore, the common intersection, O, of these three planes is equally distant from the four faces of the tetraedron; and a sphere described with O as a centre, and with a radius equal to the distance of O from any face, will be tangent to each face, and will be inscribed in the tetraedron.

54. Corollary. The six planes, bisecting the six diedral angles of a tetraedron, intersect in the same point.

SPHERICAL ANGLES.

55. Definition. The *angle of two curves* passing through the same point is the angle formed by the two tangents to the curves at that point.

This definition is applicable to any two intersecting curves in space, whether drawn in the same plane or upon a surface of any kind.

Thus, in a plane, two circumferences intersecting in a point A, make an angle equal to the angle TAT' formed by their tangents at A. In this case, the angle is also equal to the angle OAO' formed by the radii of the two circles drawn to the common point.

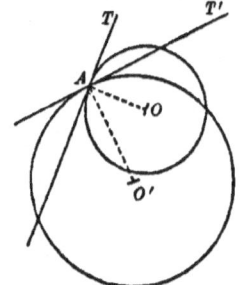

In like manner, on a sphere, the angle formed by any two intersecting curves, AB, AB', is the angle TAT', formed by the lines AT, AT', tangents to the two curves, respectively, at their common point A.

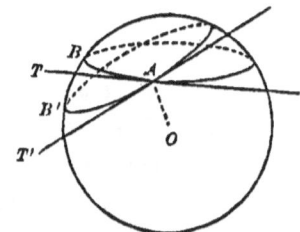

PROPOSITION XV.—THEOREM.

56. *The angle of two intersecting curves on the surface of a sphere is equal to the diedral angle between the planes passed through the centre of the sphere and the tangents to the two curves at their point of intersection.*

Let the curves, AB and AB', on the surface of a sphere whose centre is O, intersect at A, and let AT and AT' be the tangents to the two curves, respectively. Since AT and AT' do not cut the curves at A, they do not cut the surface of the sphere, and are therefore tangents to the

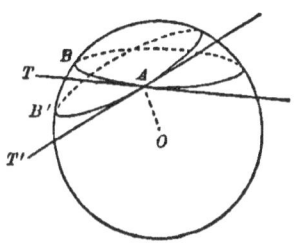

sphere. Hence they are both perpendicular to the radius OA drawn to the common point of contact, and consequently the angle $T'AT$, which is the angle of the two curves (55), measures the diedral angle of the planes OAT, OAT', passed through the radius OA and each of the tangents.

PROPOSITION XVI.—THEOREM.

57. *The angle of two arcs of great circles is equal to the angle of their planes, and is measured by the arc of a great circle described from its vertex as a pole and included between its sides (produced if necessary).*

Let AB and AB' be two arcs of great circles, AT and AT' the tangents to these arcs at A, O the centre of the sphere. The planes passing through the centre O and the tangents AT, AT', are in this case the planes of the curves AB, AB', themselves; consequently the angle BAB', or TAT', is equal to the angle of these

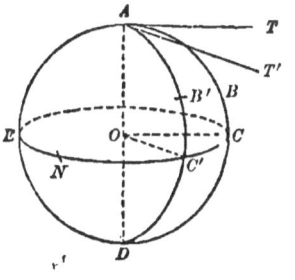

planes (56), the edge of this angle being the common diameter AOD.

Now let CC' be the arc of a great circle described from A as a pole and intersecting the arcs AB, AB' (produced if necessary), in C and C'. The radii OC and OC' are perpendicular to AO, since the arcs AC, AC', are quadrants (37); therefore, the angle COC' is also equal to the diedral angle AO, or to the angle BAB', and it is measured by the arc CC'.

58. *Corollary.* Any great circle arc AC', drawn through the pole of a given great circle CC', is perpendicular to the circumference CC'. For, the pole A being in the diameter AOD perpendicular to

the plane of CC', the plane of AC' is perpendicular to the plane of CC' (VI. 47), and hence the angle C' is a right angle.

Conversely, any great circle arc $C'A$ perpendicular to the arc CC' must pass through the pole A of CC'.

59. Scholium. If it is required to draw a great circle $B'C'$ perpendicular to a given great circle $CC'E$, through a given point B', we have only to find the pole N of the required arc by describing, from B' as a pole and at a quadrant's distance, an arc cutting $CC'E$ in N; then, from N as a pole, the perpendicular $B'C'$ can be described.

SPHERICAL POLYGONS AND PYRAMIDS.

60. Definition. A *spherical polygon* is a portion of the surface of a sphere bounded by three or more arcs of great circles, as $ABCD$.

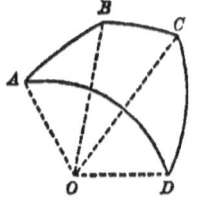

Since the planes of all great circles pass through the centre of the sphere, the planes of the sides of a spherical polygon form, at the centre O, a polyedral angle of which the edges are the radii drawn to the vertices of the polygon, the face angles are angles at the centre measured by the sides of the polygon, and the diedral angles are equal to the angles of the polygon (57).

Since in a polyedral angle each face angle is assumed to be less than two right angles, each side of a spherical polygon will be assumed to be less than a semi-circumference.

A spherical polygon is *convex* when its corresponding polyedral angle at the centre is convex (VI. 67).

A *diagonal* of a spherical polygon is an arc of a great circle joining any two vertices not consecutive.

61. Definition. A *spherical triangle* is a spherical polygon of three sides. It is called *right angled*, *isosceles*, or *equilateral*, in the same cases as a plane triangle.

62. Definition. A *spherical pyramid* is a solid bounded by a spherical polygon and the planes of the sides of the polygon; as $O-ABCD$. The centre of the sphere is the *vertex* of the pyramid; the spherical polygon is its *base*.

63. Symmetrical spherical triangles and polygons. Let ABC be a spherical triangle, and O the centre of the sphere. Drawing the radii OA, OB, OC, we form the triedral angle $O-ABC$, at the centre. The sides AB, BC, AC, of the triangle are respectively the measures of the face angles AOB, BOC, AOC, of the triedral angle; and the angles A, B, C, of the triangle are respectively equal to the diedral angles at the edges OA, OB, OC, of the triedral angle (57).

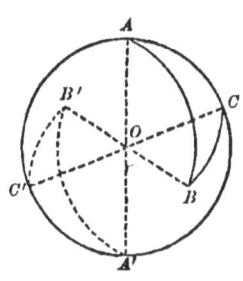

If the radii AO, BO, CO, are produced to meet the surface of the sphere in the points A', B', C', and if these points are joined by arcs of great circles $A'B'$, $B'C'$, $A'C'$, a triedral angle $O-A'B'C'$ is formed symmetrical with $O-ABC$ (VI. 68), and its corresponding spherical triangle $A'B'C'$ is symmetrical with ABC.

The spherical pyramid $O-A'B'C'$ is also symmetrical with the spherical pyramid $O-ABC$.

In the same manner, we may form two symmetrical polygons of any number of sides, and corresponding symmetrical pyramids.

64. Two symmetrical spherical triangles, or polygons, are still called symmetrical in whatever position they may be placed on the surface of the sphere. If we place the symmetrical triangles of the preceding figure with the vertices A' and B' in coincidence with their homologous vertices A and B, their third vertices C and C' will lie on opposite sides of the arc AB. In this position, it is apparent that the order of arrangement of the parts in one triangle is the reverse of that in the other, and that, in general, two symmetrical spherical triangles cannot be made to coincide by superposition.

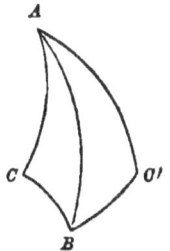

65. There is, however, one exception to the last remark, namely, the case of *symmetrical isosceles triangles*. For, if ABC is an isosceles spherical triangle and $AB = AC$, then, in its symmetrical triangle we have $A'B' = A'C'$, and consequently $AB = A'C'$, $AC = A'B'$, and since

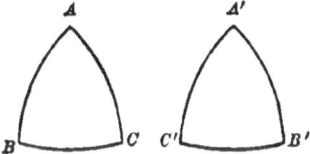

BOOK VIII. 257

the angles A and A' are equal, if AB be placed on $A'C'$, AC will fall on its equal $A'B'$ and the two triangles will coincide throughout.

66. In consequence of the relation established between polyedral angles and spherical polygons, it follows that from any property of polyedral angles we may infer an analogous property of spherical polygons.

Reciprocally, from any property of spherical polygons we may infer an analogous property of polyedral angles.

The latter is in almost all cases the more simple mode of procedure, inasmuch as the comparison of figures drawn on the surface of a sphere is nearly if not quite as simple as the comparison of plane figures.

67. *Definition.* If from the vertices of a spherical triangle as poles, arcs of great circles are described, these arcs form by their intersection a second triangle which is called the *polar triangle* of the first.

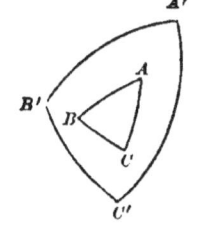

Thus, if A, B and C are the poles of the arcs of great circles, $B'C'$, $A'C'$, and $A'B'$, respectively, $A'B'C'$ is the polar triangle of ABC.

Since all great circles, when completed, intersect each other in two points, the arcs $B'C'$, $A'C'$, $A'B'$, if produced, will form three other triangles; but the triangle which is taken as the polar triangle is that whose vertex A', homologous to A, lies on the same side of the arc BC as the vertex A; and so of the other vertices.

PROPOSITION XVII.—THEOREM.

68. *If $A'B'C'$ is the polar triangle of ABC, then, reciprocally, ABC is the polar triangle of $A'B'C'$.*

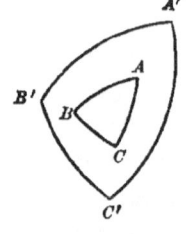

For, since A is the pole of the arc $B'C'$, the point B' is at a quadrant's distance from A; and since C is the pole of the arc $A'B'$, the point B' is at a quadrant's distance from C; therefore, B' is the pole of the arc AC (38). In the same manner, it is shown that A' is the pole of the arc BC, and C' the pole of the arc AB. Moreover, A and A' are on the same side of $B'C'$, B and B' on the same side of $A'C'$,

C and C' on the same side of $A'B'$; therefore, ABC is the polar triangle of $A'B'C'$.

PROPOSITION XVIII.—THEOREM.

69. *In two polar triangles, each angle of one is measured by the supplement of the side lying opposite to it in the other.*

Let ABC and $A'B'C'$ be two polar triangles.

Let the sides AB and AC, produced if necessary, meet the side $B'C'$ in the points b and c. The vertex A being the pole of the arc bc, the angle A is measured by the arc bc (57).

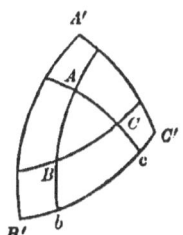

Now, B' being the pole of the arc Ac and C' the pole of the arc Ab, the arcs $B'c$ and $C'b$ are quadrants; hence we have

$$B'C' + bc = B'c + C'b = \text{a semi-circumference.}$$

Therefore bc, which measures the angle A, is the supplement of the side $B'C'$ (II. 55).

In the same manner, it can be shown that each angle of either triangle is measured by the supplement of the side lying opposite to it in the other triangle.

70. *Scholium* I. Let the angles of the triangle ABC be denoted by A, B and C, and let the sides opposite to them, namely, BC, AC and AB, be denoted by a, b and c, respectively. Let the corresponding angles and sides of the polar triangle be denoted by A', B', C', a', b' and c'. Also let both angles and sides be expressed in degrees (II. 54). Then, the preceding theorem gives the following relations:

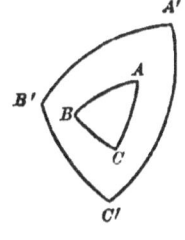

$$A + a' = B + b' = C + c' = 180°,$$
$$A' + a = B' + b = C' + c = 180°,$$

also $A - a = A' - a'$, etc.

71. *Scholium* II. Two triedral angles at the centre of the sphere, corresponding to two polar triangles on the surface, are called *supplementary triedral angles;* for, it follows from the preceding theorem, and from the relation between any spherical polygon and its corresponding polyedral angle (60), that the diedral angles of either of these triedral angles are respectively the supplements of the opposite face angles of the other.

PROPOSITION XIX.—THEOREM.

72. *Two triangles on the same sphere are either equal or symmetrical, when two sides and the included angle of one are respectively equal to two sides and the included angle of the other.*

In the triangles ABC and DEF, let the angle A be equal to the angle D, the side AB equal to the side DE, and the side AC equal to side DF.

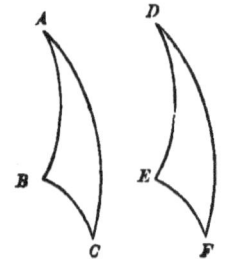

1st. When the parts of the two triangles are in the same order, ABC can be applied to DEF, as in the corresponding case of plane triangles (I. 76), and the two triangles will coincide; therefore, they are equal.

2d. When the parts of the two triangles are in inverse order, let $DE'F$ be the symmetrical triangle of DEF, and therefore having its angles and sides equal, respectively, to those of DEF. Then, in the triangle ABC and $DE'F$, we shall have the angle BAC equal to the angle $E'DF$, the side AB to the side DE', and

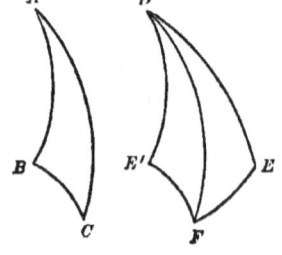

the side AC to the side DF, and these parts arranged in the same order in the two triangles; therefore, the triangle ABC is equal to the triangle $DE'F$, and consequently symmetrical with DEF.

73. *Scholium.* In this proposition, and in those which follow, the two triangles may be supposed on the same sphere, or on two equal spheres.

PROPOSITION XX.—THEOREM.

74. *Two triangles on the same sphere are either equal or symmetrical, when a side and the two adjacent angles of one are equal respectively to a side and the two adjacent angles of the other.*

For, one of the triangles may be applied to the other, or to its symmetrical triangle, as in the corresponding case of plane triangles (I. 78).

PROPOSITION XXI.—THEOREM.

75. *Two triangles on the same sphere are either equal or symmetrical, when the three sides of one are respectively equal to the three sides of the other.*

For, their corresponding triedral angles at the centre of the sphere are either equal or symmetrical (VI. 71).

PROPOSITION XXII.—THEOREM.

76. *If two triangles on the same sphere are mutually equiangular, they are also mutually equilateral; and are either equal or symmetrical.*

Let the spherical triangles M and N be mutually equiangular.

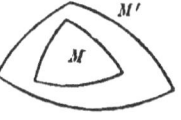 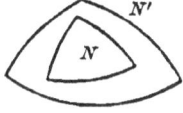

Let M' be the polar triangle of M, and N' the polar triangle of N. Since M and N are mutually equiangular, their polar triangles M' and N' are mutually equilateral (69); therefore, by the preceding proposition, the triangles M' and N' are mutually equiangular. But M' and N' being mutually equiangular, their polar triangles M and N are mutually equilateral (69). Consequently, M and N are either equal or symmetrical (75).

77. *Scholium.* It may seem to the student that the preceding property destroys the analogy which subsists between plane and spherical triangles, since two mutually equiangular plane triangles are not necessarily mutually equilateral. But in the case of spherical triangles, the equality of the sides follows from that of the angles only upon the condition that the triangles are constructed upon the same sphere or on equal spheres; if they are constructed on spheres of different radii, the homologous sides of two mutually equiangular triangles will no longer be equal, but will be proportional to the radii of the sphere; the two triangles will then be *similar*, as in the case of plane triangles.

PROPOSITION XXIII.—THEOREM.

78. *In an isosceles spherical triangle, the angles opposite the equal sides are equal.*

In the spherical triangle ABC, let $AB = AC$; then, $B = C$.

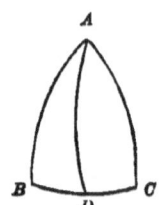

For, draw the arc AD of a great circle, from the vertex A to the middle of the base BC. The triangles ABD and ACD are mutually equilateral, and in this case are symmetrical (77); therefore $B = C$.

79. *Corollary.* Since the triangles ABD and ACD are mutually equiangular, we have the angle BAD equal to the angle CAD, and the angle ADB equal to the adjacent angle ADC; therefore, *the arc drawn from the vertex of an isosceles spherical triangle to the middle of the base is perpendicular to the base and also bisects the vertical angle.*

80. *Scholium.* This proposition and its corollary may also be proved by applying the isosceles triangle to its symmetrical triangle (65).

PROPOSITION XXIV.—THEOREM.

81. *If two angles of a spherical triangle are equal, the triangle is isosceles.*

In the triangle ABC let $B = C$; then, $AB = AC$.

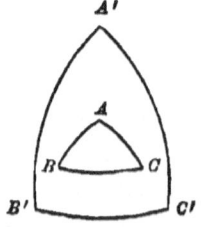

For, let $A'B'C'$ be the polar triangle of ABC. Then, the sides $A'B'$ and $A'C'$ are equal (69), and therefore the angles B' and C' are equal (78). But since the angles B' and C' are equal in the triangle $A'B'C'$, the sides AB and AC are equal in its polar triangle ABC.

PROPOSITION XXV.—THEOREM.

82. *Any side of a spherical triangle is less than the sum of the other two.*

Let ABC be a spherical triangle; then, any side, as AC, is less than the sum of the other two, AB and AC.

For, in the corresponding triedral angle formed at the centre O of the sphere, we have the angle AOC less than the sum of the angles AOB and BOC (VI. 69); and since the sides of the triangle measure these angles, respectively, we have $AC < AB + BC$.

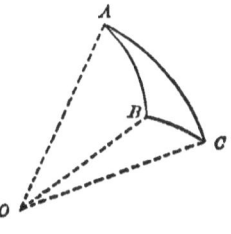

83. *Corollary.* Any side, AB, of a spherical polygon $ABCDE$ is less than the sum of all the other sides.

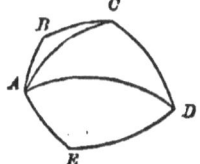

PROPOSITION XXVI.—THEOREM.

84. *In a spherical triangle, the greater side is opposite the greater angle; and conversely.*

1st. In the triangle ABC suppose $ABC > ACB$; then, $AC > AB$. For, draw the arc BD making the angle $DBC = DCB$; then, the triangle BDC is isosceles (81), and $DC = DB$. Adding DA to each of these equals we have $AC = DB + DA$. But $DB + DA > AB$ (82); therefore, $AC > AB$.

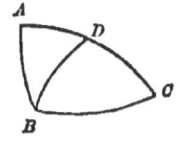

2d. Conversely, in the triangle ABC suppose $AC > AB$; then $ABC > ACB$. For, if ABC were equal to ACB, AC would be equal to AB (81), which is contrary to the hypothesis; and if ABC were less than ACB, AC would be less than AB, which is also contrary to the hypothesis; therefore, ABC must be greater than ACB.

PROPOSITION XXVII.—THEOREM.

85. *If from the extremities of one side of a spherical triangle two arcs of great circles are drawn to a point within the triangle, the sum of these arcs is less than the sum of the other two sides of the triangle.*

In the spherical triangle ABC, let the arcs BD and CD be drawn to any point D within the triangle; then, $DB + DC < AB + AC$.

For, produce BD to meet AC in E; then we have $DC < DE + EC$ (82); and adding BD to both members of this inequality, we have $DB + DC < BE + EC$. In the same manner, we prove that $BE + EC < AB + AC$; therefore, $DB + DC < AB + AC$.

PROPOSITION XXVIII.—THEOREM.

86. *The sum of the sides of a convex spherical polygon is less than the circumference of a great circle.*

For, the sum of the face angles of the corresponding polyedral angle at the centre of the sphere is less than four right angles (VI. 70).

PROPOSITION XXIX.—THEOREM.

87. *The sum of the angles of a spherical triangle is greater than two, and less than six, right angles.*

For, denoting the angles of a spherical triangle by A, B, C, and the sides respectively opposite to them in its polar triangle by a', b', c', we have (70),

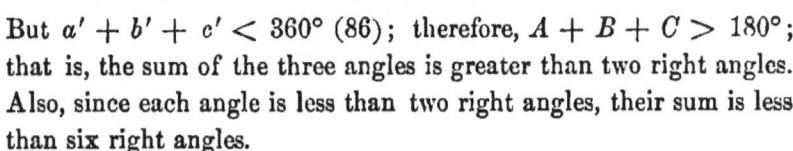

$A = 180° - a'$, $B = 180° - b'$, $C = 180° - c'$,

the sum of which is

$A + B + C = 540° - (a' + b' + c')$.

But $a' + b' + c' < 360°$ (86); therefore, $A + B + C > 180°$; that is, the sum of the three angles is greater than two right angles. Also, since each angle is less than two right angles, their sum is less than six right angles.

88. *Corollary*. A spherical triangle may have two or even three right angles; also two or even three obtuse angles.

89. *Definitions*. If a spherical triangle ABC has two right angles, B and C, it is called a *bi-rectangular triangle*; and since the sides AB and AC must each pass through the pole of BC (58), the vertex A is that pole, and therefore AB and AC are quadrants.

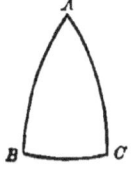

If a triangle has three right angles it is called a *tri-rectangular triangle;* each of its sides is a quadrant, and each vertex is the pole of the opposite side. Three planes passed through the centre of the sphere, each perpendicular to the other two (VI. 48), divide the surface of the sphere into eight tri-rectangular triangles, ABC, $A'BC$, etc.

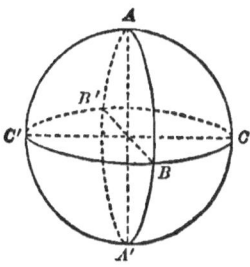

RATIO OF THE SURFACES AND VOLUMES OF SPHERICAL FIGURES.

90. *Definitions.* A *lune* is a portion of the surface of a sphere included between two semi-circumferences of great circles; as $AMBNA$.

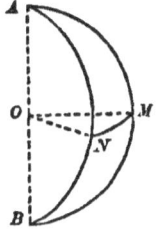

A *spherical ungula*, or *wedge*, is a solid bounded by a lune and the two semicircles which intercept the lune on the surface of the sphere; as the solid $ABMANB$. The common diameter AB, of the semicircles, is called the *edge* of the ungula; the lune is called its *base*.

91. *Definition.* The excess of the sum of the angles of a spherical triangle over two right angles is called the *spherical excess*.

If the angles of a spherical triangle ABC are denoted by A, B and C, and its spherical excess by E, and if a right angle is the unit employed in expressing the angles, we shall have

$$E = A + B + C - 2.$$

PROPOSITION XXX.—THEOREM.

92. *Two symmetrical spherical triangles are equivalent.*

Let ABC and $A'B'C'$ be two symmetrical triangles with their homologous vertices diametrically opposite to each other on the sphere. Let P be the pole of the small circle which passes through the three points A, B and C. The great circle arcs PA, PB, PC, are equal (36).

Draw the diameter POP' and the great circle arcs $P'A'$, $P'B'$, $P'C'$; these arcs being equal, respectively, to PA, PB, PC, are also equal to each other.

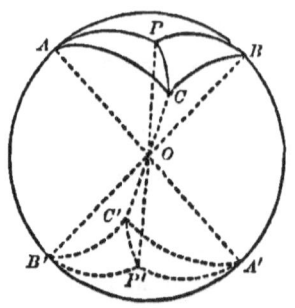

The triangles PAB, $P'A'B'$, are mutually equilateral, and also isosceles; therefore, they are superposable (65) and are equal in area. For the same reason the triangle PAC is equivalent to the triangle $P'A'C'$, and PBC is equivalent to $P'B'C'$. Therefore the triangle ABC, which is the sum of the triangles PAB, PAC and PBC, is equivalent to its symmetrical triangle $A'B'C'$ which is the sum of the triangles $P'A'B'$, $P'A'C'$ and $P'B'C'$.

If the pole P should fall without the triangle ABC, the triangle would be equivalent to the sum of two of the isosceles triangles diminished by the third; but as the same thing would occur for the symmetrical triangle, the conclusion would be the same.

93. *Corollary* I. If the arcs of two great circles, ACA', BCB', intersect on the surface of a hemisphere, the sum of the opposite triangles ACB, $A'CB'$, is equivalent to a lune whose angle is the angle ACB, formed by the great circles.

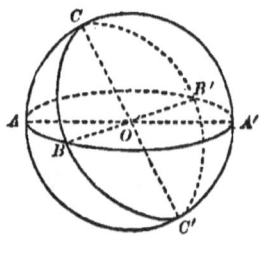

For, completing the great circle $BCB'C'$, the triangles $A'CB'$, $AC'B$, are symmetrical, and therefore equivalent. Hence, the sum of ACB and $A'CB'$ is equivalent to the sum of ACB and $AC'B$, that is, to the lune $ACBC'A$, whose angle is the angle ACB.

94. *Corollary* II. The reasoning employed in the demonstration of the theorem may be applied also to the pyramids whose bases are two symmetrical triangles. Hence, *two symmetrical spherical triangular pyramids are equivalent.*

Also by the reasoning in Corollary I. we infer that *the sum of the volumes of two spherical triangular pyramids the sum of whose bases is equivalent to a lune, is equal to the volume of the ungula whose base is that lune.*

PROPOSITION XXXI.—THEOREM.

95. *A lune is to the surface of the sphere as the angle of the lune is to four right angles.*

Let $ANBMA$ be a lune, and let MNP be the great circle whose poles are the extremities of the diameter AB.

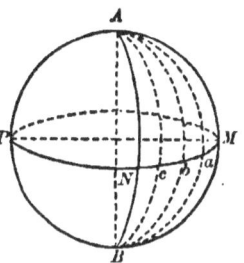

Let the circumference of the circle MNP be divided into any number of equal parts Ma, ab, etc.; and let planes be passed through the diameter AB and each of the points of division. The whole surface of the sphere will evidently be divided into equal lunes of which the given lune will contain the same number as there are parts in the arc MN. Hence, whether the number of the parts in MN and the number of the parts in the whole circumference MNP, are commensurable or incommensurable, the ratio of the lune $ANBMA$ to the surface of the sphere is the same as the ratio of the arc MN to the circumference MNP; or, since MN is the measure of the angle of the lune, and the circumference MNP is the measure of four right angles, the lune is to the surface of the sphere as the angle of the lune is to four right angles.

96. *Corollary* I. Two lunes, on the same or on equal spheres, are to each other as their angles.

97. *Corollary* II. If we denote the surface of the tri-rectangular triangle by T, the surface of the whole sphere will be $8T$ (89); therefore, denoting the surface of the lune by L and its angle by A, the unit of the angle being a right angle, we have

$$\frac{L}{8T} = \frac{A}{4}, \text{ whence } L = T \times 2A.$$

If, further, we take the tri-rectangular triangle as the unit of surface in comparing surfaces on the same sphere, we shall have

$$L = 2A;$$

that is, *a right angle being the unit of angles, and the tri-rectangular* triangle the unit of spherical surfaces, the area of a lune is expressed by twice its angle.

98. *Corollary* III. The *tri-rectangular spherical pyramid* (that whose base is the tri-rectangular triangle) being taken as the unit of volume, the same reasoning may be employed to prove that *the volume of an ungula will be expressed by twice its angle.*

PROPOSITION XXXII.—THEOREM.

99. *The area of a spherical triangle is equal to its spherical excess* (the right angle being the unit of angles and the tri-rectangular triangle the unit of areas).

For, let ABC be a spherical triangle. Complete the great circle $ABA'B'$, and produce the arcs AC and BC to meet this circle in A' and B'.

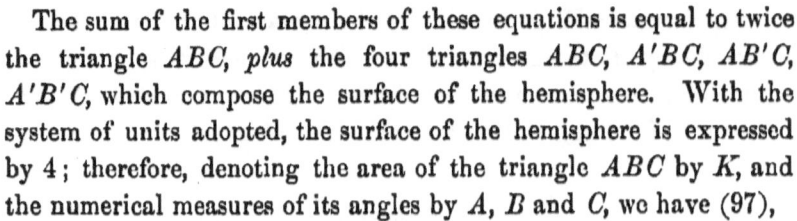

We have, by the figure,

$$ABC + A'BC = \text{lune } A,$$
$$ABC + AB'C = \text{lune } B,$$

and by (93)

$$ABC + A'B'C = \text{lune } C.$$

The sum of the first members of these equations is equal to twice the triangle ABC, *plus* the four triangles ABC, $A'BC$, $AB'C$, $A'B'C$, which compose the surface of the hemisphere. With the system of units adopted, the surface of the hemisphere is expressed by 4; therefore, denoting the area of the triangle ABC by K, and the numerical measures of its angles by A, B and C, we have (97),

$$2K + 4 = 2A + 2B + 2C,$$

whence

$$K = A + B + C - 2 = \textit{spherical excess.}$$

100. *Corollary.* The same reasoning, in connection with (94) and (98), may be employed to prove that, if V is the volume of a spherical triangular pyramid whose base is the spherical triangle ABC, and if the unit of volume is the volume of the tri-rectangular spherical pyramid, we shall have

$$V = A + B + C - 2.$$

268 GEOMETRY.

101. *Scholium.* It must not be forgotten that the preceding results are merely the expression of the *ratios* of the figures considered to the adopted units. For example, suppose the angles of a spherical triangle are given in degrees as follows: $A = 80°$, $B = 100°$, $C = 120°$; then, reducing them to the right angle as the unit,

$$K = \frac{80}{90} + \frac{100}{90} + \frac{120}{90} - 2 = \frac{4}{3},$$

therefore, the area of this triangle is $\frac{4}{3}$ of the area of the tri-rectangular triangle.

Also, the volume of the spherical pyramid of which this triangle is the base is $\frac{4}{3}$ of the volume of the tri-rectangular spherical pyramid.

Hence, also, it follows that *the volumes of two triangular spherical pyramids are to each other as the areas of their bases.*

PROPOSITION XXXIII.—THEOREM.

102. *The area of a spherical polygon is measured by the sum of its angles* minus *the product of two right angles multiplied by the number of sides of the polygon less two.*

Let $ABCDE$ be a spherical polygon. From any vertex, as A, draw the diagonals AC, AD; the polygon will be divided into as many triangles as there are sides less two. The surface of each triangle is measured by the sum of its angles *minus* two right angles; and the sum of all the angles of the triangles is equal to the sum of the angles of the polygon; therefore the surface of the polygon is measured by the sum of its angles *minus* two right angles multiplied by the number of triangles, that is, by the number of sides of the polygon less two.

103. *Corollary* I. Denoting the number of sides of the polygon by n, the sum of its angles by S, and its area by K, then, with the adopted system of units, we have

$$K = S - 2(n - 2) = S - 2n + 4.$$

104. *Corollary* II. The tri-rectangular pyramid being taken as the unit of volume, the volume of any spherical pyramid will have the

same numerical expression as the area of its base; that is, *the volume of a spherical pyramid is to the volume of the tri-rectangular pyramid as the base of the pyramid is to the tri-rectangular triangle.*

Now the volume of the tri-rectangular pyramid is one-eighth of the volume of the sphere, and the tri-rectangular triangle is one-eighth of the surface of the sphere; therefore, *the volume of a spherical pyramid is to the volume of the sphere as its base is to the surface of the sphere.*

SHORTEST LINE ON THE SURFACE OF A SPHERE BETWEEN TWO POINTS.

PROPOSITION XXXIV.—THEOREM.

105. *The shortest line that can be drawn on the surface of a sphere between two points is the arc of a great circle, less than a semi-circumference, joining the two points.*

Let AB be an arc of a great circle, less than a semi-circumference, joining any two points A and B of the surface of a sphere; and let C be any arbitrary point taken in that arc. Then we say that the shortest line from A to B, on the surface of the sphere, must pass through C.

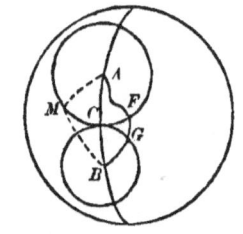

From A and B as poles, with the polar distances AC and BC, describe circumferences on the surface; these circumferences touch at C and lie wholly without each other. For, let M be any point in the circumference whose pole is A, and draw the arcs of great circles AM, BM, forming the spherical triangle AMB. We have, by (82), $AM + BM > AB$, and subtracting from the two members of this inequality the equal arcs AM and AC, we have $BM > BC$; therefore, M lies without the circumference whose pole is B.

Now let $AFGB$ be any line from A to B, on the surface of the sphere, which does not pass through the point C, and which therefore cuts the two circumferences in different points, one in F, the other in G. Whatever may be the nature of the line AF, an equal line can be drawn from A to C; for, if AC and AF be conceived to be drawn on two equal spheres having a common diameter passing through A, and therefore having their surfaces in coincidence, and if one of these spheres be turned upon the common diameter as an axis, the point A will be fixed and the point F will come into coincidence with C; the surfaces of the two spheres continuing to coincide, the line AF will then lie on the common surface between A and C. For the same reason, a line can be drawn from B to C, equal to BG. Therefore, a line can be drawn from A to B, through C, equal to the sum of AF and BG, and consequently less than any line $AFGB$ that does not pass through C. The shortest line from A to B therefore passes through C, that is, through any, or every, point in AB; consequently it must be the arc AB itself.

BOOK IX.

MEASUREMENT OF THE THREE ROUND BODIES.

THE CYLINDER.

1. *Definition.* The area of the convex, or lateral, surface of a cylinder is called its *lateral area*.

2. *Definition.* A prism is *inscribed in a cylinder* when its bases are inscribed in the bases of the cylinder.

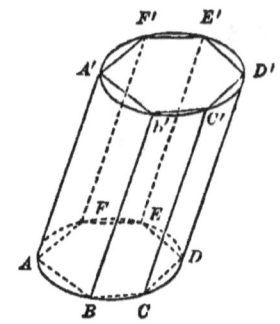

If a polygon $ABCDEF$ is inscribed in the base of a cylinder, planes passed through the sides of the polygon, parallel to the elements of the cylinder, intersect the cylinder in parallelograms, $ABB'A'$, etc. (VIII. 6), which evidently determine a prism inscribed in the cylinder.

3. *Definition.* A prism is *circumscribed about a cylinder* when its bases are circumscribed about the bases of the cylinder.

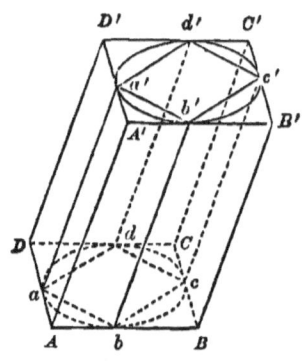

If a polygon $ABCD$ is circumscribed about the base of a cylinder, planes passed through the sides of the polygon, parallel to the elements of the cylinder, will evidently contain the elements, aa', bb', etc., drawn at the points of contact, and be tangent to the cylinder in these elements. The intersection of these planes with the plane of the upper base of the cylinder will therefore determine a polygon $A'B'C'D'$, equal to $ABCD$, circumscribed about the upper base, and a prism will be formed which is circumscribed about the cylinder.

4. Definition. A *right section* of a cylinder is a section made by a plane perpendicular to its elements; as *abcdef*.

The intersection of the same plane with an inscribed or circumscribed prism is a right section of the prism.

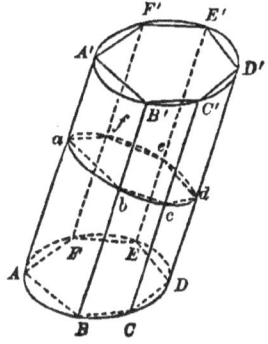

5. Definition. *Similar cylinders of revolution* are those which are generated by similar rectangles revolving about homologous sides.

PROPOSITION I.—THEOREM.

6. *A cylinder is the limit of the inscribed and circumscribed prisms, the number of whose faces is indefinitely increased.*

Let any polygon *abcd* be inscribed in the base of the cylinder *ac'*, and at the vertices of this polygon let tangents be drawn to the base of the cylinder forming the circumscribed polygon *ABCD*. Upon these polygons as bases let prisms be formed, inscribed in, and circumscribed about, the cylinder. We shall assume, as evident, that the convex surface of the cylinder is greater than that of the inscribed prism and less than that of the circumscribed prism.*

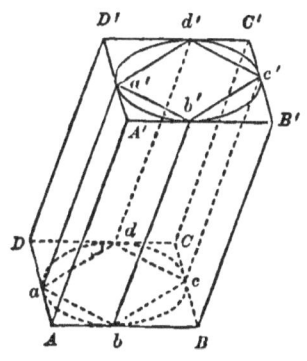

Suppose the arcs *ab*, *bc*, etc., to be bisected and polygons to be formed having double the number of sides of the first; and upon these as bases suppose prisms to be constructed, inscribed and circumscribed, as before; and let this process be repeated an indefinite number of times. The difference between the convex surface of the inscribed prism and that of the corresponding circumscribed prism will continually diminish and approach to zero as its limit. There-

---

* A proof, however, can be given analogous to that of (V. 32).

fore these convex surfaces themselves approach to the convex surface of the cylinder as their common limit.

At the same time, it is evident that the volumes of the inscribed and circumscribed prisms approach to the volume of the cylinder as their common limit.

7. *Scholium.* In the preceding demonstration, the base of the cylinder is not required to be a circle, but may be any closed convex curve. We have, however, tacitly assumed that the curve is the limit of the perimeters of the inscribed and circumscribed polygons; a principle which was rigorously proved in the case of *regular* polygons inscribed in a circle.

PROPOSITION II.—THEOREM.

8. *The lateral area of a cylinder is equal to the product of the perimeter of a right section of the cylinder by an element of the surface.*

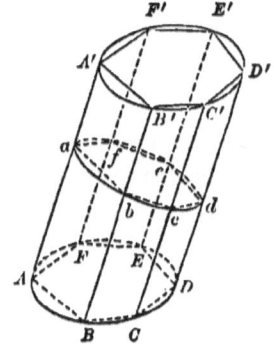

Let $ABCDEF$ be the base and AA' any element of a cylinder, and let the curve $abcdef$ be any right section of the surface. Denote the perimeter of the right section by P, the element AA' by E, and the lateral area of the cylinder by S.

Inscribe in the cylinder a prism $ABCDEFA'$ of any number of faces. The right section, $abcdef$, of this prism will be a polygon inscribed in the right section of the cylinder formed by the same plane. Denote the lateral area of the prism by s, and the perimeter of its right section by p; then, the lateral edge of the prism being equal to E, we have (VII. 16),

$$s = p \times E.$$

Let the number of lateral faces of the prism be indefinitely increased, as in the preceding proposition; then s approaches indefinitely to S as its limit, and p approaches to P; therefore, at the limit, we have (V. 31),

$$S = P \times E.$$

9. Corollary I. The lateral area of a right cylinder is equal to the product of the perimeter of its base by its altitude.

10. Corollary II. Let a cylinder of revolution be generated by the rectangle whose sides are R and H revolving about the side H. Then, R is the radius of the base, and H is the altitude of the cylinder. The perimeter of the base is $2\pi R$ (V. 40), and hence, for the lateral area S we have the expression

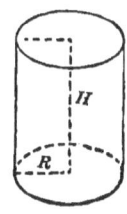

$$S = 2\pi R.H.$$

The area of each base is πR^2 (V. 43); hence the *total area* T of the cylinder of revolution, is expressed by

$$T = 2\pi R.H + 2\pi R^2 = 2\pi R(H + R).$$

11. Corollary III. Let S and s denote the lateral areas of two similar cylinders of revolution (4); T and t their total areas; R and r the radii of their bases; H and h their altitudes. The generating rectangles being similar, we have (III. 12)

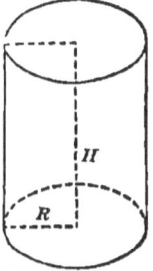
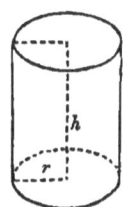

$$\frac{H}{h} = \frac{R}{r} = \frac{H+R}{h+r};$$

therefore,

$$\frac{S}{s} = \frac{2\pi RH}{2\pi rh} = \frac{R}{r} \cdot \frac{H}{h} = \frac{H^2}{h^2} = \frac{R^2}{r^2};$$

$$\frac{T}{t} = \frac{2\pi R(H+R)}{2\pi r(h+r)} = \frac{R}{r} \cdot \frac{H+R}{h+r} = \frac{H^2}{h^2} = \frac{R^2}{r^2}.$$

That is, *the lateral areas, or the total areas, of similar cylinders of revolution are to each other as the squares of their altitudes, or as the squares of the radii of their bases.*

PROPOSITION III.—PROBLEM.

12. *The volume of a cylinder is equal to the product of its base by its altitude.*

Let the volume of the cylinder be denoted by V, its base by B, and its altitude by H. Let the volume of an inscribed prism be denoted by V', and its base by B'; its altitude will also be H, and we shall have (VII. 38)

$$V' = B' \times H.$$

Let the number of faces of the prism be indefinitely increased, as in (8); then the limit of V' is V, and the limit of B' is B; therefore (V. 31),

$$V = B \times H.$$

13. *Corollary* I. Let V be the volume of a cylinder of revolution, R the radius of its base, and H its altitude; then the area of its base is πR^2 (V. 43); and therefore

$$V = \pi R^2 H.$$

14. *Corollary* II. Let V and v be the volumes of two similar cylinders of revolution; R and r the radii of their bases; H and h their altitudes; then, the generating rectangles being similar, we have

$$\frac{H}{h} = \frac{R}{r},$$

and

$$\frac{V}{v} = \frac{\pi R^2 H}{\pi r^2 h} = \frac{R^2}{r^2} \cdot \frac{H}{h} = \frac{H^3}{h^3} = \frac{R^3}{r^3};$$

that is, *the volumes of similar cylinders of revolution are to each other as the cubes of their altitudes, or as the cubes of their radii.*

THE CONE.

15. *Definition.* The area of the convex, or lateral, surface of a cone is called its *lateral area*.

16. *Definition.* A pyramid is *inscribed in a cone* when its base is inscribed in the base of the cone, and its vertex coincides with the vertex of the cone.

If a polygon *ABCD* is inscribed in the base of a cone and planes are passed through its sides and the vertex *S* of the cone, these planes intersect the convex surface of the cone in right lines (VIII. 18) and determine a pyramid inscribed in the cone.

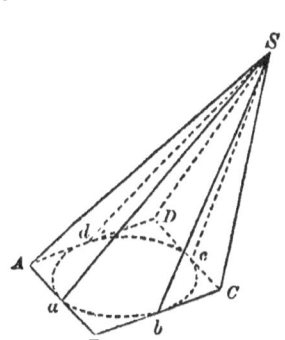

17. *Definition.* A pyramid is *circumscribed about a cone* when its base is circumscribed about the base of the cone, and its vertex coincides with the vertex of the cone.

If a polygon *ABCD* is circumscribed about the base of a cone, its points of contact with the base being *a, b, c, d*, and planes are passed through its sides and the vertex *S* of the cone, these planes will be tangent to the cone in the elements *Sa, Sb*, etc. (VIII. 21), and will determine a pyramid circumscribed about the cone.

18. *Definition.* A *truncated cone* is the portion of a cone included between its base and a plane cutting its convex surface.

When the cutting plane is parallel to the base, the truncated cone is called a *frustum of a cone;* as *ABCD–abcd*. The *altitude* of a frustum is the perpendicular distance *Tt* between its bases.

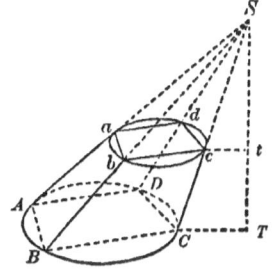

If a pyramid is inscribed in the cone, the cutting plane determines a truncated pyramid inscribed in the truncated cone; and if a pyramid is circumscribed about the cone, the cutting plane determines a truncated pyramid circumscribed about the truncated cone.

19. *Definition.* In a cone of revolution, as $S\text{-}ABC$, generated by the revolution of the right triangle SAO about the axis SO, all the elements, SA, SB, etc., are equal; and any element is called the *slant height of the cone.*

In a cone of revolution, the portion of an element included between the parallel bases of a frustum, as Aa, or Bb, is called the *slant height of the frustum.*

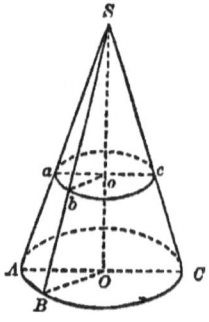

20. *Definition.* Similar cones of revolution are those which are generated by similar right triangles revolving about homologous sides.

PROPOSITION IV.—THEOREM.

21. *A cone is the limit of the inscribed and circumscribed pyramids, the number of whose faces is indefinitely increased.*

The demonstration is precisely the same as that of Proposition I., substituting a cone for a cylinder, and pyramids for prisms.

22. *Corollary.* A frustum of a cone is the limit of the frustums of the inscribed and circumscribed frustums of pyramids, the number of whose faces is indefinitely increased.

PROPOSITION V.—THEOREM.

23. *The lateral area of a cone of revolution is equal to the product of the circumference of its base by half its slant height.*

Let $S\text{-}MNPQ$ be a cone generated by the revolution of the right triangle SOM about the axis SO. Denote its lateral area by S, the circumference of its base by C, and its slant height SM by L.

Circumscribe about the base any regular polygon $ABCD$, and upon this polygon as a base construct a regular pyramid $S\text{-}ABCD$ circumscribed about the cone. Denote the

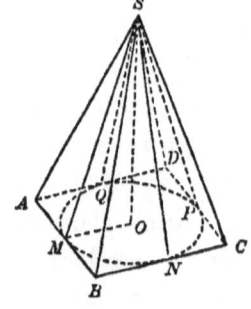

lateral area of the pyramid by s, and the perimeter of its base by p; its slant height is the same as that of the cone, since it is an element of contact, as SM or SN; therefore, we have (VII. 47),

$$s = p \times \frac{L}{2}.$$

The number of lateral faces of the pyramid being indefinitely increased, s approaches indefinitely to S, and p approaches indefinitely to C; therefore, at the limit, we have (V. 31),

$$S = C \times \frac{L}{2}.$$

24. *Corollary* I. If R is the radius of the base, we have $C = 2\pi R$ (V. 40); hence

$$S = 2\pi R \times \frac{L}{2} = \pi R L.$$

The area of the base being πR^2, the total area T of the cone is

$$T = \pi R L + \pi R^2 = \pi R (L + R).$$

25. *Corollary* II. Hence, by the same process as was employed in (11), we can prove that *the lateral areas, or the total areas, of similar cones of revolution are to each other as the squares of their slant heights, or as the squares of their altitudes, or as the squares of the radii of their bases.*

PROPOSITION VI.—THEOREM.

26. *The lateral area of a frustum of a cone of revolution is equal to the half sum of the circumferences of its bases multiplied by its slant height.*

The plane which cuts off the frustum $MNPm$, from the cone S-MNP, also cuts off from any circumscribed pyramid a frustum, as $ABCDa$, the lateral area of which is equal to the half sum of the perimeters of its bases multiplied by its slant height Mm (VII. 48). When the number of faces of the frustum of the pyramid is indefinitely increased, its lateral area approaches indefinitely to that of the frustum

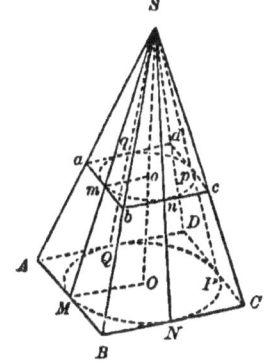

BOOK IX.

of the cone, and the perimeters of its bases approach indefinitely to the circumferences of the bases of the frustum of the cone; and the slant height Mm is common. Hence, if we express by *area Mm*, the area of the surface generated by the revolution of Mm about the axis, which is the lateral area of the frustum of the cone; and by *circ. OM*, and *circ. om*, the circumferences of the bases whose radii are OM and om; we shall have, at the limit,

$$\text{area } Mm = \tfrac{1}{2}(\text{circ. } OM + \text{circ. } om) \times Mm.$$

27. *Corollary.* Let IK be the radius of a section of the frustum equidistant from its bases; then, $IK = \tfrac{1}{2}(OM + om)$, (I. 124), and since circumferences are proportional to their radii, *circ.* $IK = \tfrac{1}{2}(\text{circ. } OM + \text{circ. } om)$; therefore,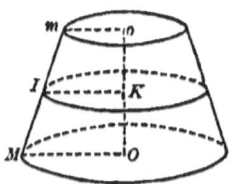

$$\text{area } Mm = \text{circ. } IK \times Mm;$$

that is, *the lateral area of a frustum of a cone of revolution is equal to the circumference of a section equidistant from its bases multiplied by its slant height.*

PROPOSITION VII.—THEOREM.

28. *The volume of any cone is equal to one-third of the product of its base by its altitude.*

Let the volume of the cone be denoted by V, its base by B, and its altitude by H.

Let the volume of an inscribed pyramid be denoted by V', and its base by B'; its altitude will also be H, and we shall have (VII. 54),

$$V' = \tfrac{1}{3} B' \times H.$$

When the number of lateral faces of the pyramid is indefinitely increased, V' approaches indefinitely to V, and B' to B; therefore, at the limit, we have

$$V = \tfrac{1}{3} B \times H.$$

29. *Corollary* I. If the cone is a cone of revolution, let R be the radius of the base, then $B = \pi R^2$, and we have

$$V = \tfrac{1}{3}\pi . R^2 . H.$$

30. *Corollary* II. Let R and r be the radii of the bases of two similar cones of revolution; H and h their altitudes, V and v their volumes; then, the generating triangles being similar, we have

$$\frac{H}{h} = \frac{R}{r},$$

and hence

$$\frac{V}{v} = \frac{\tfrac{1}{3}\pi . R^2 . H}{\tfrac{1}{3}\pi . r^2 . h} = \frac{R^2}{r^2} . \frac{H}{h} = \frac{H^3}{h^3} = \frac{R^3}{r^3};$$

that is, *similar cones of revolution are to each other as the cubes of their altitudes, or as the cubes of the radii of their bases.*

PROPOSITION VIII.—THEOREM.

31. *A frustum of any cone is equivalent to the sum of three cones whose common altitude is the altitude of the frustum, and whose bases are the lower base, the upper base, and a mean proportional between the bases of the frustum.*

Let V denote the volume of the frustum, B its lower base, b its upper base, and h its altitude.

Let V' denote the volume of an inscribed frustum of a pyramid, B' its lower base, and b' its upper base; its altitude will also be h, and we shall have (VII. 59),

$$V' = \tfrac{1}{3} h (B' + b' + \sqrt{B'b'}).$$

When the number of lateral faces of the frustum of a pyramid is indefinitely increased, V', B' and b', approach indefinitely to V, B and b, respectively; therefore, at the limit, we have

$$V = \tfrac{1}{3} h (B + b + \sqrt{Bb}),$$

which is the algebraic expression of the theorem.

BOOK IX.

32. *Corollary*. If the frustum is that of a cone of revolution, and the radii of its bases are R and r, we shall have

$$B = \pi . R^2, \qquad b = \pi . r^2, \qquad \sqrt{Bb} = \pi . Rr,$$

and consequently,

$$V = \tfrac{1}{3}\pi . h (R^2 + r^2 + Rr).$$

THE SPHERE.

33. *Definition*. A *spherical segment* is a portion of a sphere included between two parallel planes.

The sections of the sphere made by the parallel planes are the *bases* of the segment; the distance of the planes is the *altitude* of the segment.

Let the sphere be generated by the revolution of the semicircle EBF about the axis EF; and let Aa and Bb be two parallels, perpendicular to the axis. The solid generated by the figure $ABba$ is a spherical segment; the circles generated by Aa and Bb are its bases; and ab is its altitude.

If two parallels Aa and TE are taken, one of which is a tangent at E, the solid generated by the figure EAa is a spherical segment having but one base, which is the section generated by Aa. The segment is still included between two parallel planes, one of which is the tangent plane at E, generated by the line ET.

34. *Definition*. A *zone* is a portion of the surface of a sphere included between two parallel planes.

The circumferences of the sections of the sphere made by the parallel planes are the *bases* of the zone; the distance of the planes is its *altitude*.

A zone is the curved surface of a spherical segment.

In the revolution of the semicircle EBF about EF, an arc AB generates a zone; the points A and B generate the bases of the zone; and the altitude of the zone is ab.

An arc, EA, one extremity of which is in the axis, generates a zone of one base, which is the circumference described by the extremity A.

35. *Definition.* When a semicircle revolves about its diameter, the solid generated by any sector of the semicircle is called a *spherical sector.*

Thus, when the semicircle EBF revolves about EF, the circular sector COD generates a spherical sector.

The spherical sector is bounded by three curved surfaces; namely, the two conical surfaces generated by the radii OC and OD, and the zone generated by the arc CD. This zone is called the *base* of the spherical sector.

PROPOSITION IX.—LEMMA.

36. *The area of the surface generated by a straight line revolving about an axis in its plane, is equal to the projection of the line on the axis multiplied by the circumference of the circle whose radius is the perpendicular erected at the middle of the line and terminated by the axis.*

Let AB be the straight line revolving about the axis XY; ab its projection on the axis; OI the perpendicular to it, at its middle point I, terminating in the axis; then,

$$\text{area } AB = ab \times \text{circ. } OI.$$

For, draw IK perpendicular, and AH parallel to the axis. The area generated by AB is that of a frustum of a cone; hence (27),

$$\text{area } AB = AB \times \text{circ. } IK.$$

Now the triangles ABH and IOK, having their sides perpendicular each to each, are similar (III. 33), hence

$$AH \text{ or } ab : AB = IK : OI,$$

or, since circumferences are proportional to their radii,

$$ab : AB = \text{circ. } IK : \text{circ. } OI,$$

whence

$$AB \times circ.\ IK = ab \times circ.\ OI,$$

therefore,

$$area\ AB = ab \times circ.\ OI.$$

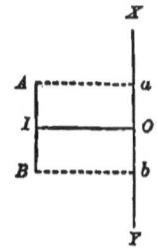

If AB is taken parallel to the axis, the result is the same, and in fact has already been proved, since in this case the surface generated is that of a cylinder whose radius is OI and whose altitude is ab (9).

PROPOSITION X.—THEOREM.

37. *The area of a zone is equal to the product of its altitude by the circumference of a great circle.*

Let the sphere be generated by the revolution of the semicircle EBF about the axis EF; and let the arc AD generate the zone whose area is required.

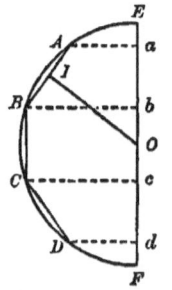

Let the arc AD be divided into any number of equal parts, AB, BC, CD. The chords AB, BC, CD, form a *regular broken line*, which differs from a portion of a regular polygon only in this, that the arc subtended by one of its sides, as AB, is not necessarily an aliquot part of the whole circumference. The sides being equidistant from the centre, a circle described with the perpendicular OI, let fall from the centre upon any side, would touch all the sides and be inscribed in the regular broken line. Drawing the perpendiculars Aa, Bb, Cc, Dd, we have by the preceding Lemma,

$$area\ AB = ab \times circ.\ OI,$$
$$area\ BC = bc \times circ.\ OI,$$
$$area\ CD = cd \times circ.\ OI,$$

the sum of which is

$$area\ ABCD = (ab + bc + cd) \times circ.\ OI,$$

or

$$area\ ABCD = ad \times circ.\ OI.$$

This being true whatever the number of sides of the regular broken line, let that number be indefinitely increased; then area $ABCD$, generated by the broken line, approaches indefinitely to the area of the zone generated by the arc AD, and *circ. OI* approaches indefinitely to *circ. OE*, or the circumference of a great circle; hence, at the limit, we have

$$\text{area of zone } AD = ad \times \text{circ. } OE,$$

which establishes the theorem.

38. *Corollary* I. Let S denote the surface of the zone whose altitude is H, the radius of the sphere being R; then,

$$S = 2\pi RH.$$

39. *Corollary* II. Zones on the same sphere, or on equal spheres, are to each other as their altitudes.

40. *Corollary* III. Let the arc AD generate a zone of a single base; its area is

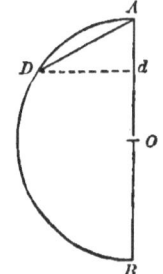

$$Ad \times 2\pi \cdot OA = \pi \cdot Ad \times AB = \pi \cdot \overline{AD}^2 \text{ (III. 47);}$$

that is, *a zone of one base is equivalent to the circle whose radius is the chord of the generating arc of the zone.*

PROPOSITION XI.—THEOREM.

41. *The area of the surface of a sphere is equal to the product of its diameter by the circumference of a great circle.*

This follows directly from the preceding proposition, since the surface of the whole sphere may be regarded as a zone whose altitude is the diameter of the sphere.

42. *Corollary* I. Let S denote the area of the surface of a sphere whose radius is R; then

$$S = 2\pi R \times 2R = 4\pi R^2;$$

that is, *the surface of a sphere is equivalent to four great circles.*

43. *Corollary* II. Let S and S' be the surfaces of two spheres whose radii are R and R'; then,

$$\frac{S}{S'} = \frac{4\pi R^2}{4\pi R'^2} = \frac{(2R)^2}{(2R')^2} = \frac{R^2}{R'^2};$$

hence, *the surfaces of two spheres are to each other as the squares of their diameters, or as the squares of their radii.*

PROPOSITION XII.—LEMMA.

44. *If a triangle revolves about an axis situated in its plane and passing through the vertex without crossing its surface, the volume generated is equal to the area generated by the base multiplied by one-third of the altitude.*

Let ABC be the triangle revolving about an axis XY passing through the vertex A; then, the volume generated is equal to the area generated by the base BC multiplied by one-third of the altitude AD.

We shall distinguish three cases:

1st. When one of the sides of the triangle, as AB, lies in the axis. (Figs. 1 and 2.)

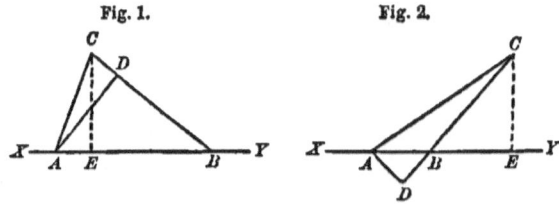

Fig. 1. Fig. 2.

Draw CE perpendicular to the axis. According as this perpendicular falls within the triangle (Fig. 1) or without it (Fig. 2), the volume generated is the sum or the difference of the cones generated by the right triangles ACE and BCE. The volumes of these cones are (29),

$$\text{vol. } ACE = \tfrac{1}{3}\pi \cdot \overline{CE}^2 \times AE,$$
$$\text{vol. } BCE = \tfrac{1}{3}\pi \cdot \overline{CE}^2 \times BE;$$

if we take their sum, we have in Fig. 1, $AE + BE = AB$; if we take their difference, we have in Fig. 2, $AE - BE = AB$; therefore, in either case,

$$\text{vol. } ABC = \tfrac{1}{3}\pi \cdot \overline{CE}^2 \times AB = \tfrac{1}{3}\pi \cdot CE \times CE \times AB;$$

or, since $CE \times AB$ is double the area of the triangle which is also expressed by $BC \times AD$ (IV. 13),

$$\text{vol. } ABC = \tfrac{1}{3}\pi \cdot CE \times BC \times AD.$$

But $\pi \cdot CE \times BC$ is the measure of the surface generated by BC (24); therefore,

$$\text{vol. } ABC = \text{area } BC \times \tfrac{1}{3} AD.$$

2d. When the triangle has only the vertex A in the axis, and the base BC when produced meets the axis in F (Fig. 3).

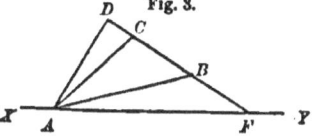
Fig. 3.

The volume generated is then the difference of the volumes generated by the triangles ACF and ABF, and, by the first case, these volumes are

$$vol.\ ACF = area\ FC \times \tfrac{1}{3}AD,$$
$$vol.\ ABF = area\ FB \times \tfrac{1}{3}AD,$$

the difference of which is

$$vol.\ ABC = (area\ FC - area\ FB) \times \tfrac{1}{3}AD = area\ BC \times \tfrac{1}{3}AD.$$

3d. When the triangle has only the vertex A in the axis, and the base BC is parallel to the axis (Figs. 4 and 5).

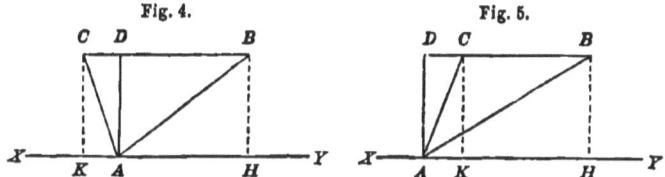

The volume generated is the sum (Fig. 4), or the difference (Fig. 5), of the volumes generated by the right triangles ABD and ACD.

Draw BH and CK perpendicular to the axis. The volume generated by the triangle ABD is the difference of the volumes of the cylinder generated by the rectangle $ADBH$ and the cone generated by the triangle ABH; therefore,

$$vol.\ ABD = \pi \cdot \overline{AD}^2 \times BD - \tfrac{1}{3}\pi \cdot \overline{AD}^2 \times BD = \tfrac{2}{3}\pi \cdot \overline{AD}^2 \times BD$$
$$= 2\pi \cdot AD \times BD \times \tfrac{1}{3}AD,$$

or, since $2\pi \cdot AD \times BD$ is the lateral area of the cylinder generated by the rectangle $AHBD$ (9),

$$vol.\ ABD = area\ BD \times \tfrac{1}{3}AD;$$

and in the same manner we have

$$vol.\ ACD = area\ CD \times \tfrac{1}{3}AD.$$

BOOK IX.

Taking the sum of these (Fig. 4), or their difference (Fig. 5), we have

$$\text{vol. } ABC = \text{area } BC \times \tfrac{1}{3} AD.$$

Therefore, in all cases, the volume generated by the triangle is equal to the area generated by its base multiplied by one-third of its altitude.

PROPOSITION XIII.—THEOREM.

45. *The volume of a spherical sector is equal to the area of the zone which forms its base multiplied by one-third the radius of the sphere.*

Let the sphere be generated by the revolution of the semicircle EBF about the axis EF; and let the circular sector AOD generate a spherical sector whose volume is required.

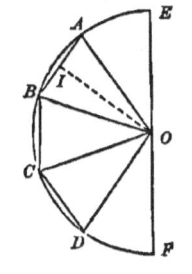

Inscribe in the arc AD a regular broken line $ABCD$, as in Proposition X., forming with the radii OA and OD a regular polygonal sector $OABCD$. Decompose this polygonal sector into triangles AOB, BOC, COD, by drawing radii to its vertices. Taking the sides AB, BC, CD, as bases, the perpendicular OI from the centre upon any side is the common altitude of these triangles.

The volume generated by the polygonal sector is the sum of the volumes generated by the triangles, and the volume generated by any triangle is equal to the area of its base multiplied by one-third of its altitude OI (44); therefore,

$$\text{vol. } OABCD = \text{area } ABCD \times \frac{OI}{3}.$$

When the number of sides of the regular polygonal sector is indefinitely increased, *vol.* $OABCD$ approaches indefinitely to the volume of the spherical sector OAD, area $ABCD$ to the area of the zone AD, and OI to the radius OA of the sphere; therefore, at the limit, we have

$$\text{vol. spherical sector } OAD = \text{zone } AD \times \tfrac{1}{3} OA;$$

which establishes the theorem.

PROPOSITION XIV.—THEOREM.

46. *The volume of a sphere is equal to the area of its surface multiplied by one-third of its radius.*

This follows directly from the preceding proposition; for, if a circular sector is increased until it becomes the semicircle which generates the sphere, the spherical sector which it generates becomes the sphere itself, and its surface becomes the surface of the sphere.

47. *Corollary* I. If V denotes the volume of a sphere whose radius is R, we have (42)

$$V = 4\pi . R^2 \times \tfrac{1}{3}R = \tfrac{4}{3}\pi . R^3.$$

Or, if D is the diameter of the sphere, whence $D^3 = (2R)^3 = 8R^3$,

$$V = \tfrac{1}{6}\pi . D^3.$$

48. *Corollary* II. *The volumes of two spheres are to each other as the cubes of their radii, or as the cubes of their diameters.*

PROPOSITION XV.—THEOREM.

49. *The solid generated by a circular segment revolving about a diameter exterior to it, is equivalent to one-sixth of the cylinder whose radius is the chord of the segment and whose altitude is the projection of that chord on the axis.*

Let $ANBIA$ be a circular segment revolving about the diameter EF, and ab the projection of the chord AB on the axis. The volume generated is the difference of the volumes generated by the circular sector AOB and the triangle AOB. Drawing OI perpendicular to AB, we have (45), (44), (38) and (36),

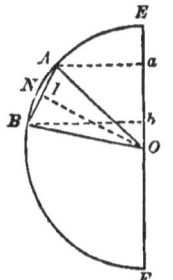

vol. sph. sector AOB = zone $AB \times \tfrac{1}{3}OA = \tfrac{2}{3}\pi . \overline{OA}^2 . ab,$
vol. triangle AOB = area $AB \times \tfrac{1}{3}OI = \tfrac{2}{3}\pi . \overline{OI}^2 . ab,$

the difference of which gives

vol. segment $ANB = \tfrac{2}{3}\pi(\overline{OA}^2 - \overline{OI}^2) \times ab.$

But $\overline{OA}^2 - \overline{OI}^2 = \overline{AI}^2 = \frac{1}{4}\overline{AB}^2$; hence

$$\text{vol. segment } ANB = \tfrac{1}{6}\pi \cdot \overline{AB}^2 \cdot ab,$$

which establishes the theorem, since $\pi \cdot \overline{AB}^2 \cdot ab$ is the volume of the cylinder whose radius is AB and whose altitude is ab (13).

PROPOSITION XVI.—THEOREM.

50. *The volume of a spherical segment is equal to the half sum of its bases multiplied by its altitude plus the volume of a sphere of which that altitude is the diameter.*

Let Aa and Bb be the radii of the bases of a spherical segment, and ab its altitude, so that the segment is generated by the revolution of the figure $ANBba$ about the axis EF.

The segment is the sum of the solid generated by the circular segment ANB and the frustum of a cone generated by the trapezoid $ABba$; hence, denoting the volume of the spherical segment by V, we have (49) and (32),

$$V = \tfrac{1}{6}\pi \cdot \overline{AB}^2 \cdot ab + \tfrac{1}{3}\pi \cdot (\overline{Bb}^2 + \overline{Aa}^2 + Bb \cdot Aa) \cdot ab.$$

Drawing AH parallel to EF, we have $BH = Bb - Aa$, and hence

$$\overline{BH}^2 = \overline{Bb}^2 + \overline{Aa}^2 - 2Bb \cdot Aa,$$

and

$$\overline{AB}^2 = \overline{AH}^2 + \overline{BH}^2 = \overline{ab}^2 + \overline{Bb}^2 + \overline{Aa}^2 - 2Bb \cdot Aa.$$

Substituting this value of $\overline{AB}^2$, we have, after reduction,

$$V = \tfrac{1}{2}(\pi \cdot \overline{Bb}^2 + \pi \cdot \overline{Aa}^2) \cdot ab + \tfrac{1}{6}\pi \cdot \overline{ab}^3,$$

which establishes the theorem, since $\pi \cdot \overline{Bb}^2$ and $\pi \cdot \overline{Aa}^2$ represent the bases of the segment, and $\tfrac{1}{6}\pi \cdot \overline{ab}^3$ is the volume of the sphere whose diameter is ab (47).

51. *Corollary.* Denoting the radii of the bases of the spherical segment by R and r, and its altitude by h, we have, for its volume,

$$V = \tfrac{1}{2}\pi(R^2 + r^2)h + \tfrac{1}{6}\pi \cdot h^3.$$

If the point A coincides with E, the upper base becomes zero, and the solid generated becomes a segment of one base. Therefore, making $r = 0$ in the above expression, the volume of a spherical segment of one base is

$$V = \tfrac{1}{2}\pi . R^2 h + \tfrac{1}{6}\pi . h^3.$$

PROPOSITION XVII.—THEOREM.

52. *The volume of a spherical pyramid is equal to the area of its base multiplied by one-third of the radius of the sphere.*

For, let v denote the volume of a spherical pyramid, and s the area of the spherical polygon which forms its base. Let V, S and R denote the volume, surface and radius of the sphere; then (VIII. 104),

$$\frac{v}{V} = \frac{s}{S}, \text{ whence } v = s \times \frac{V}{S}.$$

But $\dfrac{V}{S} = \tfrac{1}{3}R$ (46); therefore,

$$v = s \times \tfrac{1}{3}R.$$

APPENDIX I.

EXERCISES IN ELEMENTARY GEOMETRY.

EXERCISES IN ELEMENTARY GEOMETRY.

In order to make these exercises progressive as to difficulty, and to bring them fairly within the grasp of the student at the successive stages of his progress, many of them are accompanied by diagrams in which the necessary auxiliary lines are drawn, or by references to the articles in the GEOMETRY on which the exercise immediately depends, or by both. These aids are less and less freely given in the later exercises, and the student is finally left wholly to his own resources.

GEOMETRY.—BOOK I.

THEOREMS.

1. The sum of the three straight lines drawn from any point within a triangle to the three vertices, is less than the sum and greater than the half sum of the three sides of the triangle (I. 33, 66).

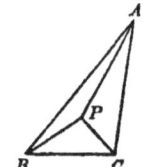

2. The medial line to any side of a triangle is less than the half sum of the other two sides, and greater than the excess of that half sum above half the third side (I. 66, 67, 112).

3. The sum of the three medial lines of a triangle is less than the *perimeter* (sum of the three sides), and greater than the semi-perimeter of the triangle.

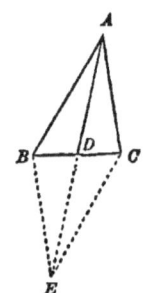

4. If from two points, A and B, on the same side of a straight line MN, straight lines, AP, BP, are drawn to a point P in that line, making with it equal angles APM and BPN, the sum of the lines AP and BP is less than the sum of any other two lines, AQ and BQ, drawn from A and B to any other point Q in MN (I. 83, 38, 66).

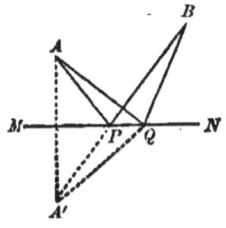

EXERCISES.

5. If from two points, A and B, on opposite sides of a straight line MN, straight lines AP, BP, are drawn to a point P in that line, making with it equal angles APN and BPN, the difference of the lines AP and BP is greater than the difference of any other two straight lines AQ and BQ, drawn from A and B to any other point Q in MN.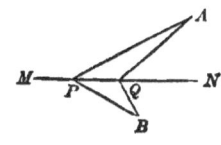

6. The three straight lines joining the middle points of the sides of a triangle divide the triangle into four equal triangles (I. 122).

7. The straight line AE which bisects the angle exterior to the vertical angle of an isosceles triangle ABC, is parallel to the base BC.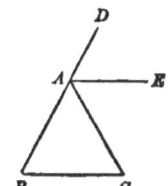

8. In any right triangle, the straight line drawn from the vertex of the right angle to the middle of the hypotenuse is equal to one-half the hypotenuse (I. 121, 38, 46).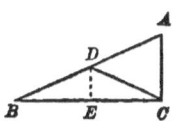

9. If one of the acute angles of a right triangle is double the other, the hypotenuse is double the shortest side (Ex. 8), (I. 69, 86, 90).

10. If ABC is any right triangle, and if from the acute angle A, AD is drawn cutting BC in E and a parallel to AC in D so that $ED = 2AB$; then, the angle DAC is one-third the angle BAC. (Ex. 8), (I. 69, 86, 49).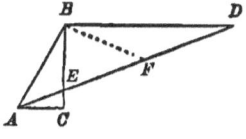

11. If BC is the base of an isosceles triangle ABC, and BD is drawn perpendicular to AC, the angle DBC is equal to one-half the angle A. (I. 73).

12. If from a *variable point* in the base of an isosceles triangle parallels to the sides are drawn, a parallelogram is formed whose perimeter is *constant*.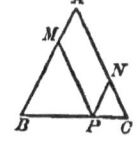

13. If from a variable point P in the base of an isosceles triangle ABC, perpendiculars, PM, PN, to the sides, are drawn, the sum of PM and PN is constant, and equal to the perpendicular from C upon AB (I. 104, 83).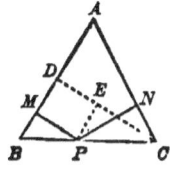

What modification of this statement is required when P is taken in BC produced?

EXERCISES. 295

14. If from any point within an equilateral triangle, perpendiculars to the three sides are drawn, the sum of these lines is constant, and equal to the perpendicular from any vertex upon the opposite side (**Ex. 12**).

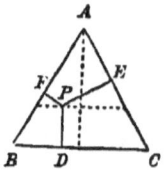

What modification of this statement is required when the point is taken *without* the triangle?

15. If ABC is an equilateral triangle, and if BD and CD bisect the angles B and C, the lines DE, DF, parallel to AB, AC, respectively, divide BC into three equal parts.

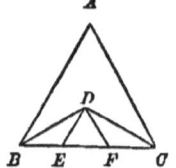

16. The locus of *all* the points which are equally distant from two intersecting straight lines consists of two perpendicular lines (I. 126, 25).

What is the locus of all the points which are equally distant from two parallel lines?

17. Let the three medial lines of a triangle ABC meet in O. Let one of them, AD, be produced to G, making $DG = DO$, and join CG. Then, the sides of the triangle OCG are, respectively, two-thirds of the medial lines of ABC (I. 134).

Also, if the three medial lines of the triangle OCG be drawn, they will be respectively equal to $\frac{1}{2} AB$, $\frac{1}{2} BC$ and $\frac{1}{2} AC$.

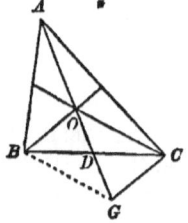

18. In any triangle ABC, if AD is drawn perpendicular to BC, and AE bisecting the angle BAC, the angle DAE is equal to one-half the difference of the angles B and C (I. 68).

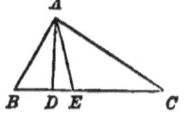

19. If BE bisects the angle B of a triangle ABC, and CE bisects the exterior angle ACD, the angle E is equal to one-half the angle A.

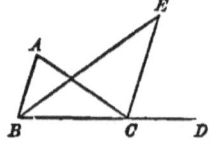

20. If from the diagonal BD of a square $ABCD$, BE is cut off equal to BC, and EF is drawn perpendicular to BD, then, $DE = EF = FC$.

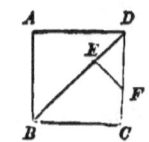

21. If E and F are the middle points of the opposite sides, AD, BC, of a parallelogram $ABCD$, the straight lines BE, DF, trisect the diagonal AC.

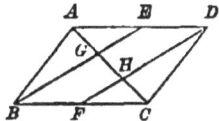

22. The sum of the four lines drawn to the vertices of a quadrilateral from any point except the intersection of the diagonals, is greater than the sum of the diagonals.

23. The straight lines joining the middle points of the adjacent sides of any quadrilateral, form a parallelogram whose perimeter is equal to the sum of the diagonals of the quadrilateral (I. 122).

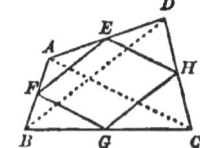

24. The intersection of the straight lines which join the middle points of opposite sides of any quadrilateral, is the middle point of the straight line which joins the middle points of the diagonals (I. 122, 108, 109).

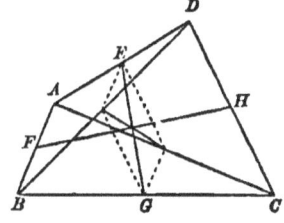

25. The four bisectors of the angles of a quadrilateral form a second quadrilateral, the opposite angles of which are supplementary.

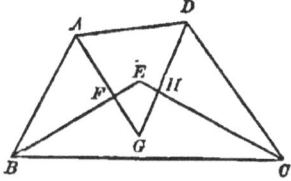

If the first quadrilateral is a parallelogram, the second is a rectangle. If the first is a rectangle, the second is a square.

26. A parallelogram is a symmetrical figure with respect to its *centre* (intersection of the diagonals), (I. 140).

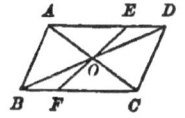

27. If in a parallelogram $ABCD$, E and G are any two symmetrical points in the sides AD, BC, and F and H any two symmetrical points in the sides AB, DC, the figure $EFGH$ is a parallelogram concentric with $ABCD$.

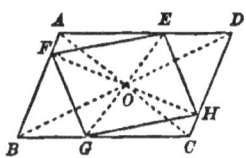

28. If the *diameters* (I. 140) EG, FH, joining symmetrical points in the opposite sides of a square $ABCD$, are perpendicular to each other, the lines joining their extremities form a second square, $EFGH$, concentric with $ABCD$.

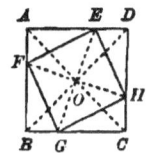

PROBLEM.

29. A billiard-ball is placed at any point P of a rectangular table $ABCD$. In what direction must it be struck to cause it to return to the same point P, after impinging successively on the four sides of the table, the ball, before and after impinging on a side, moving in lines which make equal angles with the side?

What is the length of the whole path described by the ball? Show that it is the shortest path that can be described by the ball touching the four sides and returning to the same point.

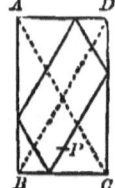

GEOMETRY.—BOOK II.

THEOREMS.

30. The circle is a symmetrical figure with respect to any diameter, or with respect to its centre.

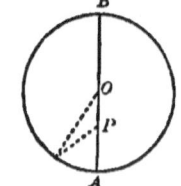

31. If P is any point within a circle whose centre is O, and $APOB$ is the diameter through P, then, AP is the least, and PB the greatest, distance from P to the circumference.

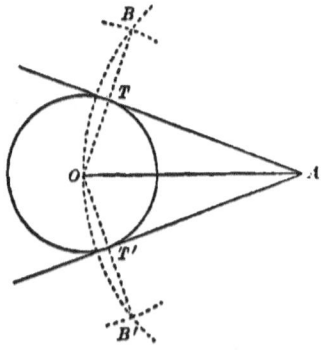

32. Prove the correctness of the following method of drawing a tangent to a given circumference O, from a given point A without it.

With radius AO and centre A, describe an arc BOB'. With centre O, and radius equal to the diameter of the given circle, describe arcs intersecting the first in B and B'. Join OB, OB', intersecting the given circumference in T, T'. Then AT, AT', are tangents.

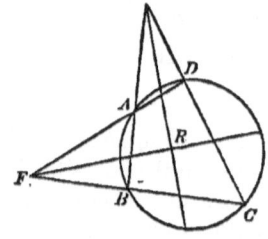

33. The bisectors of the angles contained by the opposite sides (produced) of an inscribed quadrilateral intersect at right angles.

34. If from the middle point A of an arc BC, any chords AD, AE are drawn, intersecting the chord BC in F and G, $FDEG$ is an inscriptible quadrilateral. (II. 99.)

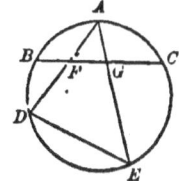

35. If $A'B'C'$ is a triangle inscribed in another triangle ABC, the three circumferences circumscribed about the triangles $AB'C'$, $BA'C'$, $CA'B'$, intersect in a common point P.

Let P be the intersection of two of the circumferences, and prove that the third must pass through P (II. 99).

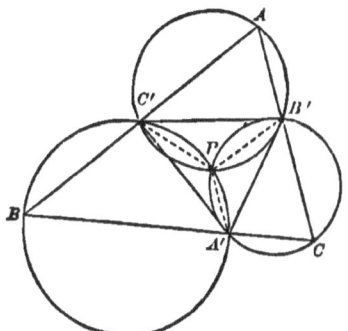

36. The perpendiculars from the angles upon the opposite sides of a triangle are the bisectors of the angles of the triangle formed by joining the feet of the perpendiculars (II. 58, 99).

37. If two circumferences are tangent internally, and the radius of the larger is the diameter of the smaller, then, any chord of the larger drawn from the point of contact is bisected by the circumference of the smaller. (II. 15).

38. If AOB is any given angle at the centre of a circle, and if BC can be drawn, meeting AO produced in C, and the circumference in D, so that CD shall be equal to the radius of the circle, then, the angle C will be equal to one-third the angle AOB.

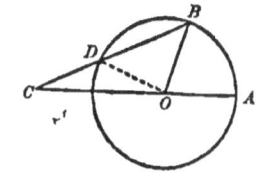

Note. There is no method known of drawing BC, under these conditions, and with the use of straight lines and circles only, AOB being *any* given angle: so that the *trisection of an angle*, in general, is a problem that cannot be solved by elementary geometry.

39. If ABC is an equilateral triangle inscribed in a circle, and P any point in the arc BC, then $PA = PB + PC$.

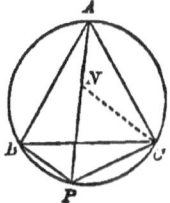

EXERCISES.

40. If a triangle ABC is formed by the intersection of three tangents to a circumference, two of which, AM and AN, are fixed, while the third BC touches the circumference at a variable point P, prove that the perimeter of the triangle ABC is constant, and equal to $AM + AN$, or $2AN$.

Also, prove that the angle BOC is constant.

41. If ABC is a triangle inscribed in a circle, and from the middle point D of the arc BC a perpendicular DE is drawn to AB; then, (II. 57), (I. 87),

$$AE = \tfrac{1}{2}(AB + AC), \quad BE = \tfrac{1}{2}(AB - AC).$$

If the perpendicular $D'E'$ is drawn from the middle point D' of the arc BAC, then

$$AE' = \tfrac{1}{2}(AB - AC), \quad BE' = \tfrac{1}{2}(AB + AC).$$

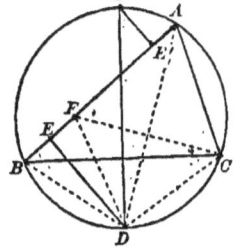

Also join AD and draw the diameter DD'; then, the angle ADD' is equal to one-half the difference of the angles ACB and ABC.

42. If two straight lines are drawn through the point of contact of two circles, the chords of the intercepted arcs are parallel.

43. Two circles are tangent internally at P, and a chord AB of the larger circle touches the smaller at C; prove that PC bisects the angle APB (II. 62).

44. If through P, one of the points of intersection of two circumferences, any two secants, APB, CPD, are drawn, the straight lines, AC, DB, joining the extremities of the secants, make a constant angle E, equal to the angle MPN formed by the tangents at P.

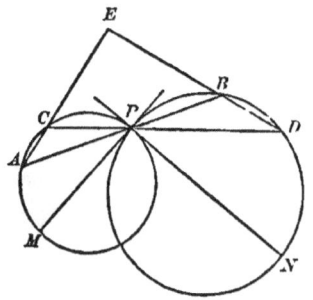

45. If through one of the points of intersection of two circumferences, a diameter of each circle is drawn, the straight line which joins the extremities of these diameters passes through the other point of intersection, is parallel to the line joining the centres, and is shorter than any other line drawn through a point of intersection and terminated by the two circumferences.

46. The feet, a, b, c, of the perpendiculars let fall from any point P in a circumference on the sides of an inscribed triangle ABC, are in a straight line.

Join ab, bc, and prove that the angle $abC =$ the angle Abc (II. 99).

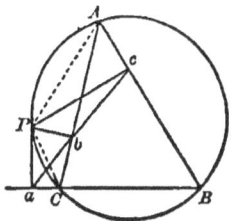

47. If equilateral triangles ABC', BCA', CAB', are constructed on the sides of any triangle ABC: 1st. The circumferences circumscribed about the equilateral triangles intersect in the same point P; 2d. The straight lines AA', BB', CC', are equal and intersect in P; 3d. The centres of the three circumferences are the vertices of an equilateral triangle $OO'O''$.

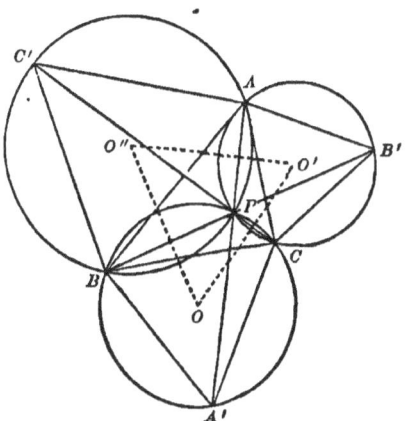

48. The inscribed and the three escribed circles of a triangle ABC being drawn, as in the figure on p. 86, let D, D', D'', D''', be the four points of contact on the same side BC. Designate the side BC, opposite to the angle A, by a, the side AC by b, and the side AB by c; and let $s = \frac{1}{2}(a+b+c)$. Prove the following properties:

$$BD'' = CD''' = s, \qquad DD'' = D'''D' = b,$$
$$BD''' = CD'' = s-a, \qquad DD''' = D'D'' = c,$$
$$BD = CD' = s-b, \qquad DD' = b-c,$$
$$BD' = CD = s-c, \qquad D'''D'' = b+c.$$

Also, let a circumference be circumscribed about the triangle ABC. Prove that this circumference bisects each of the six lines OO', OO'', OO''', $O'O''$, $O''O'''$, $O'''O'$; and that the points of bisection are the centres of circumferences that may be circumscribed about the quadrilaterals $BOCO'$, $COAO''$, $AOBO'''$, $ABO'O''$, $BCO''O'''$, $CAO'''O'$, respectively.

Finally, designating the radius of the circumscribed circle by R; the radius of the inscribed circle by r; the radii of the escribed circles by r', r'', r'''; the perpendiculars from the centre of the circumscribed circle to the three sides by p', p'', p'''; prove the following relations:

$$r' + r'' + r''' = 4R + r,$$
$$p' + p'' + p''' = R + r.$$

EXERCISES.

LOCI.

49. Find the locus of the centre of a circumference which passes through two given points.

50. Find the locus of the centre of a circumference which is tangent to two given straight lines.

51. Find the locus of the centre of a circumference which is tangent to a given straight line at a given point of that line, or to a given circumference at a given point of that circumference.

52. Find the locus of the centre of a circumference passing through a given point and having a given radius.

53. Find the locus of the centre of a circumference tangent to a given straight line and having a given radius.

54. Find the locus of the centre of a circumference of given radius, tangent externally or internally to a given circumference.

55. A straight line MN, of given length, is placed with its extremities on two given perpendicular lines, AB, CD; find the locus of its middle point P (Ex. 8).

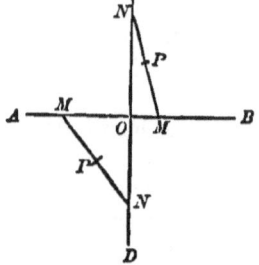

56. A straight line of given length is inscribed in a given circle; find the locus of its middle point.

57. A straight line is drawn through a given point A, intersecting a given circumference in B and C; find the locus of the middle point, P, of the intercepted chord BC.

Note the special case in which the point A is on the given circumference.

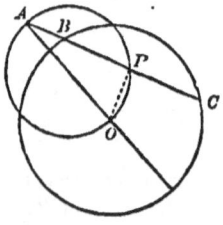

58. From any point A in a given circumference, a straight line AP of fixed length is drawn parallel to a given line MN; find the locus of the extremity P.

59. Upon a given base BC, a triangle ABC is constructed having a given vertical angle A; find the locus of the intersection of the perpendiculars from the vertices of this triangle upon the opposite sides (II. 97).

60. The angle ACB is any inscribed angle in a given segment of a circle; AC is produced to P, making CP equal to CB: find the locus of P.

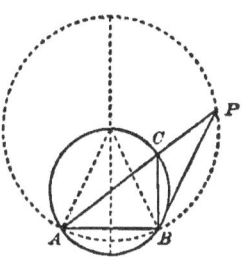

61. From one extremity A of a fixed diameter AB, any chord AC is drawn, and at C a tangent CD. From B, a perpendicular BD to the tangent is drawn, meeting AC in P. Find the locus of P.

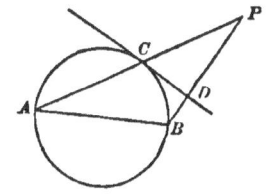

62. A triangle ABC is given, right angled at A. Any perpendicular, EF, to BC, is drawn, cutting AB in D, and CA in F. Find the locus of P, the intersection of BF and CD.

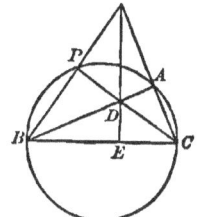

63. The base BC of a triangle is given, and the medial line BE, from B, is of a given length. Find the locus of the vertex A.

Draw AO parallel to EB. Since $BO = BC$, O is a fixed point; and since $AO = 2BE$, OA is a constant distance.

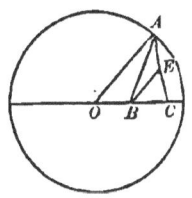

64. An angle BAC is given in position, and points B and C are taken in its sides so that $AB + AC$ shall be a given constant length. Find the locus of the centre of the circle circumscribed about the triangle ABC (Ex. 41).

Also, if the points B and C are so taken that $AB - AC$ is a given constant length, find the locus of the centre of the circle circumscribing ABC (Ex. 41).

Also find the locus of the middle point of BC.

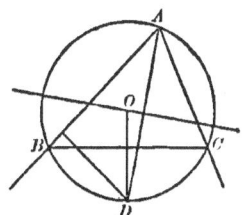

65. The base BC of a triangle ABC is given in position, and the vertical

angle A is of a given magnitude; find the loci of the centres of the inscribed and escribed circles.

In the figure, p. 86, we have the angles $BOC = 90° + \frac{1}{2} A$, $BO'C = 180° - BOC = 90° - \frac{1}{2} A$, $BO''C = BO'''C = \frac{1}{2} A$.

The loci are circumferences whose centres lie in the circumference of the circle circumscribed about ABC.

66. Find the locus of all the points the sum of the distances of each of which from two given straight lines is a given constant length (Ex. 13).

Show that the locus consists of four straight lines forming a rectangle.

67. Find the locus of all the points the difference of the distances of each of which from two given straight lines is a given constant length (Ex. 13).

Show that the locus consists of *parts* of four straight lines whose intersections form a rectangle.

68. If in Ex. 66 by *sum* is understood *algebraic sum*, and distances falling on opposite sides of the same line have opposite algebraic signs, show that Ex. 66 includes Ex. 67, and the locus consists of the *whole* of four indefinite lines whose intersections form a rectangle.

PROBLEMS.

The most useful general precept that can be given, to aid the student in his search for the solution of a problem, is the following. Suppose the problem solved, and construct a figure accordingly: study the properties of this figure, drawing auxiliary lines when necessary, and endeavor to discover the dependence of the problem upon previously solved problems. This is an *analysis* of the problem. The reverse process, or *synthesis*, then furnishes a construction of the problem. In the analysis, the student's ingenuity will be exercised especially in drawing useful auxiliary lines; in the synthesis, he will often find room for invention in combining in the most simple form the several steps suggested by the analysis.

The analysis frequently leads to the solution of a problem by the *intersection of loci*. The solution may turn upon the determination of the position of a particular point. By one condition of the problem it may appear that this required point is necessarily one of the points of a certain line; this line is a locus of the point satisfying that condition. A second condition of the problem may furnish a second locus of the point; and the point is then fully determined, being the intersection of the two loci.

Some of the following problems are accompanied by an analysis to illustrate the process.

69. A triangle ABC being given, to draw DE parallel to the base BC so that $DE = DB + EC$.

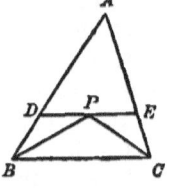

Analysis. Suppose the problem solved, and that DE is the required parallel. Since $DE = DB + EC$, we may divide it into two portions, DP and PE, respectively equal to DB and EC. Join PB, PC. Then we have the angle $DBP = DPB = PBC$; therefore, the line PB bisects the angle ABC. In the same manner it is shown that CP bisects the angle ACB. The point P, then, lies in each of the bisectors of the base angles of the triangle, and is therefore the intersection of these bisectors. Hence we derive the following construction.

Construction. Bisect the angles B and C by straight lines. Through the intersection P of the bisectors, draw the line DPE parallel to BC. This line satisfies the conditions. For we have, by the construction, the angle $DBP = PBC = BPD$; therefore, $PD = DB$; and in the same manner, $PE = EC$; hence, $DE = DB + EC$.

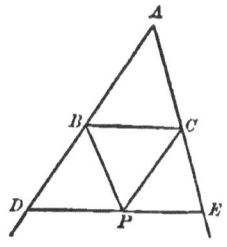

We have, however, tacitly assumed that DE is to be drawn so as to cut the sides of the triangle ABC between the vertex and the base. Suppose it drawn cutting AB and AC produced. Then the same analysis shows that the point P is found by bisecting the exterior angles CBD, BCE. Thus the problem has two solutions, if the position of DE is not limited to one side of the base BC.

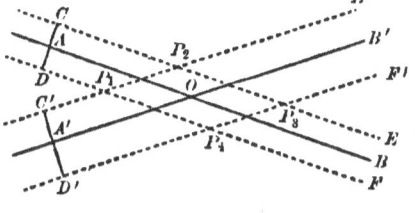

70. To determine a point whose distances from two given intersecting straight lines, AB, $A'B'$, are given.

Analysis. The locus of all the points which are at a given distance from AB consists of two parallels to AB, CE and DF, each at the given distance from AB. The locus of all the points at a given distance from $A'B'$ consists of two parallels, $C'E'$ and $D'F'$, each at the given distance from $A'B'$. The required point must be in both loci, and therefore in their intersection. There are in this case four intersections of the loci, and the problem has four solutions.

Construction. At any point of AB, as A, erect a perpendicular CD, and make $AC = AD =$ the given distance from AB; through C and D draw parallels to AB. In the same manner, draw parallels to $A'B'$ at the given distance $A'C' = A'D'$. The intersection of the four parallels determines the four points P_1, P_2, P_3, P_4, each of which satisfies the conditions.

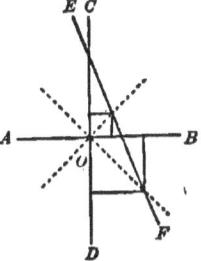

71. Given two perpendiculars, AB and CD, intersecting in O. To construct a square, one of whose angles shall coincide with one of the right angles at O, and the vertex of the opposite angle of the square shall lie on a given straight line EF. (Two solutions.)

72. In a given rhombus, $ABCD$, to inscribe a square $EFGH$. (Ex. 71.)

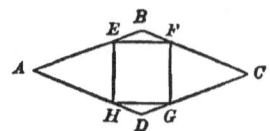

73. In a given straight line, to find a point equally distant from two given points without the line.

74. To construct a square, given its diagonal.

75. In a given square, to inscribe a square of given magnitude.

76. From two given points on the same side of a given straight line, to draw two straight lines meeting in a given straight line and making equal angles with it. (Ex. 4.)

77. Through a given point P within a given angle, to draw a straight line, terminated by the sides of the angle, which shall be bisected at P.

78. Given two straight lines which cannot be produced to their intersection, to draw a third which would pass through their intersection and bisect their contained angle.

79. Through a given point, to draw a straight line making equal angles with two given straight lines.

80. Given the middle point of a chord in a given circle, to draw the chord.

81. To draw a tangent to a given circle which shall be parallel to a given straight line.

82. In the prolongation of any given chord AB of a circle, to find a point P, such that the tangent PT, drawn from it to the circle, shall be of a given length.

83. To draw a tangent to a given circle, such that its segment intercepted between the point of contact and a given straight line shall have a given length.

In general, there are four solutions. Show when there will be but three solutions, and when but two; also, when no solution is possible.

84. Through a given point within or without a given circle, to draw a straight line, intersecting the circumference, so that the intercepted chord shall have a given length. (Two solutions.)

85. Through a given point, to draw a straight line, intersecting two given circumferences, so that the portion of it intercepted between the circumferences shall have a given length. (Two solutions.)

86. In a given circle, inscribe a chord of a given length which produced shall be tangent to another given circle.

87. Construct an angle of 60°, one of 120°, one of 30°, one of 150°, one of 45° and one of 135°.

88. To find a point within a given triangle, such that the three straight lines drawn from it to the vertices of the triangle shall make three equal angles with each other.

When will the problem be impossible?

89. Construct a parallelogram, given, 1st, two adjacent sides and one diago-

nal; 2d, one side and the diagonals; 3d, the diagonals and the angle they contain.

90. Construct a triangle, given the base, the angle opposite to the base, and the altitude.

Analysis. Suppose BAC to be the required triangle. The side BC being fixed in position and magnitude, the vertex A is to be determined. One locus of A is an arc of a segment, described upon AB, containing the given angle. Another locus of A is a straight line MN drawn parallel to BC, at a distance from it equal to the given altitude. Hence the position of A will be found by the intersection of these two loci, both of which are readily constructed.

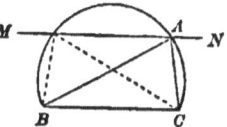

Limitation. If the given altitude were greater than the perpendicular distance from the middle of BC to the arc BAC, the arc would not intersect the line MN, and there would be no solution possible.

The limits of the data, within which the solution of any problem is possible, should always be determined.

91. Construct a triangle, given the base, the medial line to the base, and the angle opposite to the base.

92. Construct a triangle, given the base, an angle at the base, and the sum or difference of the other two sides.

Analysis. On the sides of the given angle, B, take BC = given base, and BD = given sum or difference of the sides. Join CD. The problem is reduced to drawing, from C, a line CA, which shall cut BD, or BD produced, in a point A, so that CA shall be equal to AD, which is obviously effected by making the angle DCA = the angle ADC.

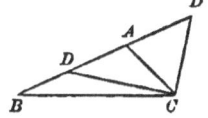

If, when the difference of AB and AC is given, AC is to be the greater side, $BD = AC - AB$ is to be taken in AB produced through B.

93. Construct a triangle, given the base, the angle opposite to the base, and the sum or difference of the other two sides.

Analysis. Suppose ABC is the required triangle. First, when the sum of AB and AC is given, produce BA to D, making $BD = AB + AC$. Join CD. The angle D is one-half of BAC, and is therefore known. Hence the following construction. Make an angle BDC equal to one-half the given angle. Take DB = given sum of sides. From B as centre, and with radius equal to the given base, describe an arc cutting DC in C. Draw CA, making the angle DCA = the angle BDC.

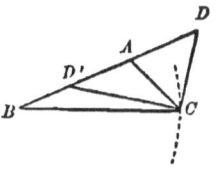

Secondly, when the difference of AB and AC is given; take $BD' = AB - AC$, and join CD'. The angle $AD'C$ is one-half the supplement of BAC, and hence the construction can readily be found.

This problem can also be solved by an application of Ex. 41.

94. Construct a triangle, given the base, the sum or difference of the other two sides, and the difference of the angles at the base.

Analysis. In the preceding figure, the angle $BCD' = \frac{1}{2}(ACB - ABC)$, and $BCD = 90° + BCD'$; and hence a construction can readily be found.

95. Construct a triangle, given, 1st, two angles and the sum of two sides; 2d, two angles and the perimeter.

96. Construct a triangle, given,

> 1st. Two sides and one medial line
> 2d. One side and two medial lines;
> 3d. The three medial lines.

See Exercise 17.

97. Construct a triangle, given an angle, the bisector of that angle, and the perpendicular from that angle to the opposite side.

98. Construct a triangle, given the middle points of its sides.

99. Construct a triangle, given the feet of the perpendiculars from the angles on the opposite sides. (Ex. 36.)

100. Construct a triangle, given the perimeter, one angle, and one altitude.

101. Construct a triangle, given an angle, together with the medial line and the perpendicular from that angle to the opposite side.

102. Construct a triangle, given the base, the sum or the difference of the other two sides, and the radius of the inscribed circle. (Ex. 48.)

103. Construct a triangle, given the centres of the three escribed circles.

104. Construct a triangle having its vertices on two given concentric circumferences, its angles being given.

105. Divide a given arc into two parts such that the sum of their chords shall be a given length. (Ex. 41.)

106. Construct a square, given the sum or the difference of its diagonal and side. (Ex. 20.)

107. With a given radius, describe a circumference, 1st, tangent to two given straight lines; 2d, tangent to a given straight line and to a given circumference; 3d, tangent to two given circumferences; 4th, passing through a given point and tangent to a given straight line; 5th, passing through a given point and tangent to a given circumference; 6th, having its centre on a given straight line, or a given circumference, and tangent to a given straight line, or to a given circumference. (Exs. 52, 53, 54.)

108. Describe a circumference, 1st, tangent to two given straight lines, and touching one of them at a given point (Exs. 50, 51); 2d, tangent to a given circumference at a given point and tangent to a given straight line; 3d, tangent to a given straight line at a given point and tangent to a given circumference (Ex. 51); 4th, passing through a given point and tangent to a given straight line at a given point; 5th, passing through a given point and tangent to a given circumference at a given point.

109. Draw a straight line equally distant from three given points.

When will there be but three solutions, and when an indefinite number of solutions?

110. Describe a circumference equally distant from four given points; the distance from a point to the circumference being measured on a radius, or radius produced.

In general there are four solutions. If three of the given points are in a straight line, one of the four circumferences becomes a straight line.

111. An angle A is given in position, and a point P in its plane. It is required to draw a straight line through P, intersecting the sides of the angle and forming a triangle ABC of a given perimeter. (Ex. 40.)

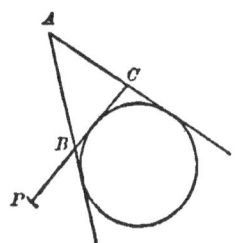

112. With a given point as a centre, describe a circle which shall be divided by a given straight line into segments containing given angles.

113. Through a given point without a given circle, draw a secant, so that the portion of it without the circle shall be equal to the portion within. (Ex. 96.)

114. Inscribe a straight line MN, of given length, between two given straight lines AB, CD, and parallel to a given straight line EF.

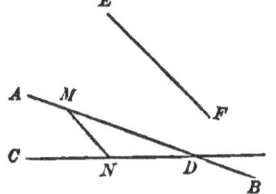

115. Inscribe a straight line of given length between two given circumferences, and parallel to a given straight line.

116. Through P, one of the points of intersection of two circumferences, draw a straight line, terminated by the circumferences, which shall be bisected in P.

117. Through one of the points of intersection of two circumferences, draw a straight line, terminated by the circumferences, which shall have a given length.

118. Given two parallels AB, CD, and two other parallels $A'B'$, $C'D'$, inclined to the first; through a given point P, in their plane, draw a straight line such that the portion of it intercepted between the parallels AB, CD, shall be equal to the portion of it intercepted between the parallels $A'B'$, $C'D'$. (Ex. 77.)

119. From two given points, A, B, on the same side of a given straight line MN, draw straight lines, meeting in a point P of MN, so that the angle APM shall be equal to double the angle BPN.

Analysis. The solution of Exercise 76, suggests the possible advantage of employing the symmetrical point of one of the given points. Let B' be the symmetrical point of B with respect to MN (I. 135). Join $B'P$ and produce it toward E. Then, since $APM = 2BPN = 2B'PN = 2MPE$, $B'PE$ bisects the angle APM. Therefore, the problem is reduced to finding a point

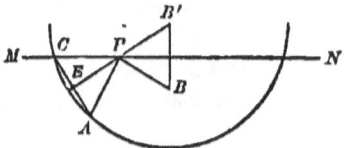

P in MN such that $B'PE$ shall bisect the angle APM. With B' as centre describe an arc through A, cutting MN in C. The perpendicular $B'E$ upon AC, evidently intersects MN in the required point.

120. With the vertices of a given triangle as centres, describe three circumferences each of which shall be tangent to the other two. (Four solutions.) (Ex. 48.)

121. Construct a quadrilateral, given its four sides and the straight line joining the middle points of two opposite sides. (Ex. 24.)

122. Construct a pentagon, given the middle points of its sides.

The middle points of all the diagonals can be determined by the principle of Ex. 23.

123. Find a point in a given straight line, such that tangents drawn from it to two given circumferences shall make equal angles with the line. (Four solutions.) Compare Ex. 76.

124. If a figure is moved in a plane, it may be brought from one position to any other, by revolving it about a certain fixed point (that is, by causing each point of the figure to move in the circumference of a circle whose centre is the fixed point). Find that point, for two given positions of the figure.

GEOMETRY.—BOOK III.

THEOREMS.

125. If three parallels AA', BB', CC', intercept on two straight lines AC, $A'C'$, the segments AB and BC, or $A'B'$ and $B'C'$, in a given ratio $m : n$, that is, if

$$AB : BC = A'B' : B'C' = m : n;$$

then, $(m + n) \cdot BB' = n \cdot AA' + m \cdot CC'$.
(III. 25, 10.)

126. In any triangle ABC, if from the vertex A, AE is drawn to the circumference of the circumscribed circle, and AD to the base BC, making the angles CAE and BAD equal to each other, then (III. 25),

$$AB \times AC = AD \times AE.$$

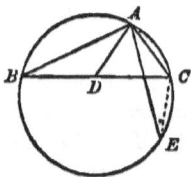

127. From the preceding theorem, deduce as a corollary the following: In any triangle ABC, if from the vertex A, AE is drawn bisecting the angle A, meeting the circumference of the circumscribed circle in E and the base BC in D, then

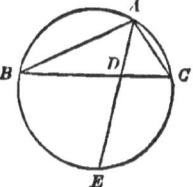

$$AB \times AC = AD \times AE.$$

Also deduce (III. 65).

128. If $ABCD$ is a given parallelogram, and AN a variable straight line drawn through A cutting BC in M and CD in N; then, the product $BM.DN$ is constant. (III. 25.)

129. If $ABCD$ is any parallelogram, and from any point P in the diagonal AC (or in AC produced) PM is drawn cutting BA in M, BC in N, AD in M' and DC in N'; then, $PM.PN = PM'.PN'$. (III. 25.)

130. If a square $DEFG$ is inscribed in a right triangle ABC, so that a side DE coincides with the hypotenuse BC (the vertices F and G being in the sides AC and AB); then, the side DE is a mean proportional between the segments BD and EC of the hypotenuse. (III. 25.)

131. If a straight line AB is divided at C and D so that $AB.AD = \overline{AC}^2$, and if from A any straight line AE is drawn equal to AC; then, EC bisects the angle DEB. (III. 10, 32, 23.)

132. If a, b, c, denote the three perpendiculars from the three vertices of a triangle upon any straight line MN in its plane, and p the perpendicular from the intersection of the three medial lines of the triangle upon the same straight line MN; then, (Ex. 125,)

$$p = \frac{a+b+c}{3}.$$

133. If ABC and $A'BC$ are two triangles having a common base BC and their vertices in a line AA' parallel to the base, and if any parallel to the base cuts the sides AB and AC in D and E, and the sides $A'B$ and $A'C$ in D' and E'; then $DE = D'E'$.

134. If two sides of a triangle are divided proportionally, the straight lines drawn from corresponding points of section to the opposite angles intersect on the line joining the vertex of the third angle and the middle of the third side.

135. The difference of the squares of two sides of any triangle is equal to the difference of the squares of the projections of these sides on the third side. (III. 48.)

136. If from any point in the plane of a polygon, perpendiculars are drawn to all the sides, the two sums of the squares of the alternate segments of the sides are equal. (Ex. 135.)

137. If O is the centre of the circle circumscribed about a triangle ABC, and P is the intersection of the perpendiculars from the angles upon the opposite sides; the perpendicular from O upon the side BC is equal to one-half the distance AP. (I. 132), (III. 25, 30.)

138. In any triangle, the centre of the circumscribed circle, the intersection of the medial lines, and the intersection of the perpendiculars from the angles upon the opposite sides, are in the same straight line; and the distance of the first point from the second is one-half the distance of the second from the third.

139. If d denotes the distance of a point P from the centre of a circle, and r the radius; and if any straight line drawn through P cuts the circumference in the points A and B; then, the product $PA.PB$ is equal to $r^2 - d^2$ if P is within the circle, and to $d^2 - r^2$ if P is without the circle.

140. In any quadrilateral, the sum of the squares of the diagonals is equal to twice the sum of the squares of the straight lines joining the middle points of the opposite sides. (III. 64) and (Ex. 23.)

141. In any quadrilateral $ABCD$ inscribed in a circle, the product of the diagonals is equal to the sum of the products of the opposite sides; that is,

$$AC.DB = AD.BC + AB.DC.$$

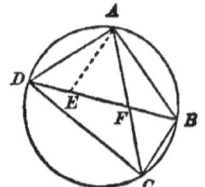

(Make the angle $DAE = BAC$, and prove that $AD.BC = AC.DE$, and $AB.DC = AC.BE$.)

142. In an inscribed quadrilateral $ABCD$, if F is the intersection of the diagonals AC and BD; then

$$\frac{AD.AB}{CB.CD} = \frac{AF}{FC} \qquad \text{(III. 65.)}$$

143. In an inscribed quadrilateral $ABCD$,

$$\frac{AD.AB + CB.CD}{BA.BC + DA.DC} = \frac{AC}{BD}.$$

144. In an inscribed quadrilateral, the product of the perpendiculars let fall from any point of the circumference upon two opposite sides is equal to the product of the perpendiculars let fall from the same point upon the other two sides. (III. 65.)

145. If from a point P in a circumference, any chords PA, PB, PC, are drawn, and any straight line MN parallel to the tangent at P, cutting the chords (or chords produced) in a, b, c; then, the products $PA.Pa$, $PB.Pb$, $PC.Pc$, are equal.

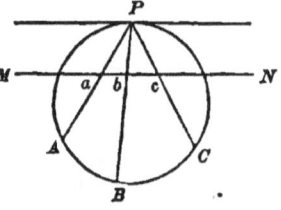

146. If two tangents are drawn to a circle at the extremities of a diameter, the portion of any third tangent intercepted between them is divided at its point of contact into segments whose product is equal to the square of the radius.

147. If on a diameter of a circle two points are taken equally distant from the centre, the sum of the squares of the distances of any point of the circumference from these two points is constant.

148. If a point P on the circumference of a circle is taken as the centre of a second circle, and any tangent is drawn to the second circle cutting the first in M and N; then, the product $PM.PN$ is constant.

149. The perpendicular from any point of a circumference upon a chord is a mean proportional between the perpendiculars from the same point upon the tangents drawn at the extremities of the chord.

150. If AB is the chord of a quadrant of a circle whose centre is O, and any chord MN parallel to AB cuts the radii OA, OB in P and Q; then

$$\overline{MP}^2 + \overline{PN}^2 = \overline{AB}^2. \qquad \text{(III. 48) and (Ex. 139.)}$$

151. If $ABCD$ is any parallelogram, and any circumference is described passing through A and cutting AB, AC, AD, in the points F, G, H, respectively; then

$$AF.AB + AH.AD = AG.AC.$$

152. In any isosceles triangle, the square of one of the equal sides is equal to the square of any straight line drawn from the vertex to the base plus the product of the segments of the base.

153. If A and B are the centres of two circles which touch at C, and P is a point at which the angles APC and BPC are equal, and if from P tangents PD and PE are drawn to the two circles; then,

$$PD.PE = \overline{PC}^2. \qquad \text{(III. 21 and 66.)}$$

154. If two circles touch each other, secants drawn through their point of contact and terminating in the two circumferences are divided proportionally at the point of contact.

155. If two circles are tangent externally, the portions of their common tangent included between the points of contact is a mean proportional between the diameters of the circles.

156. Two circles are tangent internally at A, and from any point P in the circumference of the exterior circle a tangent PM is drawn to the interior circle; prove that the ratio $PA : PM$ is constant.

157. If two circles intersect in the points A and B, and through A any secant CAD is drawn terminated by the circumferences at C and D, the straight lines BC and BD are to each other as the diameters of the circles.

158. If a fixed circumference is cut by any circumference which passes through two fixed points, the common chord passes through a fixed point.

159. Two chords AB and CD, perpendicular to each other, intersect in a point P either within or without the circle, and the line OP is drawn from the centre O. Prove that if D is the diameter of the circle,

$$\overline{PA}^2 + \overline{PB}^2 + \overline{PC}^2 + \overline{PD}^2 = D^2,$$

and
$$\overline{AB}^2 + \overline{CD}^2 + 4\overline{OP}^2 = 2D^2.$$

EXERCISES.

160. If any number of circumferences pass through the same two points, and if through one of these points any two straight lines are drawn, the corresponding segments of these lines intercepted between the circumferences are proportional.

161. If a triangle ABC is inscribed in a circle, and from the vertex A, AD and AE are drawn parallel to the tangents at B and C respectively and cutting the base BC in D and E; then

$$BD.DE = \overline{AD}^2 = \overline{AE}^2,$$

$$BD : DE = \overline{AB}^2 : \overline{AC}^2.$$

162. Let AB be a given straight line. At A erect the perpendicular $AD = AB$; in BA produced take $AO = \tfrac{1}{2} AB$; with centre O and radius OD describe a circumference, cutting AB and AB produced in C and C'; prove that AB is divided in extreme and mean ratio, internally at C, and externally at C'.

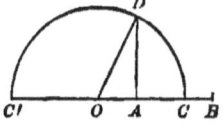

163. If a rhombus $ABCD$ is circumscribed about a circle, any tangent MN determines on two adjacent sides AB, AD, two segments BM, DN, whose product is constant.

164. If in a semicircle whose diameter is AB, any two chords AC and BD are drawn intersecting in P, then

$$\overline{AB}^2 = AC.AP + BD.BP.$$

165. If O is the intersection of the three medial lines of a triangle ABC, prove the relations

$$\overline{AB}^2 + \overline{AC}^2 + \overline{BC}^2 = 3(\overline{OA}^2 + \overline{OB}^2 + \overline{OC}^2),$$

$$\overline{BC}^2 + 3\,\overline{OA}^2 = \overline{AC}^2 + 3\,\overline{OB}^2 = \overline{AB}^2 + 3\,\overline{OC}^2.$$

166. If O is the intersection of the three medial lines of a triangle ABC, and P any point in the plane of the triangle; then,

$$\overline{PA}^2 + \overline{PB}^2 + \overline{PC}^2 = \overline{OA}^2 + \overline{OB}^2 + \overline{OC}^2 + 3\,\overline{PO}^2.$$

167. If R, r, r', r'', r''', are respectively the radii of the circumscribed, the inscribed, and the three escribed circles in any triangle, and if d, d', d'', d''', are respectively the distances from the centre of the circumscribed circle to the centres of the inscribed and escribed circles, prove the relations

$$R^2 = d^2 + 2Rr = d'^2 - 2Rr' = d''^2 - 2Rr'' = d'''^2 - 2Rr''',$$

$$R^2 = \frac{d^2 + d'^2 + d''^2 + d'''^2}{12}.$$

LOCI.

168. From a fixed point O, a straight line OA is drawn to any point in a given straight line MN, and divided at P in a given ratio $m : n$ (i. e., so that $OP : PA = m : n$); find the locus of P.

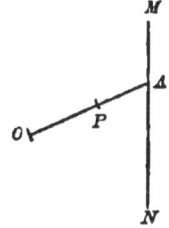

169. From a fixed point O, a straight line OA is drawn to any point in a given circumference, and divided at P in a given ratio; find the locus of P.

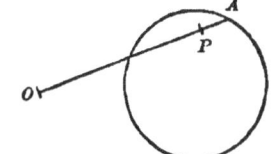

170. Find the locus of a point whose distances from two given straight lines are in a given ratio. (The locus consists of two straight lines.)

171. Find the locus of the points which divide the various chords of a given circle into segments (external or internal) whose product is equal to a given constant, k^2. (III. 56, 59.)

172. Find the locus of a point the sum of whose distances from two given straight lines is equal to a given constant k.

173. Find the locus of a point the difference of whose distances from two given straight lines is equal to a given constant k.

174. Find the locus of a point such that the sum of the squares of its distances from two given points is equal to a given constant, k^2. (III. 62.)

175. Find the locus of a point such that the difference of the squares of its distances from two given points is equal to a given constant k^2. (III. 62.)

176. If A, B and C are three given points in the same straight line, find the locus of a point P at which the angles APB and BPC are equal. (Ex. 131.)

177. Through A, one of the points of intersection of two given circles, any secant is drawn cutting the two circumferences in the points B and C; find the locus of the middle point of BC.

178. Through A, one of the points of intersection of two given circles, any secant is drawn cutting the two circumferences in the points B and C, and on this secant AP is laid off equal to the sum of AB and AC; find the locus of P.

179. From a given point O, any straight line OA is drawn to a given straight line MN, and divided at P (internally or externally) so that the product $OA \cdot OP$ is equal to a given constant; find the locus of P. (Ex. 145.)

180. From a given point O in the circumference of a given circle, any chord OA is drawn and divided at P (internally or externally) so that the product $OA \cdot OP$ is equal to a given constant; find the locus of P.

EXERCISES.

181. From a given point O, any straight line OA is drawn to a given straight line MN, and OP is drawn making a given angle with OA, and such that OP is to OA in a given ratio; find the locus of P.

With the same construction, if OP is so taken that the product $OP.OA$ is equal to a given constant; find the locus of P.

182. From a given point O, any straight line OA is drawn to a given circumference, and OP is drawn making a given angle with OA, and such that OP is to OA in a given ratio; find the locus of P.

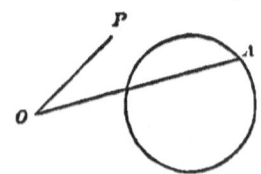

With the same construction, if OP is so taken that the product $OP.OA$ is equal to a given constant; find the locus of P.

183. One vertex of a triangle whose angles are given is fixed, while the second vertex moves on the circumference of a given circle; what is the locus of the third vertex?

184. Given a circle O and a point A; find the locus of the point P such that the distance PA is equal to the tangent from P to the circle O.

185. Find the locus of a point from which two given circles are seen under equal angles.

Note. The angle under which a circle is seen from a point is the angle contained by the two tangents from that point.

186. Find the locus of a point, such that the sum of the squares of its distances from the vertices of a given triangle is equal to the square of a given line. (Ex. 166.)

187. From any point A within a given circle, two straight lines AM and AN are drawn perpendicular to each other, intersecting the circumference in M and N; find the locus of the middle point of the chord MN.

PROBLEMS.

188. To divide a given straight line into three segments, A, B and C, such that A and B shall be in the ratio of two given straight lines M and N, and B and C shall be in the ratio of two other given straight lines P and Q.

189. Through a given point, to draw a straight line so that the portion of it intercepted between two given straight lines shall be divided at the point in a given ratio.

190. Through a given point, to draw a straight line so that the distances from two other given points to this line shall be in a given ratio. (Two solutions.)

191. Through a given point P, to draw a straight line cutting two given parallels in M and N, so that the distances AM and BN of the points of intersection from two given points A and B on these parallels shall be in a given ratio.

EXERCISES.

192. To determine a point whose distances from three given points shall be proportional to three given straight lines. (III. 79.)

193. To determine a point whose distances from three given indefinite straight lines shall be proportional to three given straight lines. (Ex. 170.)

194. Given two straight lines which cannot be produced to their intersection, to draw a straight line through a given point which would, if sufficiently produced, pass through the unknown point of intersection of the given lines. (III. 35.)

195. In a given triangle ABC to draw a parallel EF to the base BC, intersecting the sides AB and AC in E and F respectively, so that $BE + CF = BC$; or so that $BE - CF = BC$. (III. 19, 21.)

196. In a given triangle ABC, to inscribe a square $DEFG$. (Exs. 71 and 133.)

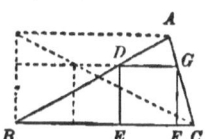

197. In a given triangle ABC, to inscribe a parallelogram $DEFG$, such that the adjacent sides DE and DG shall be in a given ratio and contain a given angle. (Remark, that the solution of this problem includes that of the preceding.)

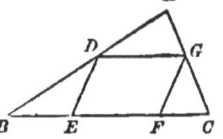

198. Construct a triangle, given its base, the ratio of the other two sides, and one angle. (III. 79.)

199. To determine a point in a given arc of a circle, such that the chords drawn from it to the extremities of the arc shall have a given ratio.

200. To find a point P in the prolongation of a given chord CD of a given circle, such that the sum of the two tangents PA and PB, drawn from it to the circle, shall be equal to the entire secant PC.

201. To divide a given straight line into two segments, such that the sum of their squares shall be equal to the square of a given straight line.

202. Given an obtuse-angled triangle; it is required to draw a straight line from the vertex of the obtuse angle to the opposite side, the square of which shall be equal to the product of the segments into which it divides that side.

203. Through a given point P to draw a straight line intersecting a given circumference in two points A and B, so that PA shall be to PB in a given ratio.

204. Given two circumferences intersecting in A; to draw through A a secant, BAC, such that AB shall be to AC in a given ratio.

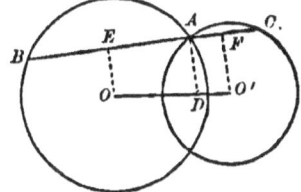

205. Given two circumferences intersecting in A; to draw through A a secant ABC, such that the product $AB.AC$ shall be equal to the square of a given line.

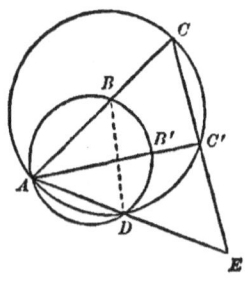

Construction. Produce the common chord AD, and take E so that $AD.AE=$ the square of the given line (III. 71). Make the angle AEC equal to the angle inscribed in the segment ABD, and let EC cut the circumference in C and C'. Join AC and AC'. Either of these lines satisfies the conditions of the problem.

206. To describe a circumference passing through two given points A and B, and tangent to a given circumference O.

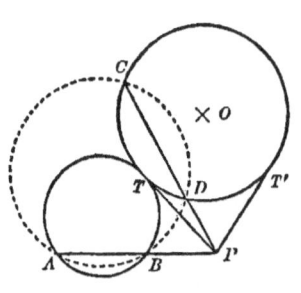

Analysis. Suppose ATB is the required circumference tangent to the given circumference at T, and $ACDB$ any circumference passing through A and B and cutting the given circumference in C and D. The common chords AB and CD, and the common tangent at T, all pass through a common point P (Ex. 158); from which a simple construction may be inferred. There are two solutions, given by the two tangents that can be drawn from P.

207. To describe a circumference passing through two given points and tangent to a given straight line. (Two solutions.)

208. To describe a circumference passing through a given point and tangent to two given straight lines.

209. To describe a circumference passing through a given point P, and tangent to a given straight line MN and to a given circumference O.

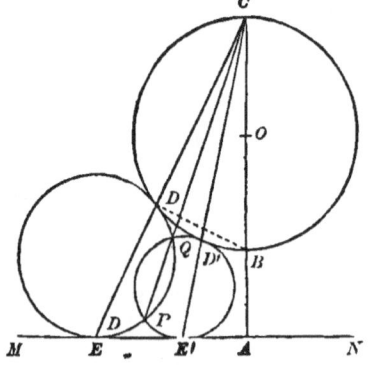

Analysis. Suppose EPD is one of the circumferences which satisfy the conditions, passing through P, touching MN at E and the circumference O at D. Through the centre O, draw $COBA$ perpendicular to MN; join CP meeting the circumference EPD in Q; also join CE. It can be proved that CE passes through D, and that

$$CP.CQ = CE.CD = CA.CB;$$

the point Q is therefore determined, and the problem is reduced to that of Ex. 206 or 207. The point Q may be taken either in PC or in PC produced through C, and thus there will be obtained, in all, four solutions.

210. To describe a circumference passing through a given point A and tangent to two given circumferences, O and O'.

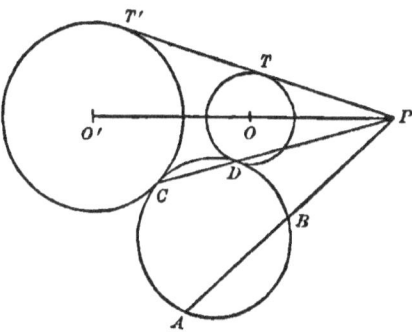

Analysis. If $ACDB$ is one of the circumferences satisfying the conditions, we can show that the line CD, joining the points of contact with the given circles, passes through P, the intersection of the line of centres, OO', with a common tangent TT', and that $PC.PD = PT.PT'$. Hence, joining PA, we have $PA.PB = PT.PT'$, and PB is known; or B is a known point on the required circumference. The problem is thus reduced to Ex. 206. By employing also the internal common tangent, we find, in all, four solutions.

211. To describe a circle tangent to two given straight lines and to a given circle.

This is reducible to Ex. 208. If both the given straight lines intersect the given circle, there may be eight solutions.

212. To describe a circle tangent to two given circles, and to a given straight line.

This is reducible to Ex. 209. There may be eight solutions.

213. To describe a circle tangent to three given circles.

This is reducible to Ex. 210. There may be eight solutions.

*214. To describe a circumference which shall bisect three given circumferences.

*215. To construct a triangle, given its base in position and magnitude, and the sum (or the difference) of the other two sides, the locus of the vertex opposite the base being a given straight line.

*216. To construct a triangle, given the product of two sides, the medial line to the third side, and the difference of the angles adjacent to the third side.

*217. To construct a triangle, similar to a given triangle, and having its three vertices on the circumferences of three given concentric circles.

The same problem, substituting three parallel straight lines for the three circumferences.

*218. In a given circle, to inscribe a triangle, such that

1st. Its base shall be parallel to a given straight line, and its other two sides shall pass through two given points in that line; or,

2d. Its base shall be parallel to a given straight line, and its other two sides shall pass through two given points not in that line; or,

3d. Its three sides shall pass through three given points.

---

* Exercises 214 to 218 are intended only for the most advanced students.

GEOMETRY.—BOOK IV.

THEOREMS.

219. Two triangles which have an angle of the one equal to the supplement of an angle of the other are to each other as the products of the sides including the supplementary angles. (IV. 22.)

220. Prove, geometrically, that the square described upon the sum of two straight lines is equivalent to the sum of the squares described on the two lines *plus* twice their rectangle.

Note. By the "rectangle of two lines" is here meant the rectangle of which the two lines are the adjacent sides.

221. Prove, geometrically, that the square described upon the difference of two straight lines is equivalent to the sum of the squares described on the two lines *minus* twice their rectangle.

222. Prove, geometrically, that the rectangle of the sum and the difference of two straight lines is equivalent to the difference of the squares of those lines.

223. Prove, geometrically, that the sum of the squares on two lines is equivalent to twice the square on half their sum plus twice the square on half their difference.

Or, the sum of the squares on the two segments of a line is equivalent to twice the square on half the line plus twice the square on the distance of the point of section from the middle of the line.

224. The area of a triangle is equal to the product of half its perimeter by the radius of the inscribed circle; that is, if a, b and c denote the sides opposite the angles A, B and C respectively, r the radius of the inscribed circle, S the area, and

$$s = \text{semi-perimeter} = \tfrac{1}{2}(a+b+c),$$

then

$$S = sr.$$

Also, if r', r'', r''', denote the radii of the three escribed circles, prove, by Ex. 48 with the figure of (II. 95), that

$$\frac{r}{r''} = \frac{s-b}{s}, \qquad rr'' = (s-a)(s-c),$$

and hence the following expressions for S, r, r', r'', r''',

$$S = \sqrt{s(s-a)(s-b)(s-c)},$$

$$r = \frac{S}{s}, \quad r' = \frac{S}{s-a}, \quad r'' = \frac{S}{s-b}, \quad r''' = \frac{S}{s-c}.$$

Also prove that

$$S = \sqrt{rr'r''r'''}.$$

225. The area of a triangle is equal to the product of its three sides divided by four times the radius of the circumscribed circle; that is, denoting this radius by R,

$$S = \frac{abc}{4R}.$$

(IV. 13) and (III. 65.)

226. The area of a triangle is equal to the product of the radius of the circumscribed circle by the semi-perimeter of the triangle formed by joining the feet of the perpendiculars drawn from the vertices of the given triangle to the opposite sides.

227. The area of the triangle formed with the three medial lines of a given triangle is three-fourths of the area of that triangle. (IV. 20) and (Ex. 17.)

228. The two opposite triangles, formed by joining any point in the interior of a parallelogram to its four vertices, are together equivalent to one-half the parallelogram.

229. The triangle formed by joining the middle point of one of the non-parallel sides of a trapezoid to the extremities of the opposite side, is equivalent to one-half the trapezoid.

230. The figure formed by joining consecutively the four middle points of the sides of any quadrilateral is equivalent to one-half the quadrilateral.

231. If through the middle point of each diagonal of any quadrilateral a parallel is drawn to the other diagonal, and from the intersection of these parallels straight lines are drawn to the middle points of the four sides, these straight lines divide the quadrilateral into four equivalent parts.

232. Two quadrilaterals are equivalent if their diagonals are equal, each to each, and contain equal angles.

233. If in a rectangle $ABCD$ we draw the diagonal AC, inscribe a circle in the triangle ABC, and from its centre draw OE and OF perpendicular to AD and DC, respectively, the rectangle OD will be equivalent to one-half the rectangle $ABCD$.

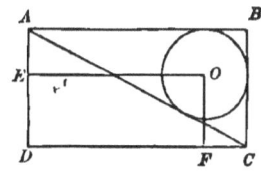

234. Let ABC be any triangle, and upon the sides AB, AC, construct parallelograms AD, AF, of any magnitude or form. Let their exterior sides DE, FG meet in M; join MA, and upon BC construct a parallelogram BK, whose side BH is equal and parallel to MA. Then the parallelogram BK is equivalent to the sum of the parallelograms AD and AF.

From this, deduce the Pythagorean Theorem. (IV. 25.)

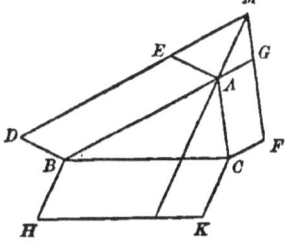

EXERCISES. 321

235. Upon the sides of any triangle ABC, construct squares AD, AG, BH; join EF, GH, DK; from A draw AP perpendicular to BC, and produce it to Q, making $AQ = BC$; join BQ, CQ, BG, CD, and from D and G, draw DM, GN, perpendicular to BC. Prove the following properties:

1st. The triangles AEF, CGH, DKB, are each equivalent to ABC.

2d. $DM + GN = BC$.

3d. BQ is perpendicular to CD, and CQ to BG.

4th. CD and BG intersect on the perpendicular AP.

5th. The lines AQ and EF bisect each other at R.

6th. EF, GH, DK, are respectively equal to twice the medial lines of the triangle ABC.

236. If three straight lines Aa, Bb, Cc, drawn from the vertices of a triangle ABC to the opposite sides, pass through a common point O within the triangle, then

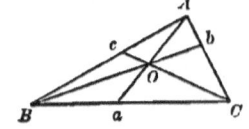

$$\frac{Oa}{Aa} + \frac{Ob}{Bb} + \frac{Oc}{Cc} = 1.$$

What modification of this statement is necessary if the point O is without the triangle?

237. If from any point O within a triangle ABC, any three straight lines, Oa, Ob, Oc, are drawn to the three sides, and through the vertices of the triangle three straight lines Aa', Bb', Cc', are drawn parallel respectively to Oa, Ob, Oc, then

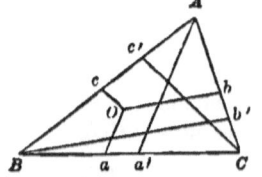

$$\frac{Oa}{Aa'} + \frac{Ob}{Bb'} + \frac{Oc}{Cc'} = 1.$$

What modification of this statement is necessary if the point O is taken without the triangle?

Deduce the preceding theorem from this.

238. If from the vertices of a triangle ABC, three straight lines, AA', BB', CC', are drawn to the opposite sides (or these sides produced), each equal to a given line L, and from any point O within the triangle, Oa, Ob,

Oc, are drawn parallel respectively to AA', BB', CC', and terminating in the same sides, then, the sum of Oa, Ob and Oc is equal to the given length L.

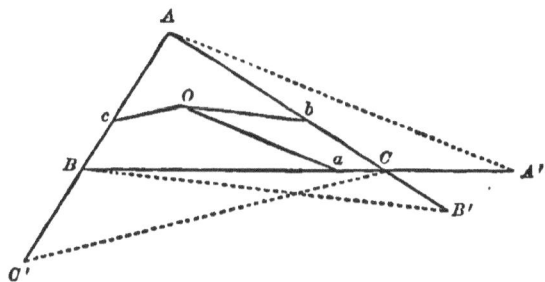

239. If a, b, c and d denote the four sides of any quadrilateral, m and n its diagonals, and S its area, then

$$S = \tfrac{1}{4}\sqrt{(2mn + a^2 - b^2 + c^2 - d^2)(2mn - a^2 + b^2 - c^2 + d^2)}.$$

If the quadrilateral is inscribed in a circle, this formula becomes

$$S = \sqrt{(p-a)(p-b)(p-c)(p-d)},$$

in which $p = \tfrac{1}{2}(a + b + c + d)$.

If the quadrilateral is such that it can be circumscribed about a circle and also inscribed in a circle, then the formula becomes

$$S = \sqrt{abcd}.$$

PROBLEMS.

240. To construct a triangle, given one angle, the side opposite to that angle, and the area (equal to that of a given square).

241. To construct a triangle, given its angles and its area.

242. To construct a triangle, given one angle, the medial line from one of the other angles, and the area.

243. To construct a triangle, given its area, the radius of the inscribed circle, and the radius of one of the escribed circles; or, given its area and the radii of two escribed circles. (Exercises 48 and 224.)

244. Given any triangle, to construct an isosceles triangle of the same area, whose vertical angle is an angle of the given triangle.

245. Given any triangle, to construct an equilateral triangle of the same area.

246. Given the three straight lines EF, GH and DK, in the figure of Exercise 235, to construct the triangle ABC.

247. Bisect a given triangle by a parallel to one of its sides.

Or, more generally, divide a given triangle into two or more parts proportional to given lines, by parallels to one of its sides.

EXERCISES.

248. Bisect a triangle by a straight line drawn through a given point in one of its sides.

249. Through a given point, draw a straight line which shall form with two given intersecting straight lines a triangle of a given area.

Remark that the area and an angle being known, the product of the sides including that angle is known. (IV. 22.)

250. Bisect a trapezoid by a parallel to its bases.

251. Inscribe a rectangle of a given area in a given circle.

252. Inscribe a trapezoid in a given circle, knowing its area and the common length of its inclined sides. (See Ex. 229.)

253. Given three points, A, B and C, to find a fourth point P, such that the areas of the triangles APB, APC, BPC, shall be equal. (Four solutions.)

254. Given three points, A, B and C, to find a fourth point P, such that the areas of the triangles APB, APC, BPC, shall be proportional to three given lines L, M, N. (Four solutions.)

See Exercise 170.

GEOMETRY.—BOOK V.

THEOREMS.

255. An equilateral polygon inscribed in a circle is regular.

256. An equilateral polygon circumscribed about a circle is regular if the number of its sides is *odd*.

257. An equiangular polygon inscribed in a circle is regular if the number of its sides is *odd*.

258. An equiangular polygon circumscribed about a circle is regular.

259. The area of the regular inscribed hexagon is three-fourths of that of the regular circumscribed hexagon.

260. The area of the regular inscribed hexagon is a mean proportional between the areas of the inscribed and circumscribed equilateral triangles.

261. A plane surface may be entirely covered (as in the construction of a pavement) by *equal* regular polygons of either three, four, or six sides.

262. A plane surface may be entirely covered by a combination of squares and regular octagons having the same side, or by dodecagons and equilateral triangles having the same side.

263. The area of a regular inscribed octagon is equal to that of a rectangle whose adjacent sides are equal to the sides of the inscribed and circumscribed squares.

264. The area of a circle is a mean proportional between the areas of any two similar polygons, one of which is circumscribed about the circle and the other isoperimetrical with the circle. (*Galileo's Theorem.*)

265. Two diagonals of a regular pentagon, not drawn from a common vertex, divide each other in extreme and mean ratio.

266. If $a =$ the side of a regular pentagon inscribed in a circle whose radius is R, then,
$$a = \frac{R}{2}\sqrt{10 - 2\sqrt{5}}.$$

267. If $a =$ the side of a regular octagon inscribed in a circle whose radius is R, then,
$$a = R\sqrt{2 - \sqrt{2}}.$$

268. If $a =$ the side of a regular dodecagon inscribed in a circle whose radius is R, then,
$$a = R\sqrt{2 - \sqrt{3}}.$$

269. If $a =$ the side of a regular pentedecagon inscribed in a circle whose radius is R, then,
$$a = \frac{R}{4}\left(\sqrt{10 + 2\sqrt{5}} + \sqrt{3} - \sqrt{15}\right).$$

270. If $d =$ the diagonal of a regular pentagon inscribed in a circle whose radius is R, then,
$$d = \frac{R}{2}\sqrt{10 + 2\sqrt{5}}.$$

271. If $a =$ the side of a regular polygon inscribed in a circle whose radius is R, and $A =$ the side of the similar circumscribed polygon, then,
$$A = \frac{2aR}{\sqrt{(4R^2 - a^2)}}, \qquad a = \frac{2AR}{\sqrt{(4R^2 + A^2)}}.$$

272. If $a =$ the side of a regular polygon inscribed in a circle whose radius is R, and $a' =$ the side of the regular inscribed polygon of double the number of sides, then,
$$a'^2 = R\left(2R - \sqrt{4R^2 - a^2}\right), \qquad a^2 = \frac{a'^2(4R^2 - a'^2)}{R^2}.$$

273. If AB and CD are two perpendicular diameters in a circle, and E the middle point of the radius OC, and if EF is taken equal to EA, then OF is equal to the side of the regular inscribed decagon, and AF is equal to the side of the regular inscribed pentagon.

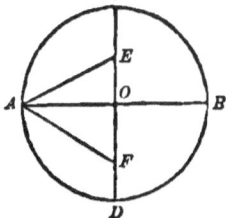

Corollary. If $a =$ the side of a regular pentagon and $a' =$ the side of a regular decagon, inscribed in a circle whose radius is R, then,
$$a^2 - a'^2 = R^2.$$

EXERCISES.

274. In two circles of different radii, angles at the centres subtended by arcs of equal length are to each other inversely as the radii.

275. From any point within a regular polygon of n sides, perpendiculars are drawn to the several sides; prove that the sum of these perpendiculars is equal to n times the apothem. (V. 22.)

What modification of this statement is required if the point is taken without the polygon?

276. If perpendiculars are dropped from the vertices of a regular polygon upon any diameter of the circumscribed circle, the sum of the perpendiculars which fall on one side of this diameter is equal to the sum of those which fall on the opposite side.

277. If n is the number of sides of a regular polygon inscribed in a circle whose radius is R, and a point P is taken such that the sum of the squares of its distances from the vertices of the polygon is equal to a given quantity k^2, the locus of P is the circumference of a circle, concentric with the given circle, whose radius r is determined by the relation

$$R^2 + r^2 = \frac{k^2}{n}.$$

(III. 52 and 53), (Ex. 276.)

PROBLEMS.

278. Divide a given circle into a given number of equivalent parts, by concentric circumferences.

Also, divide it into a given number of parts proportional to given lines, by concentric circumferences.

279. A circle being given, to find a given number of circles whose radii shall be proportional to given lines, and the sum of whose areas shall be equal to the area of the given circle.

280. In a given equilateral triangle, inscribe three equal circles tangent to each other and to the sides of the triangle.

Determine the radius of these circles in terms of the side of the triangle.

281. In a given circle, inscribe three equal circles tangent to each other and to the given circle.

Determine the radius of these circles in terms of the radius of the given circle.

GEOMETRY.—BOOK VI.

THEOREMS.

282. If a straight line AB is parallel to a plane MN, any plane perpendicular to the line AB is perpendicular to the plane MN.

283. If a plane is passed through one of the diagonals of a parallelogram, the perpendiculars to this plane from the extremities of the other diagonal are equal.

284. If the intersections of a number of planes are parallel, all the perpendiculars to these planes, drawn from a common point in space, lie in one plane.

285. If the projections of a number of points on a plane are in a straight line, these points are in one plane.

286. If each of the projections of a line AB upon two intersecting planes is a straight line, the line AB is a straight line.

287. Let A and B be two points, and M and N two planes. If the sum of the two perpendiculars from the point A upon the planes M and N is equal to the sum of those from B upon these planes, this sum is the same for every other point in the straight line AB. (Ex. 125.)

Extend the theorem to any number of planes.

288. Let A, B and C be three points, and M and N two planes. If the sum of the two perpendiculars from each of the points A, B and C, upon the planes M and N, is the same for the three points, it will be the same for every point in the plane ABC. (Ex. 287.)

Extend the theorem to any number of planes.

289. A plane passed through the middle point of the common perpendicular to two straight lines in space (VI. 63), and parallel to both these lines, bisects every straight line joining a point of one of these lines to a point of the other.

290. In any triedral angle, the three planes bisecting the three diedral angles, intersect in the same straight line.

291. In any triedral angle, the three planes passed through the edges, perpendicular to the opposite faces respectively, intersect in the same straight line.

292. In any triedral angle, the three planes passed through the edges and the bisectors of the opposite face angles respectively, intersect in the same straight line.

293. In any triedral angle, the three planes passed through the bisectors of the face angles, and perpendicular to these faces respectively, intersect in the same straight line.

294. If through the vertex of a triedral angle, three straight lines are drawn, one in the plane of each face and perpendicular to the opposite edge, these three straight lines are in one plane.

LOCI.

295. Find the locus of the points in space which are equally distant from two given points.

296. Locus of the points which are equally distant from two given planes; or whose distances from two given planes are in a given ratio. (Compare Ex. 170.)

297. Locus of the points which are equally distant from two given straight lines in the same plane.

298. Locus of the points which are equally distant from three given points.

299. Locus of the points which are equally distant from three given planes.

300. Locus of the points which are equally distant from three given straight lines in the same plane.

301. Locus of the points which are equally distant from the three edges of a triedral angle. (Ex. 293.)

302. Locus of the points in a given plane which are equally distant from two given points out of the plane.

303. Locus of the points which are equally distant from two given planes, and at the same time equally distant from two given points. (Exs. 295 and 296.)

304. Locus of a point in a given plane such that the straight lines drawn from it to two given points out of the plane make equal angles with the plane. (III. 79.)

305. Locus of a point such that the sum of its distances from two given planes is equal to a given straight line.

306. Locus of a point such that the difference of the squares of its distances from two given points is equal to a given constant.

307. Locus of a point in a given plane such that the difference of the squares of its distances from two given points is equal to a given constant.

308. A straight line of a given length moves so that its extremities are constantly upon two given perpendicular but non-intersecting straight lines; what is the locus of the middle point of the moving line?

309. Two given non-intersecting straight lines in space are cut by an indefinite number of parallel planes, the two intersections of each plane with the given lines are joined by a straight line, and each of these joining lines is divided in a given ratio $m : n$; what is the locus of the points of division?

PROBLEMS.

In the solution of problems in space, we assume—1st, that a plane can be *drawn* passing through three given points (or two intersecting straight lines) and its intersections with given straight lines or planes determined—and 2d, that a perpendicular to a given plane can be drawn at a given point in the plane, or from a given point without it; and the solution of a problem will consist, not in giving a graphic construction, but in determining the *conditions* under which the proposed problem is solved by the application of these elementary problems. The graphic solution of problems belongs to *Descriptive Geometry.*

310. Through a given straight line, to pass a plane perpendicular to a given plane.

311. Through a given point, to pass a plane perpendicular to a given straight line.

312. Through a given point, to pass a plane parallel to a given plane.

313. To determine that point in a given straight line which is equidistant from two given points not in the same plane with the given line.

314. To find a point in a plane which shall be equidistant from three given points in space.

315. Through a given point in space, to draw a straight line which shall cut two given straight lines not in the same plane.

316. Given a straight line AB parallel to a plane M; from any point A in AB, to draw a straight line AP, of a given length, to the plane M, making the angle BAP equal to a given angle.

317. Through a given point A in a plane, to draw a straight line AT in that plane, which shall be at a given distance PT from a given point P without the plane.

318. A given straight line AB meets a given plane at the point A; to draw through A a straight line AP in the given plane, making the angle BAP equal to a given angle.

319. Through a given point A, to draw to a given plane M a straight line which shall be parallel to a given plane N and of a given length.

320. Through a given point A, to draw to a given plane M a straight line which shall be parallel to a given plane N and make a given angle with the plane M.

321. Given two straight lines, CD and EF, not in the same plane, and AB any third straight line in space; to draw a straight line PQ from AB to EF which shall be parallel to CD.

322. Given two straight lines AB and CD, not in the same plane; to draw a straight line PQ from AB to CD which shall make a given angle with AB.

323. Given two straight lines, AB and CD, not in the same plane, to find a point in AB at a given perpendicular distance from CD.

324. Through a given point, to draw a straight line which shall meet a given straight line and the circumference of a given circle not in the same plane.

325. In a given plane and through a given point of the plane, to draw a straight line which shall be perpendicular to a given line in space. (VI. 62.)

326. In a given plane, to determine a point such that the sum of its distances from two given points on the same side of the plane shall be a minimum.

327. In a given plane, to determine a point such that the difference of its distances from two given points on opposite sides of the plane shall be a maximum.

328. To cut a given polyedral angle of four faces by a plane so that the section shall be a parallelogram.

EXERCISES. 329

GEOMETRY.—BOOK VII.

THEOREMS.

329. The volume of a triangular prism is equal to the product of the area of a lateral face by one-half the perpendicular distance of that face from the opposite edge.

330. In any quadrangular prism, the sum of the squares of the twelve edges is equal to the sum of the squares of its four diagonals *plus* eight times the square of the line joining the common middle points of the diagonals taken two and two.

Deduce (VII. 20) from this.

331. Of all quadrangular prisms having equivalent surfaces, the cube has the greatest volume.

332. The lateral surface of a pyramid is greater than the base.

333. At any point in the base of a regular pyramid a perpendicular to the base is erected which intersects the several lateral faces of the pyramid, or these faces produced. Prove that the sum of the distances of the points of intersection from the base is constant.

(See Ex. 275.)

334. In a tetraedron, the planes passed through the three lateral edges and the middle points of the edges of the base intersect in a straight line. The four straight lines so determined, by taking each face as a base, meet in a point which divides each line in the ratio 1 : 4.

Note. This point is the *centre of gravity* of the tetraedron.

335. The perpendicular from the centre of gravity of a tetraedron to any plane is equal to the arithmetical mean of the four perpendiculars from the vertices of the tetraedron to the same plane. (Ex. 125.)

336. In any tetraedron, the straight lines joining the middle points of the opposite edges meet in a point and bisect each other in that point.

337. The plane which bisects a diedral angle of a tetraedron divides the opposite edge into segments which are proportional to the areas of the adjacent faces.

338. Any plane passing through the middle points of two opposite edges of a tetraedron divides the tetraedron into two equivalent solids.

339. If one of the triedral angles of a tetraedron is *tri-rectangular* (*i. e.*, composed of three right angles), the square of the area of the face opposite to it is equal to the sum of the squares of the areas of the three other faces.

340. If a, b, c, d, are the perpendiculars from the vertices of a tetraedron upon the opposite faces, and a', b', c', d', the perpendiculars from any point within the tetraedron upon the same faces respectively, then,

$$\frac{a'}{a} + \frac{b'}{b} + \frac{c'}{c} + \frac{d'}{d} = 1.$$

341. If $ABCD$ is any tetraedron, and O any point within it; and if the straight lines AO, BO, CO, DO, are produced to meet the faces in the points a, b, c, d, respectively; then

$$\frac{Oa}{Aa} + \frac{Ob}{Bb} + \frac{Oc}{Cc} + \frac{Od}{Dd} = 1.$$

342. The volume of a truncated triangular prism is equal to the product of the area of its lower base by the perpendicular upon the lower base let fall from the intersection of the medial lines of the upper base.

343. The volume of a truncated parallelopiped is equal to the product of the area of its lower base by the perpendicular from the centre of the upper base upon the lower base.

344. The volume of a truncated parallelopiped is equal to the product of a right section by one-fourth the sum of its four lateral edges. (VII. 62.)

345. The altitude of a regular tetraedron is equal to the sum of the four perpendiculars let fall from any point within it upon the four faces.

346. Any plane passed through the centre of a parallelopiped divides it into two equivalent solids.

PROBLEMS.

347. To cut a cube by a plane so that the section shall be a regular hexagon.

348. Given three indefinite straight lines in space which do not intersect, to construct a parallelopiped which shall have three of its edges on these lines.

349. A parallelopiped is given in position, and a straight line in space; to pass a plane through the line which shall divide the parallelopiped into two equivalent solids.

350. To find two straight lines in the ratio of the volumes of two given cubes.

351. Within a given tetraedron, to find a point such that planes passed through this point and the edges of the tetraedron shall divide the tetraedron into four equivalent tetraedrons.

352. To pass a plane, either through a given point, or parallel to a given straight line, which shall divide a given tetraedron into two equivalent solids.

353. Find the difference between the volume of the frustum of a pyramid and the volume of a prism of the same altitude whose base is a section of the frustum parallel to its bases and equidistant from them.

The difference may be expressed in the form $\frac{h}{12}(\sqrt{B} - \sqrt{b})^2$, if B and b are the areas of the bases, and h the altitude of the frustum.

GEOMETRY.—BOOKS VIII AND IX.

THEOREMS.

354. If through a fixed point, within or without a sphere, three straight lines are drawn perpendicular to each other, intersecting the surface of the sphere, the sum of the squares of the three intercepted chords is constant. Also, the sum of the squares of the six segments of these chords is constant.

355. If three radii of a sphere, perpendicular to each other, are projected upon any plane, the sum of the squares of the three projections is equal to twice the square of the radius of the sphere. (Ex. 339.)

356. If two circles revolve about the line joining their centres, a common tangent to the two circles generates the surface of a common tangent cone to the two spheres generated by the circles. The vertex of the cone generated by an external common tangent may be called an *external* vertex, and that of the cone generated by an internal common tangent may be called an *internal* vertex. These terms being premised, prove the following theorem:

If three spheres of different radii are placed in any position in space, and the six common tangent cones, external and internal, are drawn to these spheres taken two and two, 1st, the three external vertices are in a straight line; 2d, each external vertex lies in the same straight line with two internal vertices.

357. The volumes of a cone of revolution, a sphere, and a cylinder of revolution, are proportional to the numbers 1, 2, 3, if the bases of the cone and cylinder are each equal to a great circle of the sphere, and their altitudes are each equal to a diameter of the sphere.

358. An *equilateral cylinder* (of revolution) is one a section of which through the axis is a square. An *equilateral cone* (of revolution) is one a section of which through the axis is an equilateral triangle. These definitions premised, prove the following theorems:

I. The total area of the equilateral cylinder inscribed in a sphere is a mean proportional between the area of the sphere and the total area of the inscribed equilateral cone. The same is true of the volumes of these three bodies.

II. The total area of the equilateral cylinder circumscribed about a sphere is a mean proportional between the area of the sphere and the total area of the circumscribed equilateral cone. The same is true of the volumes of these three bodies.

359. If h is the altitude of a segment of one base in a sphere whose radius is r, the volume of the segment is equal to $\pi h^2 (R - \frac{1}{3} h)$.

360. The volumes of polyedrons circumscribed about the same sphere are proportional to their surfaces.

LOCI.

361. Locus of the points in space which are at a given distance from a given straight line.

362. Locus of the points which are at the distance a from a point A, and at the distance b from a point B.

363. Locus of the centres of the spheres which are tangent to three given planes.

364. Locus of a point in space the ratio of whose distances from two fixed points is a given constant.

365. Locus of the centres of the sections of a given sphere made by planes passing through a given straight line.

366. Locus of the centres of the sections of a given sphere made by planes passing through a given point.

367. Locus of a point in space the sum of the squares of whose distances from two fixed points is a given constant. (Ex. 174.)

368. Locus of a point in space the difference of the squares of whose distances from two fixed points is a given constant. (Ex. 175.)

PROBLEMS.

369. To cut a given sphere by a plane passing through a given straight line so that the section shall have a given radius.

370. To construct a spherical surface with a given radius, 1st, passing through three given points; 2d, passing through two given points and tangent to a given plane, or to a given sphere; 3d, passing through a given point and tangent to two given planes, or to two given spheres, or to a given plane and a given sphere; 4th, tangent to three given planes, or to three given spheres, or to two given planes and a given sphere, or to a given plane and two given spheres.

371. Through a given point on the surface of a sphere, to draw a great circle tangent to a given small circle.

372. To draw a great circle tangent to two given small circles.

373. At a given point in a great circle, to draw an arc of a great circle which shall make a given angle with the first.

374. To find the ratio of the volumes of two cylinders whose convex areas are equal.

375. To find the ratio of the convex areas of two cylinders whose volumes are equal.

376. To find the ratio of the volumes generated by a rectangle revolving successively about its two adjacent sides.

APPENDIX II.

INTRODUCTION TO MODERN GEOMETRY.

INTRODUCTION TO MODERN GEOMETRY.

TRANSVERSALS.

1. *Definition.* Any straight line cutting a system of lines is called a *transversal.*

PROPOSITION I.—THEOREM.

2. *If a transversal cuts the sides of a triangle (produced if necessary), the product of three non-adjacent segments of the sides is equal to the product of the other three segments.*

Let ABC be the given triangle, and abc the transversal. When the transversal cuts a side produced, as the side BC at a, the segments are the distances, aB, aC, of the point of section from the extremities of the line (III. 22). The segments aB, bC, cA, having no extremity in common, are non-adjacent.

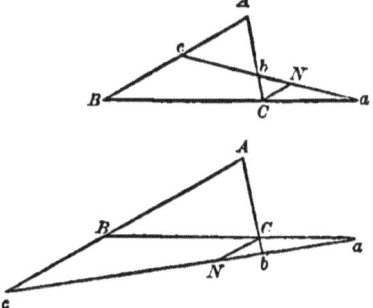

Draw CN parallel to AB. By similar triangles, we have

$$\frac{aB}{aC} = \frac{cB}{NC}, \qquad \frac{bC}{bA} = \frac{NC}{cA},$$

and multiplying,

$$\frac{aB \times bC}{aC \times bA} = \frac{cB}{cA},$$

whence,

$$aB \times bC \times cA = aC \times bA \times cB.$$

3. *Corollary.* Conversely, *if three points are taken on the sides of a triangle (one of the points, or all three, lying in the sides produced), so that the product of three non-adjacent segments of the sides is equal to the product of the other three, then the three points lie in the same straight line.*

Let a, b, c, be so taken on the sides of the triangle ABC, that the relation [I] is satisfied. Join ab, and let ab produced be supposed to cut

AB in a point which we shall call c'. Then by the above theorem, we have

$$aB \times bC \times c'A = aC \times bA \times c'B,$$

and since by hypothesis we also have

$$aB \times bC \times cA = aC \times bA \times cB,$$

there follows, by division,

$$\frac{c'A}{cA} = \frac{c'B}{cB},$$

which can evidently be true only when c' coincides with c; that is, the three points a, b and c are in the same straight line.

4. *Scholium.* The principle in the corollary often serves to determine, in a very simple manner, whether three points lie in the same straight line.

For example, take the following theorem:

The middle points of the three diagonals of a complete quadrilateral are in a straight line.

A *complete quadrilateral* is the figure formed by four straight lines intersecting in six points, as $ABCDEF$. The line EF is called the *third diagonal*.

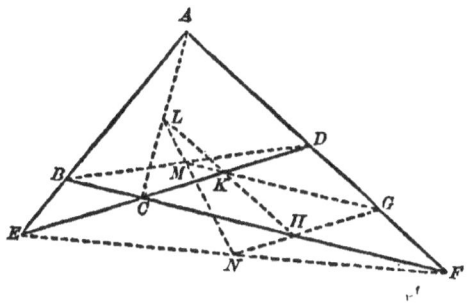

Let L, M, N, be the middle points of the three diagonals. Let G, H, K, be the middle points of the sides of the triangle FDC. The sides of the triangle GHK pass through the points LMN, respectively (I. 121 and 122). The line ABE, considered as a transversal of the triangle CDF, gives

$$AD.BF.EC = AF.BC.ED.$$

Dividing each factor of this equation by 2, and observing that we have $\frac{1}{2}AD = LK$, $\frac{1}{2}BF = MG$, etc. (I. 121), we find

$$LK.MG.NH = LH.MK.NG;$$

therefore, the points L, M, N, lying in the sides of the triangle GHK, satisfy the condition of the preceding corollary, and are in a straight line.

PROPOSITION II.—THEOREM.

5. *Three straight lines, drawn through the vertices of a triangle and any point in its plane, divide the sides into segments such that the product of three non-adjacent segments is equal to the product of the other three.*

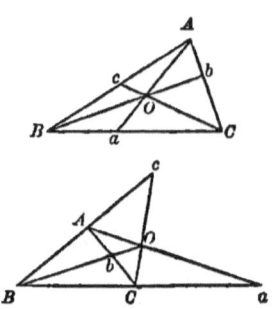

Let ABC be the triangle, and O any point in its plane, through which Aa, Bb, Cc, are drawn.

The triangle ACa is cut by the transversal Bb; hence, by (2),

$$aB.bC.AO = BC.bA.aO;$$

and the triangle ABa, cut by the transversal Cc, gives

$$BC.aO.cA = aC.AO.cB.$$

Multiplying these equations together, and omitting the common factors, we obtain

$$aB.bC.cA = aC.bA.cB.$$

6. *Corollary.* Conversely, *if three points are taken on the sides of a triangle (all the points being on the sides themselves or two on the sides produced), so that the product of three non-adjacent segments of the sides is equal to the product of the other three, the straight lines joining these points with the opposite vertices of the triangle meet in one point.*

The proof is similar to that of (3).

7. *Scholium.* The principle of this corollary often serves to determine whether three straight lines meet in a point. For example, if Aa, Bb, Cc, are the bisectors of the angles of the triangle ABC, we have by (III. 21),

$$\frac{aB}{AB} = \frac{aC}{AC}, \qquad \frac{bC}{BC} = \frac{bA}{AB}, \qquad \frac{cA}{AC} = \frac{cB}{BC},$$

and the product of these equalities, omitting the common denominator $AB \times BC \times AC$, is

$$aB.bC.cA = aC.bA.cB;$$

therefore, *the three bisectors of the angles of a triangle pass through the same point.*

With the same facility, it can be shown that *the straight lines joining the points of contact of the inscribed circle with the opposite vertices of the triangle meet in a point;* that *the three perpendiculars from the vertices of a triangle to the opposite sides meet in a point;* and that *the three medial lines of a triangle meet in a point.*

ANHARMONIC RATIO.

8. *Definition.* If four points are taken in a straight line, the quotient obtained by dividing the ratio of the distances of the first two from the third by the ratio of the distances of the first two from the fourth, is called the *anharmonic ratio* of the four points.

Thus the anharmonic ratio of the four points A, B, C, D, is

$$\frac{CA}{CB} : \frac{DA}{DB},$$

which for brevity is denoted by $[ABCD]$.

In applying the definition the points may be taken in any order we please, but the adopted order is always to be indicated in the notation. Thus, the same points, considered in the order A, C, B, D, give the anharmonic ratio

$$[ACBD] = \frac{BA}{BC} : \frac{DA}{DC}.$$

9. *The anharmonic ratio of four points is not changed in value when two of the points are interchanged, provided the other two are interchanged at the same time.*

Thus

$$[ABCD] = \frac{CA}{CB} : \frac{DA}{DB} = \frac{CA.DB}{CB.DA}.$$

$$[BADC] = \frac{DB}{DA} : \frac{CB}{CA} = \frac{CA.DB}{CB.DA}.$$

$$[CDAB] = \frac{AC}{AD} : \frac{BC}{BD} = \frac{AC.BD}{BC.AD}.$$

$$[DCBA] = \frac{BD}{BC} : \frac{AD}{AC} = \frac{AC.BD}{BC.AD}.$$

Therefore, $[ABCD] = [BADC] = [CDAB] = [DCBA]$. There are then four different ways in which the same anharmonic ratio can be expressed.

There are, in all, twenty-four ways in which the four letters may be written, and therefore four points give rise to six anharmonic ratios differing in value. Three of these six are the reciprocals of the other three.

10. *Definition.* A system of straight lines diverging from a point is called a *pencil;* each diverging line is called a *ray;* and the point from which they diverge is called the *vertex* of the pencil.

PROPOSITION III.—THEOREM.

11. *If a pencil of four rays is cut by a transversal, any anharmonic ratio of the four points of intersection is constant for all positions of the transversal.*

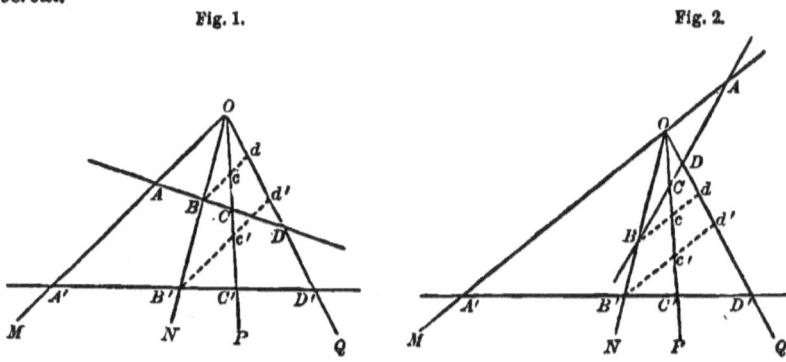

Fig. 1.   Fig. 2.

Let $O\text{-}MNPQ$ be the pencil; and let $ABCD$, $A'B'C'D'$, be any two positions of a transversal; then

$$[ABCD] = [A'B'C'D'].$$

For, drawing Bcd parallel to OM, we have by similar triangles,

$$\frac{CA}{CB} = \frac{OA}{cB}, \qquad \frac{DA}{DB} = \frac{OA}{dB}.$$

Dividing the first of these equations by the second, we have

$$[ABCD] = \frac{dB}{cB}.$$

Drawing $B'c'd'$ parallel to OM, we have in the same manner,

$$[A'B'C'D'] = \frac{d'B'}{c'B'}.$$

The second members of these two equations being equal (III. 35), we have

$$[ABCD] = [A'B'C'D'].$$

It is important to observe that the preceding demonstration applies when the transversals cut one or more of the rays on opposite sides of the vertex, as in Fig. 2.

12. *Definition.* The *anharmonic ratio of a pencil* of four rays is the anharmonic ratio of the four points on these rays determined by any transversal. Thus, $[ABCD]$, $[ACBD]$, etc., are the anharmonic ratios of the pencil formed by the rays OM, ON, OP, OQ, in the preceding figure. To

distinguish the pencil in which the ratio is considered, the letter at the vertex is prefixed to the ratio; thus, [*O.ABCD*], [*O.ACBD*], etc.

13. The angles of a pencil are the six angles which the rays, taken two and two, form with each other. It follows from the preceding proposition that the values of the anharmonic ratios in two pencils will be equal, if the angles of the pencils are equal, each to each.

14. *Definition.* The *anharmonic ratio of four fixed points A, B, C, D, on the circumference of a circle*, is the anharmonic ratio of the pencil formed by joining the four points to any variable point *O* on the circumference.

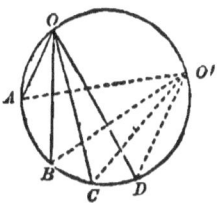

PROPOSITION IV.—THEOREM.

15. *The anharmonic ratio of four fixed points on the circumference of a circle is constant.*

For, the angles of the pencil remain the same for all positions, *O*, *O'*, etc., of its vertex, on the circumference. (II. 58.)

16. *Definition.* If four fixed tangents to a circle are cut by a fifth (variable) tangent, the anharmonic ratio of the four points of intersection is called *the anharmonic ratio of the four tangents.*

PROPOSITION V.

17. *The anharmonic ratio of four fixed tangents to a circle is constant.*

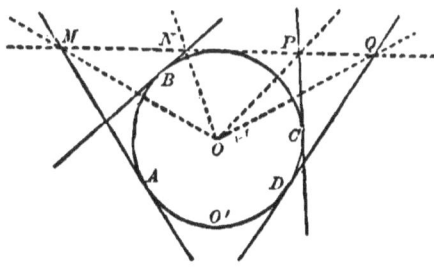

For, let four tangents, touching the circle *O* at the points *A*, *B*, *C*, *D*, be intersected by any fifth tangent in *M*, *N*, *P*, *Q*. The pencil formed by the rays *OM*, *ON*, *OP*, *OQ*, will have constant angles for all positions of the variable tangent, since (as the reader can readily prove) the angle *MON* will be measured by one-half of the fixed arc *AB*, the angle *NOP* by one-half of the arc *BC*, and the angle *POQ* by one-half of the arc *CD*. The angles of the pencil being constant, the anharmonic ratio [*O.MNPQ*] is constant.

18. *Corollary.* The *anharmonic ratio of four tangents to a circle is equal to the anharmonic ratio of the four points of contact.*

For, if any point *O'* in the circumference be joined to *A*, *B*, *C*, *D*, the pencil formed will have the same angles as the pencil formed by the rays *OM*, *ON*, *OP*, *OQ*, since these angles will also be measured by one-half the arcs *AB*, *BC*, *CD*, respectively.

PROPOSITION VI.

Theorem.

20. *When two pencils $O-ABCD$, $O'-A'B'C'D'$, have the same anharmonic ratio and a homologous ray OA common, the intersections b, c, d, of the other three pairs of homologous rays, are in a straight line.*

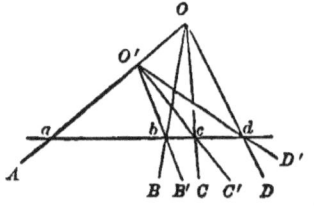

For, let the straight line joining b and c meet OA in a, and let the points in which it meets OD and $O'D'$ be called δ and δ', respectively. By hypothesis we have (11),

$$[abc\delta] = [abc\delta'],$$

which can be true only when δ and δ' coincide; but δ and δ' being on the different lines OD and $O'D'$ can coincide only when they are identical with their intersection d. Therefore, a, b, c, d, are in a straight line.

22. *Corollary.* If one of the anharmonic ratios of a pencil is equal to one of those of a second pencil, the remaining anharmonic ratios of the

Theorem.

21. *When two right-lined figures of four points A, B, C, D, and A, B', C', D', have the same anharmonic ratio and a homologous point A common, the straight lines joining the other three pairs of homologous points meet in the same point O.*

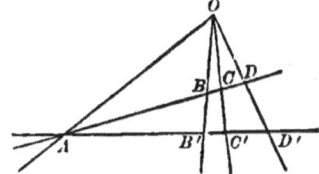

For, let O be the point of meeting of BB' and CC'; draw OA and OD', and let the point in which OD' meets AD be called δ. Then we have (11),

$$[AB'C'D'] = [ABC\delta],$$

and, by hypothesis,

$$[AB'C'D'] = [ABCD];$$

hence,

$$[ABC\delta] = [ABCD].$$

Therefore δ coincides with D, and the straight line DD' also passes through the point O.

23. *Corollary.* If one of the anharmonic ratios of a system of four points is equal to one of those of a second system, the remaining anhar-

first pencil are equal to those of the second, each to each.

For, let the pencils be placed so as to have a common homologous ray. Since one of the ratios has the same value in both pencils, the intersections of the other three pairs of homologous rays lie in a straight line, which is a common transversal; and then any two corresponding anharmonic ratios in the two pencils will be equal to that of the four points on the common transversal (11), and therefore equal to each other.

monic ratios of the two systems are equal, each to each.

For, let the two systems be placed so as to have a common homologous point. Since one of the anharmonic ratios has the same value in both systems, the straight lines joining the other three pairs of homologous points meet in a point; and then any two corresponding anharmonic ratios in the two systems are equal, being determined in the same pencil (11).

PROPOSITION VII.

Theorem.

24. *If two triangles, ABC, $A'B'C'$, are so situated that the three straight lines, AA', BB', CC', joining their corresponding vertices, meet in a point, O, the three intersections, a, b, c, of their corresponding sides, are in a straight line.*

Theorem.

25. *If two triangles, ABC, $A'B'C'$, are so situated that the three intersections, a, b, c, of their corresponding sides are in a straight line, the three straight lines, AA', BB', CC', joining their corresponding vertices, meet in a point, O.*

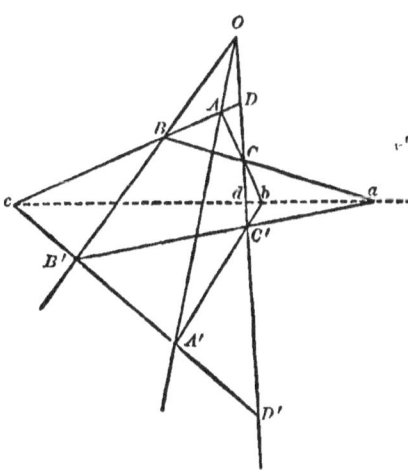

For, let BA and $B'A'$ meet OC in D and D', and suppose Oc to be drawn. The pencil Oc, OB, OA,

For, let the straight line abc meet CC' in d. Taking C and C' as the vertices of pencils having the com-

OC, intersected by the transversals cD and cD', gives

$$[cBAD] = [cB'A'D'],$$

or considering these ratios as belonging to pencils whose vertices are C and C', respectively,

$$[C.cBAD] = [C'.cB'A'D'].$$

These pencils having a common anharmonic ratio and a common ray CC', the intersections a, b, c, of the other three pairs of homologous rays are in a straight line (20).

mon transversal ac, we have, identically,

$$[C.cdba] = [C'.cdba].$$

The first pencil being cut by the transversal $cBAD$, and the second by the transversal $cB'A'D'$, the preceding equation gives (11),

$$[cDAB] = [cD'A'B'].$$

The two systems, c, B, A, D, and c, B', A', D', having a common anharmonic ratio and a common homologous point c, the lines BB', AA', DD' (or CC'), meet in the same point O (21).

26. *Definition.* Two triangles ABC, $A'B'C'$, which satisfy the conditions of the preceding two theorems, are called *homological;* the point O is called the *centre of homology;* the line abc is called the *axis of homology.*

PROPOSITION VIII.

Theorem.

27. In any hexagon $ABCDEF$ inscribed in a circle, the intersections,

Theorem.

28. In any hexagon $ABCDEF$ circumscribed about a circle, the three

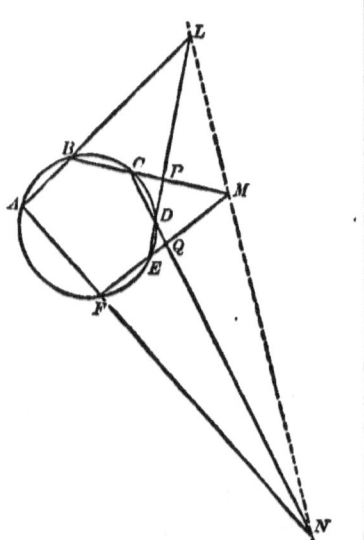
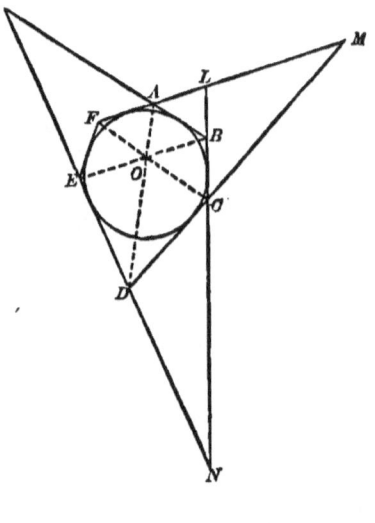

L, M, N, of the three pairs of opposite sides, are in a straight line.

For, considering two pencils formed by joining B and F as vertices, with A, C, D and E, we have (15),
$$[B.ACDE] = [F.ACDE].$$
Cutting the first pencil by the transversal $LPDE$, and the second pencil by the transversal $NQDC$, the preceding equation gives (11),
$$[LPDE] = [NCDQ].$$
Since the two systems of points $LPDE$ and $NCDQ$ have a common anharmonic ratio and a common homologous point D, the lines LN, PC, EQ, joining the other three pairs of homologous points, meet in a common point M (21). Therefore L, M, N, are in the same straight line.

This theorem is due to Pascal.

29. *Corollary 1.* If the vertex D is brought nearer and nearer to the vertex C, the side CD will approach to the tangent at C; therefore, when the point D is finally made to coincide with C, the theorem will still apply to the resulting pentagon if we substitute the tangent at C for the side CD. The theorem then takes the following form.

In any pentagon $ABCEF$ inscribed

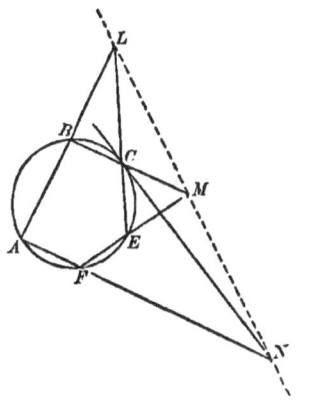

diagonals, AD, BE, CF, joining the opposite vertices, intersect in the same point.

For, regarding AB, BC, CD and EF, as fixed tangents cut by the tangent ED in P, N, D, E, and by the tangent FA in A, L, M, F, we have (17),
$$[PNDE] = [ALMF],$$
or, considering these anharmonic ratios as belonging to pencils whose vertices are B and C, respectively,
$$[B.PNDE] = [C.ALMF].$$
These two pencils, having a common anharmonic ratio and a common homologous ray LN, the intersections, A, D, O, of the other three pairs of homologous rays are in a straight line (20). Therefore AD, BE and CF, meet in the same point O.

This theorem is due to Brianchon.

30. *Corollary 1.* If a vertex C is brought into the circumference, the sides BC and CD will become a single line touching the circle at C. The theorem will still apply to the resulting pentagon if we regard the point of contact of this side as the vertex of a circumscribed hexagon. The theorem then takes the following form.

In any pentagon $ABDEF$ circumscribed about a circle, the line joining

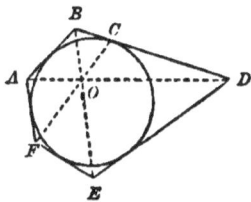

a vertex and the point of contact of the opposite side, and the diagonals joining the other non-consecutive vertices meet in a point.

in a circle, the intersection *N* of a side with the tangent drawn at the opposite vertex, and the intersections *L*, *M*, of the other non-consecutive sides are three points in a straight line.

By the same process we can reduce the hexagon to a quadrilateral and finally to a triangle; whence the following corollaries.

31. *Corollary II.* In any quadrilateral inscribed in a circle, if tangents are drawn at two consecutive vertices, the point of intersection of each of them with the side passing through the point of contact of the other, and the intersection of the other two sides, are three points in a straight line.

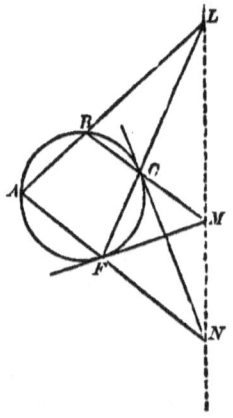

32. *Corollary II.* In any quadrilateral circumscribed about a circle, if we take the points of contact of two

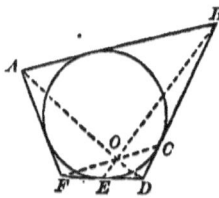

adjacent sides, and join the point of contact of each side with the vertex on the other side, and if the remaining two vertices are joined, the three straight lines so drawn meet in a point.

33. *Corollary III.* In any quadri-

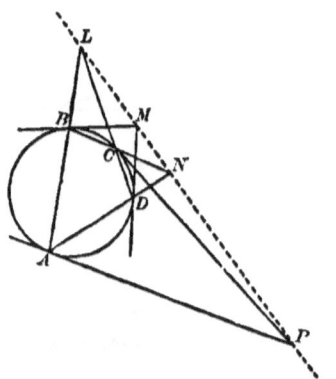

34. *Corollary III.* In any quadri-

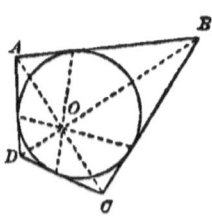

lateral circumscribed about a circle, the straight lines joining the points of contact of opposite sides, and the diagonals, are four straight lines which meet in a point.

lateral inscribed in a circle, the intersections of the tangents drawn at opposite vertices and the intersections of the opposite sides are four points in a straight line.

35. *Corollary* IV. In any triangle inscribed in a circle, the intersections

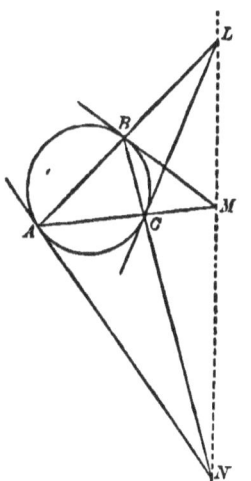

36. *Corollary* IV. In any triangle circumscribed about a circle, the straight lines joining the point of con-

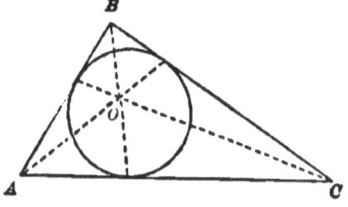

tact of each side with the opposite vertex meet in a point.

of each side with the tangent drawn at the opposite vertex are in a straight line.

37. *Scholium.* Pascal's Theorem (27) may be applied to the figure $ABCDEFA$, formed by joining any six points of the circumference by consecutive straight lines in any order whatever, a figure which may still be called a hexagon (non-convex), but which for distinction has been called a hexagram.

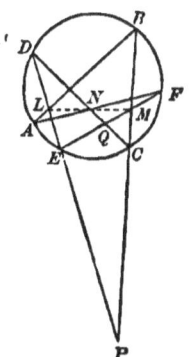

The demonstration (27) applies to this figure, word for word.

Brianchon's Theorem (28) may also be extended to a circumscribed hexagram, formed by six tangents at any six points taken in any assumed order of succession.

MODERN GEOMETRY.

HARMONIC PROPORTION.

38. *Definition.* Four points $A, B, C, D,$ are called four *harmonic points* when their anharmonic ratio $[ABCD]$ is equal to unity; that is, when

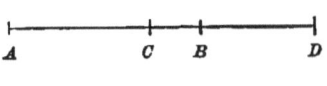

$$\frac{CA}{CB} \div \frac{DA}{DB} = 1, \quad \text{or} \quad \frac{CA}{CB} = \frac{DA}{DB},$$

which agrees with the definition of harmonic points in (III. 76).

39. *Definition.* A *harmonic pencil* is a pencil of four rays whose anharmonic ratio is equal to unity; that is, a pencil O, which determines upon any transversal a system of four harmonic points A, B, C, D. From (11) it follows that if one transversal of a pencil is divided harmonically, all other transversals of the pencil are also divided harmonically.

The points A and B are called conjugate points with respect to C and D, that is, they divide the distance CD harmonically; and C and D are called conjugate points with respect to A and B, that is, they divide the distance AB harmonically (III. 76). In like manner, the rays OA and OB are called *conjugate rays* with respect to the rays OC and OD, and are said to *divide the angle COD harmonically;* and the rays OC and OD are *conjugate rays* with respect to OA and OB, and *divide the angle AOB harmonically.*

PROPOSITION IX.—THEOREM.

40. *If a straight line AB is divided harmonically at the points C and D, the half of AB is a mean proportional between the distances of its middle point O from the conjugate points C and D; that is,* $\overline{OB}^2 = OC \cdot OD.$

For, the harmonic proportion,

$$CA : CB = DA : DB,$$

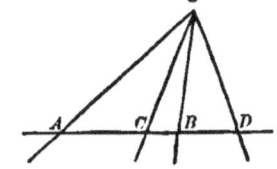

gives, by composition and division (III. 10) and (III. 9),

$$\frac{CA - CB}{2} : \frac{CA + CB}{2} = \frac{DA - DB}{2} : \frac{DA + DB}{2},$$

or, $\qquad OC : OB = OB : OD.$ [1]

41. *Corollary.* Conversely, *if we have given AB and its middle point O, and if C and D are so taken that* $\overline{OB}^2 = OC \cdot OD$, *then, A, B, C and D are four harmonic points.*

For, the proportion [1] gives

$$OB + OC : OB - OC = OD + OB : OD - OB;$$

that is, $\qquad CA : CB = DA : DB.$ [2]

42. *Scholium.* The three straight lines AC, AB, AD, are in harmonic progression. For, the harmonic proportion [2] may be written thus,

$$AC : AD = AB - AC : AD - AB,$$

or, AC, AB, AD, are such that the first is to the third as the difference between the first and second is to the difference between the second and third; that is, they are in harmonic progression, according to the definition commonly given in algebra.

Of three straight lines AC, AB, AD, in harmonic progression, the second AB is called a *harmonic mean* between the *extremes* AC and AD.

PROPOSITION X.—THEOREM.

43. *In a complete quadrilateral, each diagonal is divided harmonically by the other two.*

Let $ABCDEF$ be a complete quadrilateral (4), and L, M, N, the intersections of its three diagonals. In the triangle AEF, the transversal DBM gives (2),

$$DF.BA.ME = DA.BE.MF,$$

and since the three lines AL, FB, ED, pass through the common point C, we have by (5),

$$DF.BA.LE = DA.BE.LF.$$

Dividing one of these equations by the other, we have

$$\frac{ME}{LE} = \frac{MF}{LF}, \quad \text{or} \quad \frac{ME}{MF} = \frac{LE}{LF}.$$

therefore, EF is divided harmonically at M and L. Hence, if AM be joined, the four rays AM, AE, AL, AF, will form a harmonic pencil; consequently M, N, B, D, are also four harmonic points, or the diagonal BD is divided harmonically at M and N. Finally, if FN be joined, the four rays FM, FB, FN, FD, will form a harmonic pencil; consequently L, N, C, A, are four harmonic points, or the diagonal AC is divided harmonically at L and N.

POLE AND POLAR IN THE CIRCLE.

44. *Definition.* If through a fixed point P in the plane of a circle (either without the circle, Fig. 1, or within it, Fig. 2), we draw a secant and determine on this secant the point Q the harmonic conjugate of P with respect to the points of intersection C and D, the locus of Q, as the secant turns about P, is called the *polar* of the point P, and P is called the *pole* of this locus, with respect to the circle.

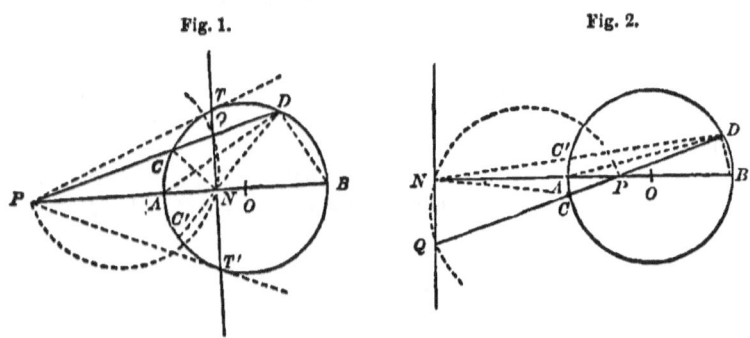

Fig. 1.   Fig. 2.

PROPOSITION XI.—THEOREM.

45. *The polar of a given point with respect to a circle is a straight line perpendicular to the diameter drawn through the given point.*

Let P be the given point (Figs. 1 and 2), O the centre of the circle, C and D the points in which any secant drawn through P cuts the circumference, Q the harmonic conjugate of P with respect to C and D. Draw QN perpendicular to the diameter AB which passes through P. Draw DN meeting the circumference in C'. Join CN, DA, DB.

Since PNQ is a right angle, the circumference described upon PQ as a diameter passes through N, and since CD is divided harmonically at P and Q, the line NP bisects the angle CNC' (III. 79 and 23); therefore the arcs AC and AC' are equal. Hence the line DA bisects the angle PDN, and DB, perpendicular to DA, bisects the angle exterior to PDN; therefore PN is divided harmonically at A and B (III. 79), or N is the harmonic conjugate of P with respect to A and B. Consequently N is a fixed point, and Q is always in the perpendicular to the diameter AB, erected at N; that is, QN is the polar of P.

46. *Corollary* I. Hence, *to construct the polar of a given point P,* with respect to a given circle, find on the diameter AB drawn through P the harmonic conjugate N of P with respect to A and B, and draw NQ perpendicular to that diameter; then NQ is the polar of P.

47. *Corollary* II. *To find the pole of a given straight line NQ,* draw a diameter AB perpendicular to the given line intersecting it in N, and on this diameter take P the harmonic conjugate of N with respect to A and B; then P is the pole of NQ.

48. *Corollary* III. Since AB is divided harmonically at P and N, and $OA = \frac{1}{2} AB$, we have (40),

$$\overline{OA}^2 = OP.ON,$$

hence, *the radius is a mean proportional between the distances of the polar and its pole from the centre of the circle.*

This principle may be used to determine the point N from P, or P from N, instead of the methods of (46) and (47).

49. *Corollary* IV. *When the point P is without the circle, its polar is the line TT' joining the points of contact of the tangents drawn from P.* For the secant PCD turning about P approaches the tangent PT as its limit (II. 28); and at the limit, the points C and D and hence also Q (which is always between C and D) all coincide with T. Therefore T and T' are points of the polar.

50. *Corollary* V. *The polar of a point on the circumference is the tangent at that point.* For, as the point P approaches the circumference, the point N also approaches the circumference (since $OP.ON = \overline{OA}^2$); and when OP becomes equal to OA, ON also becomes equal to OA.

PROPOSITION XII.—THEOREM.

51. 1st. *The polars of all the points of a straight line pass through the pole of that line.* 2d. *The poles of all the straight lines which pass through a fixed point are situated on the polar of that point.*

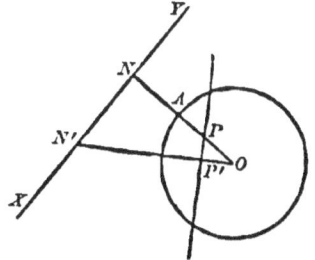 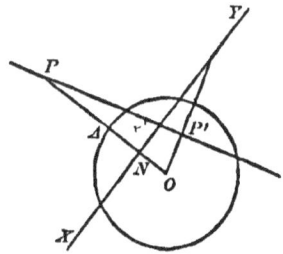

Let XY be any straight line, P its pole with respect to the circle O, and N' any point of the line. Drawing OPN, which is perpendicular to XY, we have $OP.ON = \overline{OA}^2$ (48). Let PP' be drawn perpendicular to ON'; then, the similar triangles OPP', ONN', give

$$OP'.ON' = OP.ON = \overline{OA}^2,$$

therefore, PP' is the polar of N' (48). Hence, 1st, the polar of any point N' of the line XY passes through the pole P of that line; 2d, the pole P

of any straight line XY which passes through the point N' is situated on the polar PP'' of that point.

52. *Corollary.* *The pole of a straight line is the intersection of the polars of any two of its points. The polar of any point is the straight line joining the poles of any two straight lines passing through that point.*

PROPOSITION XIII.—THEOREM.

53. *If through a fixed point P, in the plane of a circle, any two secants PCD, $PC'D'$, are drawn, and their intersections with the circumference are joined by chords CC', DD', CD', $C'D$, the locus of the intersections, M and N, of these chords, is the polar of the fixed point P.*

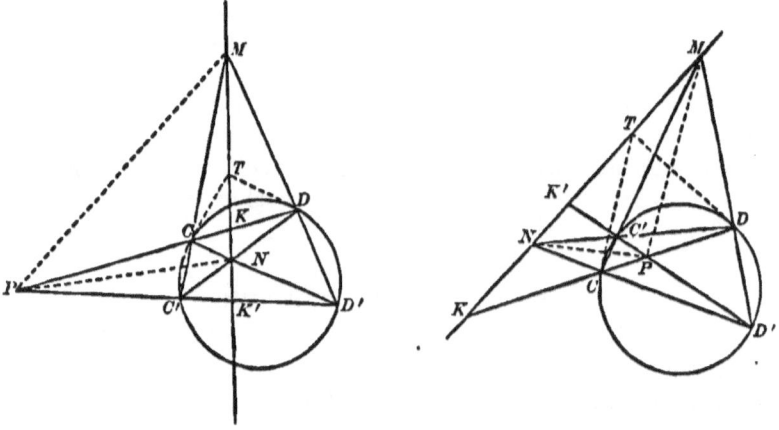

For, let K and K' be the points in which CD and $C'D'$ intersect MN. Then, considering the complete quadrilateral $MCC'NDD'$, the systems $PCKD$, $PC'K'D'$, are harmonic (43); therefore K and K' are on the polar of P (44); that is, MN is the polar of P.

54. *Corollary* I. The secants NCD', $NC'D$, being drawn through N, the line PM is the polar of N; and in like manner PN is the polar of M; therefore, *in any quadrilateral, $CC'D'D$, inscribed in a circle, the intersection N of the diagonals and the intersections M and P of the opposite sides, form a triangle MNP each vertex of which is the pole of the opposite side.*

55. *Corollary* II. As the transversal $PC'D'$ approaches to PCD, the secants MC, MD, approach to the tangents at C and D as their limits; therefore, at the limit, the tangents at C and D intersect on the polar of P. Hence, *if through a fixed point P in the plane of a circle any secant PCD is drawn, and tangents CT and DT to the circle are drawn at the points of intersection, the locus of the intersection T of these tangents is the polar of the fixed point P.*

56. *Corollary* III. From the last property it follows, that if we draw tangents to the circle at the vertices of the inscribed quadrilateral $CC'DD'$,

the complete circumscribed quadrilateral thus formed will have for its diagonals the three indefinite sides of the triangle *MNP*. Hence, *in any complete quadrilateral circumscribed about a circle, the three diagonals form a triangle each vertex of which is the pole of the opposite side.*

57. *Corollary* IV. Combining (54) and (56), we arrive at the following proposition: *If at the vertices of an inscribed quadrilateral, tangents to the circle are drawn forming a circumscribed quadrilateral, then,* 1st, *the interior diagonals of the two quadrilaterals intersect in the same point and form a harmonic pencil;* 2d, *the third diagonals of the completed quadrilaterals are situated on the same straight line, and their extremities are four harmonic points.*

RECIPROCAL POLARS.

58. *Definition.* From (51) it follows that if the points M, N, P, Q, are the poles of the sides of a polygon $ABCD$, then the points A, B, C, D, are the poles of the sides of the polygon $MNPQ$. Each of the two polygons thus related is called the *reciprocal polar* of the other, with respect to the circle, which receives the name of *auxiliary circle*.

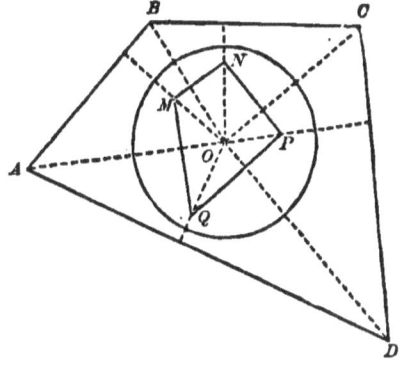

It will be observed that either of the two reciprocal polars may be derived from the other by either of two processes. If the polygon $ABCD$ is given, the polygon $MNPQ$ may be derived from it, 1st, by taking the poles M, N, P, Q, of the *sides* of the given polygon as the *vertices* of the derived polygon, or 2d, the *polars* MN, NP, PQ, QM of the *vertices* of the given polygon may be taken as the *sides* of the derived polygon. In like manner, if the polygon $MNPQ$ is given, the polygon $ABCD$ may be derived from it by either of these processes.

59. *Method of reciprocal polars.* Since the relation between two reciprocal polars is such that for each *line* of one figure there exists a corresponding *point* in the other, and reciprocally, any theorem in relation to the *lines* or *points* of one figure may be converted at once into a theorem in relation to the *points* or *lines* of the other. This is called *reciprocating* the theorem. The fecundity of this method is especially proved in its application to the theory of curves which do not belong to elementary geometry; but we can give some simple illustrations of the nature of the method by applying it to rectilinear figures.

The student will have no difficulty in showing that the theorems which we have placed against each other, in Proposition VIII., are reciprocal theorems. Thus, the reciprocal polar of an inscribed hexagon being the circumscribed hexagon formed by drawing tangents at the vertices of the first

MODERN GEOMETRY. 353

(49), (50), we can immediately infer Brianchon's Theorem (28) from Pascal's Theorem (27); for, the diagonals joining opposite vertices of the circumscribed hexagon will be the polars of the intersections of opposite sides of the inscribed hexagon (56), and therefore pass through the pole of the straight line in which these intersections lie (51). Similarly, the theorem of Pascal may be directly inferred from that of Brianchon.

The three following propositions are of frequent use in deducing reciprocal theorems.

PROPOSITION XIV.—THEOREM.

60. *The angle contained by two straight lines is equal to the angle contained by the straight lines joining their poles to the centre of the auxiliary circle.*

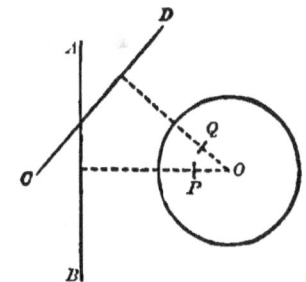

For, the poles P and Q of two straight lines AB and CD are situated respectively on the perpendiculars let fall from the centre of the auxiliary circle upon the lines AB and CD (45).

PROPOSITION XV.—THEOREM.

61. *The ratio of the distances of any two points from the centre of the auxiliary circle is equal to the ratio of the distances of each point from the polar of the other.* (Salmon's Theorem.)

Let P and Q be the points, AB and CD their polars, PF the distance of P from CD, and QE the distance of Q from AB, and O the centre of the auxiliary circle. Draw OPM and OQN, which will be parallel to QE and PF respectively; draw PH perpendicular to ON and QK perpendicular to OM. If R is the radius of the circle, we have $R^2 = OP.OM = OQ.ON$ (48), whence

$$\frac{OP}{OQ} = \frac{ON}{OM}.$$

The similar triangles POH and QOK give

$$\frac{OP}{OQ} = \frac{OH}{OK};$$

therefore (III. 12),

$$\frac{OP}{OQ} = \frac{ON + OH}{OM + OK} = \frac{HN}{KM} = \frac{PF}{QE}.$$

PROPOSITION XVI.—THEOREM.

62. *The anharmonic ratio of four points in a straight line is equal to that of the pencil formed by the four polars of these points.*

For, the pencil formed by joining the four points to the centre of the circle has the same angles as the pencil formed by their polars (60), and these pencils have equal anharmonic ratios (13).

PROPOSITION XVII.—PROBLEM.

63. *It is a known theorem that the three perpendiculars from the vertices of a triangle to the opposite sides meet in a point; it is required to determine its reciprocal theorem by the method of reciprocal polars.*

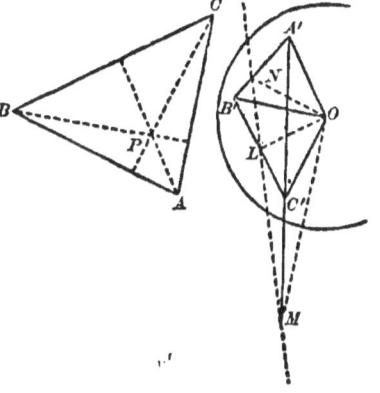

Let the perpendiculars from the angles upon the opposite sides of the triangle ABC meet in P. Let $A'B'C'$ be the reciprocal polar triangle of ABC, A' being the pole of BC, B' the pole of AC, and C' the pole of AB. The pole of the perpendicular AP is a point L on the line $B'C'$, since $B'C'$ is the polar of A (51); the pole of BP is a point M on $A'C'$, and the pole of CP is a point N on $A'B'$. The direct theorem being that the three lines AP, BP, CP, meet in a point, the reciprocal theorem will be that their poles L, M, N, are in a straight line, the polar of P; but we must express the reciprocal theorem in relation to the triangle $A'B'C'$. Now joining OL, OM, ON, and OA', OB', OC', the angle $A'OL$ is a right angle, by (60); and so also $B'OM$ and $C'ON$ are right angles. Hence, the reciprocal theorem may be expressed as follows:

If from any point O in the plane of a triangle $A'B'C'$, straight lines OA', OB', OC', are drawn to its vertices, the lines OL, OM, ON, drawn at O perpendicular respectively to the lines OA', OB', OC', meet the sides respectively opposite to the corresponding vertices in three points, L, M, N, which are in a straight line.

RADICAL AXIS OF TWO CIRCLES.

64. *Definition.* If through a point P, in the plane of a circle, a straight line is drawn cutting the circumference in the points A and B, the product of the segments into which the chord AB is divided at P, namely, the product $PA.PB$, is constant; that is, independent of the direction of the secant (III. 61). This constant product, depending upon the position of the point P with respect to the circle, is called *the power of the point with respect to the circle.*

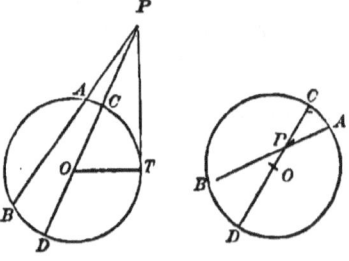

If we consider especially the secant PCD, drawn through P and the centre O of the circle, and designate the distance PO by d and the radius of the circle by r, we have, when the point P is without the circle, $PC = d - r$, $PD = d + r$, and hence the power of the point is expressed by the product $(d - r)(d + r)$ or $d^2 - r^2$.

If the point P is within the circle, the absolute values of PC and PD are $r - d$ and $r + d$; but the segments PC and PD lying in opposite directions with respect to P, are conceived to have opposite algebraic signs, so that the product $PC.PD$ must be negative; hence the power of the point P is expressed by the product $-(r-d)(r+d) = -(r^2 - d^2) = d^2 - r^2$. Thus, *in all cases, whether the point is without or within the circle, its power is expressed by the square of its distance from the centre diminished by the square of the radius.*

65. *When the point P is without the circle, its power is equal to the square of the tangent to the circle drawn from that point.*

When the point is on the circumference, its power is zero.

PROPOSITION XVIII.—THEOREM.

66. *The locus of all the points whose powers with respect to two given circles are equal, is a straight line perpendicular to the line joining the centres of the circles, and dividing this line so that the difference of the squares of the two segments is equal to the difference of the squares of the radii.*

Let O and O' be the centres of the two circles whose radii are r and r'; let P be any point whose distances from O and O' are d and d', then the powers of P with respect to the two circles are $d^2 - r^2$ and $d'^2 - r'^2$, and these being equal, by hypothesis, we have $d^2 - r^2 = d'^2 - r'^2$, whence $d^2 - d'^2 = r^2 - r'^2$. Now, drawing PX perpendicular to OO', we have from the right triangles POX, $PO'X$,

$$\overline{OX}^2 - \overline{O'X}^2 = \overline{PO}^2 - \overline{PO'}^2 = d^2 - d'^2 = r^2 - r'^2;$$

therefore, the quantity $\overline{OX}^2 - \overline{O'X}^2$, being equal to $r^2 - r'^2$, is constant, and X is a fixed point. Hence the point P is always in the perpendicular

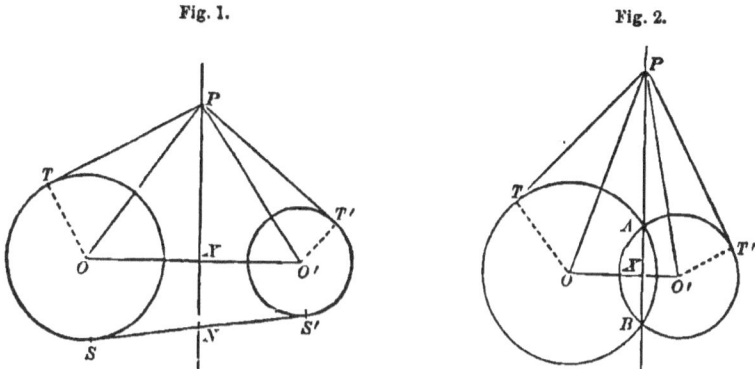

Fig. 1.   Fig. 2.

to OO' erected at the fixed point X; that is, this perpendicular is the locus of P.

67. *Definition.* The locus, PX, of the points whose powers with respect to two given circles O and O' are equal, is called the *radical axis* of the two circles.

68. *Corollary* I. If the two circles have no point in common, the radical axis does not intersect either of them. Fig. 1.

If the circles intersect, the power of each of the points of intersection is equal to zero; therefore, each of these points is a point in the radical axis; hence, *in the case of two intersecting circles, their common chord is their radical axis.* Fig. 2.

If the circles touch each other, either externally or internally, their common tangent at the point of contact is their radical axis.

69. *Corollary* II. From (65) and (67) it follows that *the tangents PT, PT', drawn to the two circles from any point of the radical axis without the circles, are equal.*

Hence, if SS' is a common tangent to the two circles, intersecting the radical axis in N, we have $NS = NS'$. Therefore, the radical axis can be constructed by joining the middle points of any two common tangents.

PROPOSITION XIX.—THEOREM.

70. *The radical axes of a system of three circles, taken two and two, meet in a point.*

Let O, O', O'', be the given circles. Designate the radical axis of O' and O'' by X, that of O and O'' by X', and that of O and O' by X''. The three centres not being in the same straight line, the axes X and X', perpendicular to the intersecting lines OO'' and $O''O'$, will meet in a certain point V. This point will have equal powers with respect to O' and O'', and with respect to O and O'', consequently it will also have equal

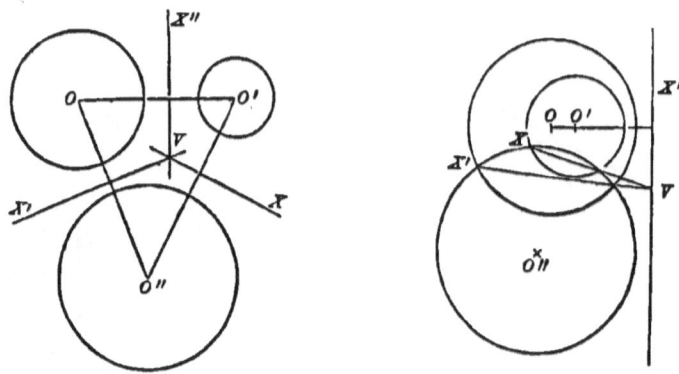

powers with respect to O and O', and is therefore a point in their radical axis X''.

71. Definition. The point in which the radical axes of a system of three circles meet is called the *radical centre* of the system.

If the three centres of the circles are in a straight line, the three axes are parallel, and the radical centre is at an infinite distance.

72. Definition. Two circles O and O' intersect *orthogonally*, that is, *at right angles*, when their tangents at the point of intersection are at right angles, or, which is the same thing, when their radii, OT, $O'T$, drawn to the common point, are at right angles.

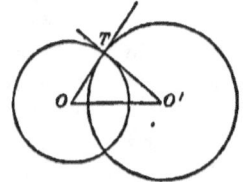

Denoting OO' by d, and the radii by r and r', we have in the right triangle OTO', $d^2 - r^2 = r'^2$; hence, *when two circles intersect orthogonally, the square of the radius of either is equal to the power of its centre with respect to the other circle.*

PROPOSITION XX.—THEOREM.

73. *The radical axis of two given circles is the locus of the centres of a system of circles which intersect both the given circles orthogonally; and the line joining the centres of the given circles is the common radical axis of all the circles of that system.*

Let P be the centre of any circle which cuts the two given circles O and O' orthogonally; then, by (72), the powers of the point P with respect to the two circles are each equal to the square of the radius of the circle P, that is, equal to each other; therefore, the centre P is in the radical axis of the two given circles.

Again, let P and Q be the centres of any two of the circles which cut both

O and O' orthogonally. Since the circle O cuts the two circles P and Q orthogonally, its centre O lies in the radical axis of P and Q; and for the

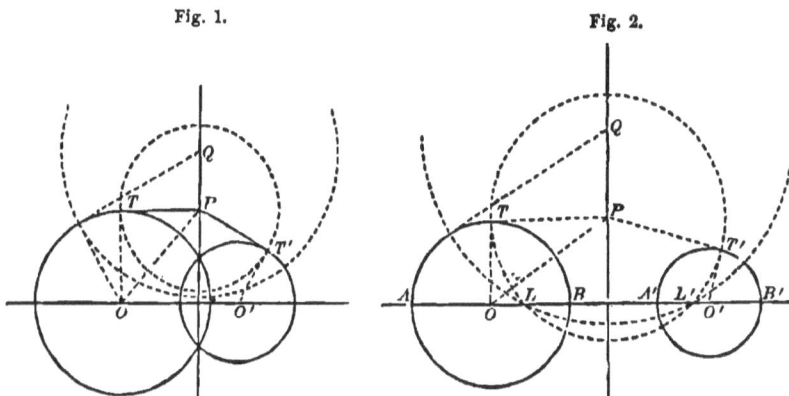

Fig. 1.   Fig. 2.

same reason the centre O' lies in the radical axis of P and Q; therefore, the line OO' is that radical axis, and is consequently the common radical axis of all the circles which cut both O and O' orthogonally.

74. *Scholium.* When the given circles intersect, Fig. 1, the radius of any one of the circles P, Q, etc., is evidently less than the distance of its centre from OO', and therefore no one of these circles cuts OO'.

But when the circles have no point in common, Fig. 2 (whether one circle is wholly without the other, as in Fig. 2, or wholly within the other), all the circles, P, Q, etc., cut the line OO'; and since OO' is their common radical axis, it is their common chord; therefore, these circles all pass through two fixed points L and L' in the line OO'.

Also, since OT is a tangent to the circle P, we have $OL.OL' = \overline{OT}^2 = \overline{OB}^2$; therefore, the diameter AB is divided harmonically at L and L' (41). For a like reason, $A'B'$ is divided harmonically at L and L'.

CENTRES OF SIMILITUDE OF TWO CIRCLES.

75. *Definition.* If the straight line joining the centres of two circles is divided externally and internally in the ratio of the corresponding radii, the points of section are called, respectively, *the external* and *the internal centres of similitude* of the two circles.

PROPOSITION XXI.—THEOREM.

76. *If in two circles two parallel radii are drawn, one in each circle, the straight line joining their extremities intersects the line of centres in the external centre of similitude if the parallel radii are in the same direction, and in the internal centre of similitude if these radii are in opposite directions.*

For, OA and $O'A'$ being any two parallel radii in the same direction, and

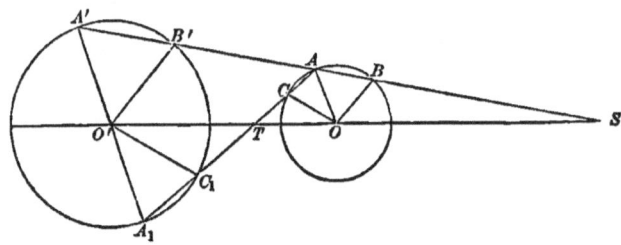

the line $A'A$ intersecting the line of centres in S, the similar triangles SOA, $SO'A'$, give
$$SO : SO' = OA : O'A',$$
and therefore, by the definition (75), S is the external centre of similitude.

Also, OA and $O'A_1$ being parallel radii in opposite directions, and the line AA_1 intersecting the line of centres in T, the similar triangles TOA, $TO'A_1$, give
$$TO : TO' = OA : O'A_1,$$
and therefore T is the internal centre of similitude.

77. *Corollary* I. It is easily shown that, conversely, *if any transversal is drawn through a centre of similitude, the radii drawn to the points in which it cuts the circumferences will be parallel, two and two.*

Of the four points in which the transversal cuts the circumferences, two points at the extremities of parallel radii, as A and A', or B and B', are called *homologous points;* and two points at the extremities of non-parallel radii, as A and B', or B and A', are called *anti-homologous points.*

78. *Corollary* II. Hence, if a transversal drawn through a centre of similitude is a tangent to one of the circles it is also a tangent to the other; so that *when one circle is wholly without the other, the centres of similitude are the intersections of the pairs of external and internal common tangents, respectively.*

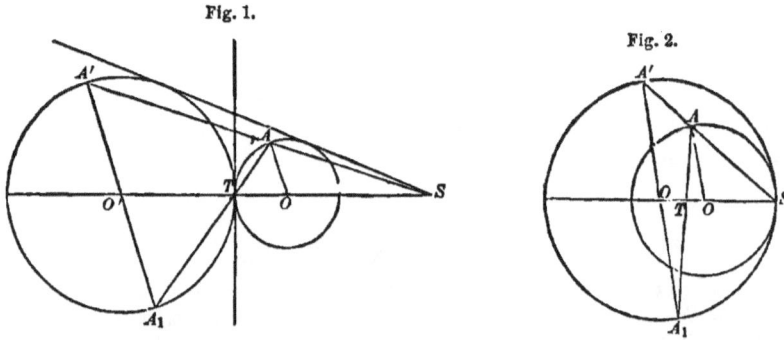

If the circles touch each other externally (Fig. 1), the point of contact is their internal centre of similitude. If they touch internally (Fig. 2), the point of contact is their external centre of similitude.

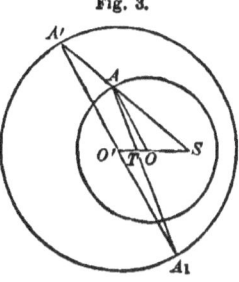

Fig. 3.

If one circle is wholly within the other (Fig. 3), both centres of similitude lie within both circles.

79. *Corollary* III. *The distances* (as SA and SA', or TA and TA_1, etc.) *of a centre of similitude from two homologous points are to each other as the radii of the circles.*

80. *Corollary* IV. Since we have

$$SO : SO' = TO : TO',$$

the line OO' is divided harmonically at S and T; that is, *the centres of two circles and their two centres of similitude are four harmonic points.*

PROPOSITION XXII.—THEOREM.

81. *The product of the distances of a centre of similitude of two circles from two anti-homologous points is constant.*

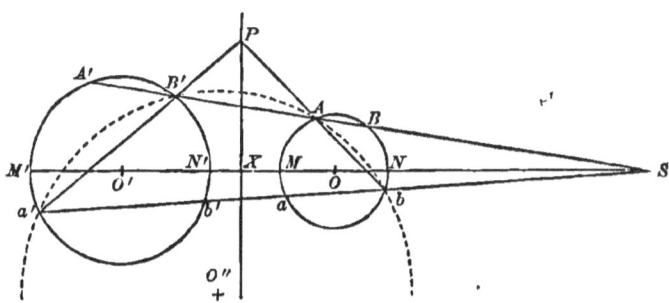

Let a transversal through the centre of similitude S intersect the circumferences O and O' in the homologous points A, A', and B, B'. The line of centres intersects the circumferences in the homologous points M, M', and N, N', respectively. Hence, by (79),

$$\frac{ON}{O'N'} = \frac{SA}{SA'} = \frac{SB}{SB'} = \frac{SM}{SM'} = \frac{SN}{SN'};$$

from which equations we deduce

$$SA \cdot SB' = SA' \cdot SB = \frac{SM}{SM'} \times SA' \cdot SB'.$$

But $SA' \cdot SB' = SM' \cdot SN'$ (III. 58); therefore we have

$$SA \cdot SB' = SA' \cdot SB = SM \cdot SN' = SM' \cdot SN.$$

The products $SM \cdot SN'$, $SM' \cdot SN$, are constant; therefore, the products $SA \cdot SB'$, $SA' \cdot SB$, are constant.

82. *Corollary* I. Hence, if A and B' are anti-homologous points of one secant drawn through S, and b and a' are anti-homologous points of a second secant, we have

$$SA \cdot SB' = Sb \cdot Sa';$$

therefore, *the four points A, B', a', b, lie on the circumference of a circle O''.*

83. *Corollary* II. The chords Ab, $a'B'$, joining pairs of anti-homologous points in the two given circles, may be called *anti-homologous chords*.

The chord Ab is the radical axis of the circles O and O''; the chord $a'B'$ is the radical axis of the circles O' and O'' (68); and these intersect the radical axis PX of the circles O and O' in the same point P (70). Hence, *pairs of anti-homologous chords in two circles intersect on the radical axis of the circles.*

84. *Corollary* III. If the secant Sa' approaches indefinitely to SA', the anti-homologous chords $a'A'$, bB, approach indefinitely to the tangents at A' and B. Hence, at the limit, we infer that *the tangents at two anti-homologous points in two circles intersect on the radical axis.*

PROPOSITION XXIII.—THEOREM.

85. *Three circles being given, and considered when taken two and two as forming three pairs of circles; then, 1st. The straight lines joining the centre of each circle and the internal centre of similitude of the other two meet in a point; 2d. The external centres of similitude of the three pairs of circles are in a straight line; 3d. The external centre of similitude of any pair and the internal centres of similitude of the other two pairs are in a straight line.*

Let O, O', O'', be the given circles; S and T the external and internal centres of similitude of O' and O''; S' and T' those of O and O''; S'' and T'' those of O and O'. Let R, R' and R'' denote the radii of the three circles.

362 MODERN GEOMETRY.

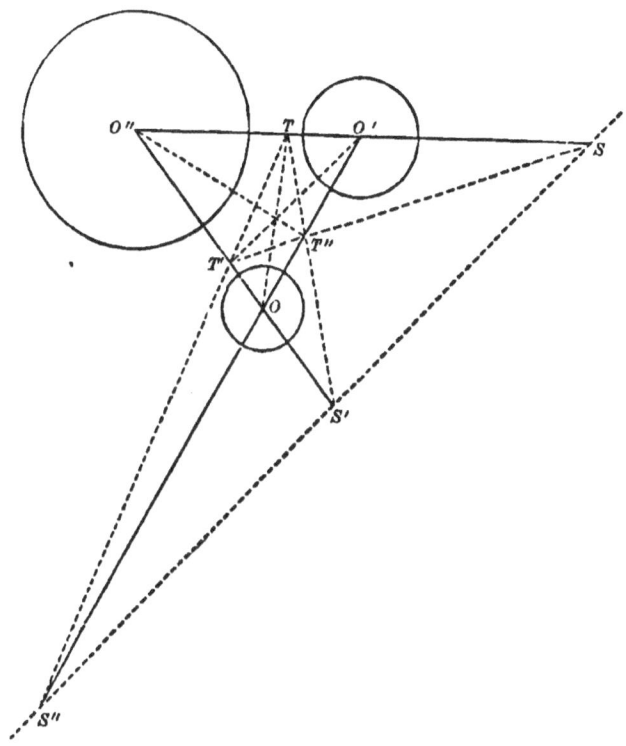

1st. By the definition (75) we have

$$\frac{T''O}{T''O'} = \frac{R}{R'}, \quad \frac{TO'}{TO''} = \frac{R'}{R''}, \quad \frac{T'O''}{T'O} = \frac{R''}{R},$$

the product of which equations gives

$$\frac{T''O . TO' . T'O''}{T''O' . TO'' . T'O} = \frac{R . R' . R''}{R' . R'' . R} = 1,$$

or $\qquad T''O . TO' . T'O'' = T''O' . TO'' . T'O;$

therefore, in the triangle $OO'O''$, the three straight lines OT, $O'T'$, $O''T''$, meet in a point (6).

2d. By the definition we also have

$$\frac{S''O}{S''O'} = \frac{R}{R'}, \quad \frac{SO'}{SO''} = \frac{R'}{R''}, \quad \frac{S'O''}{S'O} = \frac{R''}{R},$$

whence, by multiplying these equations together,

$$S''O . SO' . S'O'' = S''O' . SO'' . S'O;$$

MODERN GEOMETRY. 363

therefore, the points S, S', S'', being in the sides (produced) of the triangle $OO'O''$, are in a straight line (3).

3d. The product of the equations

$$\frac{T''O}{T''O'} = \frac{R}{R'}, \quad \frac{TO'}{TO''} = \frac{R'}{R''}, \quad \frac{S'O''}{S'O} = \frac{R''}{R},$$

gives $\quad T''O \cdot TO' \cdot S'O'' = T''O' \cdot TO'' \cdot S'O$;

therefore, the points T, T'', S', are in a straight line (3). In the same manner it is shown that T, T', S'', are in a straight line; and T', T'', S, are in a straight line.

86. *Definition.* The straight line $SS'S''$, on which the three external centres of similitude lie, is called the *external axis of similitude* of the three circles; and the lines $ST''T'$, $S'T''T$, $S''T'T$, are called the three *internal axes of similitude*.

PROPOSITION XXIV.—THEOREM.

87. *If a variable circle touch two fixed circles, the chord of contact passes through their external centre of similitude when the contacts are of the same kind (both external or both internal), and through their internal centre of similitude when the contacts are of different kinds.*

Let the circle C touch the circles O and O' in A and B'; and let the chord of contact AB' cut the two circles again in B and A'. The lines OC, $O'C$, pass through the points of contact. Drawing the radii OB, $O'A'$,

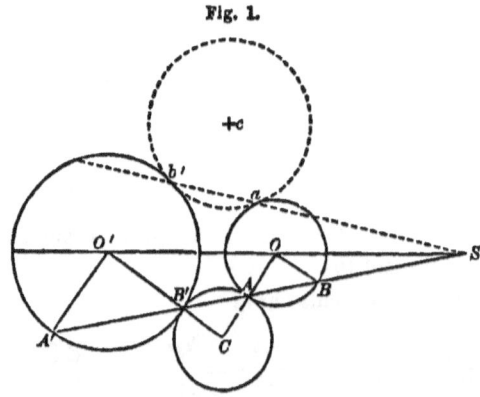

Fig. 1.

the isosceles triangles CAB', OAB, $O'A'B'$, are similar; consequently the radii OA and $O'A'$ are parallel. Therefore (76), the chord AB' passes through the external centre of similitude S, when the contacts are of the same kind (Fig. 1), and through the internal centre of similitude T, when the contacts are of a different kind (Fig. 2).

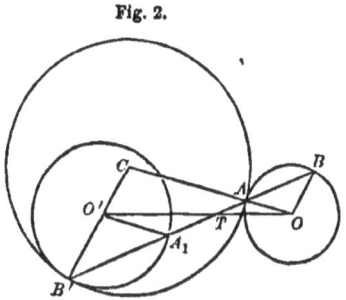

Fig. 2.

88. *Corollary.* *If two variable circles C and c touch two fixed circles O and O', their radical axis passes through the external centre of similitude of the fixed circles when the contacts of each of the two circles are of the same kind, and through their internal centre of similitude when these contacts are of different kinds.*

For, the four points of contact A, B', b', a, (Fig. 1), lie on the circumference of a circle (82) which may be designated as the circle Q. The chord AB' is the radical axis of the circle Q and the circle C; the chord ab' is the radical axis of the circle Q and the circle c; and these two axes meet the radical axis of the circles C and c in the same point (70), that is, in the point S (87). The proof is similar when the contacts are of different kinds.

PROPOSITION XXV.—THEOREM.

89. *The radical axis of two circles which touch three given circles is an axis of similitude of the three given circles.*

Let the circles M and N (figure on next page) touch the three given circles O, O', O'', the contacts of each of the two circles being all of the same kind, that is, all internal in the case of the circle M, and all external in the case of the circle N. Let S, S', S'', be the three external centres of similitude of the given circles taken in pairs, so that $SS'S''$ is their external axis of similitude (86).

Since the circles M and N touch the two given circles O' and O'', and the contacts are of the same kind in each case, the radical axis of M and N passes through S (88). For the same reason, it passes through S' and through S''. Therefore $SS'S''$ is the radical axis of the circles M and N.

In the same manner it may be shown that if each of the two circles M and N has like contacts with the pair of circles O' and O'', but unlike contacts with the other two pairs (that is, if M touches both O' and O'' internally and O externally, and N touches both O' and O'' externally and O internally), the radical axis of M and N is the internal axis of similitude which passes through the external centre of similitude, S, of the circles O' and O''.

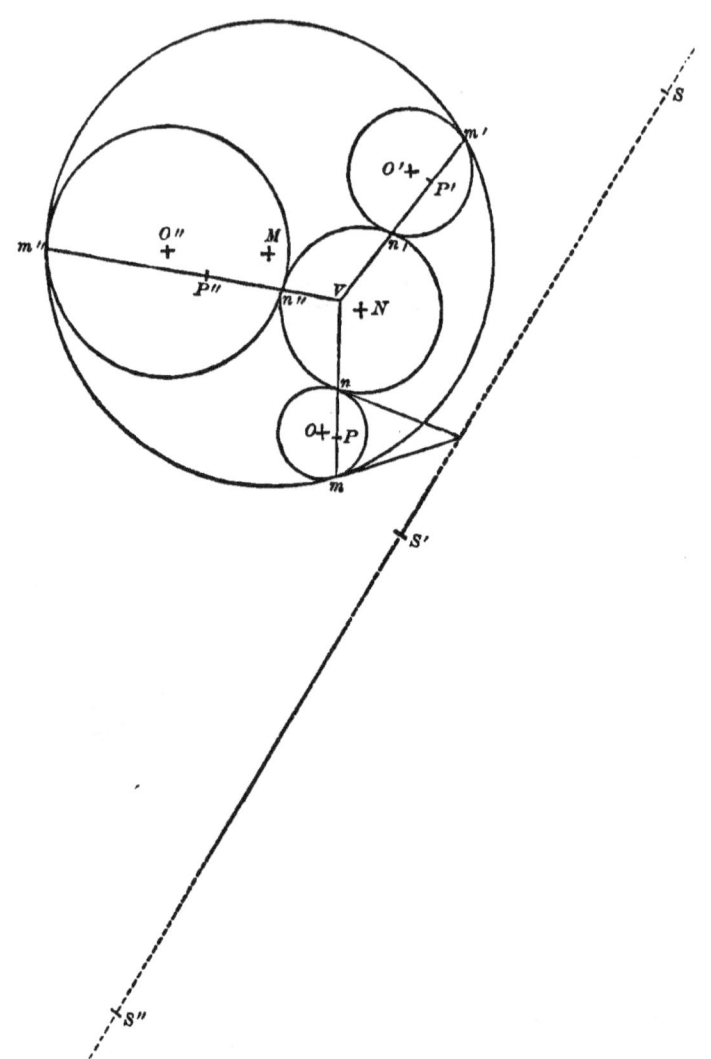

90. Scholium. There are in general eight different circles which can be drawn tangent to three given circles, and these eight circles exist in pairs the four radical axes of which are the four axes of similitude of the three given circles.

91. Corollary I. *When two circles M and N touch three given circles O, O', O'', the three chords of contact mn, $m'n'$, $m''n''$, meet in a point V, which is a centre of similitude of the two circles M and N and the radical centre of the three given circles O, O', O''.*

For, since the circle O touches the circles M and N, and the contacts are of different kinds, the chord of contact mn passes through the internal centre of similitude V of M and N (87); and for the same reason the chords $m'n'$ and $m''n''$ pass through V.

Also, since the two circles O and O' touch the two circles M and N, and the contacts are of different kinds, the radical axis of O and O' passes through the internal centre of similitude, V, of M and N (88). For the same reason, the radical axis of O' and O'', and the radical axis of O'' and O, pass through V. Therefore, V is the radical centre of the three circles O, O', O''.

92. *Corollary* II. *The pole of the radical axis of the circles M and N, with reference to any one of the three given circles, lies in the chord of contact of that circle.* Thus, in the case represented in the figure, the pole of $SS'S''$ with respect to the circle O is a point P lying in the chord of contact mn.

For, let R be the point of meeting of the tangents to the circle O drawn at m and n. These tangents are equal and touch the circles M and N; therefore, the point R is on the radical axis of M and N, that is, upon the line $SS'S''$. But mn is the polar of the point R with respect to the circle O (49), and therefore the pole of $SS'S''$ with respect to the circle O is a point P on the chord of contact mn (51).

PROPOSITION XXVI.—PROBLEM.

93. *To describe a circle tangent to three given circles.*

As remarked in (90), there are in general eight solutions of this problem. The solutions may all be brought under two cases: viz.—

1st. A pair of circles can be found one of which will touch all the given circles internally, and the other will touch all the given circles externally.

2d. A pair of circles can be found one of which will touch the first of the given circles internally and the other two externally, and the other will touch the first externally and the other two internally.

By taking each of the given circles successively as the "first," this second case gives six circles, thus making, in all, the eight solutions.

The principles developed in the preceding proposition furnish the following simple and elegant solution of the problem, first given by GERGONNE.*

Let O, O', O'' (preceding figure) be the three given circles. Let $SS'S''$ be their external axis of similitude and V their radical centre. Find the poles P, P', P'', of $SS'S''$ with respect to each of the given circles, and draw VP, VP', VP'', intersecting the three circles in the points m and n, m' and n', m'' and n'', respectively. The circumference described through the three points m, m', m'', will touch the three given circles internally; and the circumference described through the three points n, n', n'', will touch the three given circles externally.

By substituting successively each internal axis of similitude for $SS'S''$, we obtain the other three pairs of circles.

---

* *Annales de Mathématiques*, t. IV.

94. *Scholium*. This general solution embraces the solution of ten distinct problems, special cases of the general problem, in which one or more of the given circles may be reduced to points (that is, circles of infinitely small radius) or to straight lines (that is, circles of infinitely great radius).

EXERCISES.

1. If L and L' are two fixed straight lines and O a fixed point, and if through O any two straight lines OAA', OBB', are drawn cutting L in A and B and L' in A' and B', find the locus of the intersection of the lines AB' and $A'B$ (43).

2. If the three sides of a triangle pass through three fixed points which are in a straight line, and two vertices of the triangle move on two fixed straight lines, the third vertex moves on a straight line which passes through the intersection of the two fixed lines (25).

3. If the three vertices of a triangle move on three fixed straight lines which meet in a point, and two sides of the triangle pass through two fixed points, the third side passes through a fixed point which is in a straight line with the other two (24).

4. If Q is any point in the polar of a point P with respect to a given circle, the circle described upon PQ as a diameter cuts the given circle orthogonally (48).

5. Let the polars of any point P, with respect to two given circles O and O', intersect in Q. Then, the circle described upon PQ as a diameter cuts both the given circles orthogonally, and its centre is on the radical axis of the given circles.

6. Describe a circumference which shall pass through a given point and cut two given circles orthogonally.

7. The polars of any point in the radical axis of two circles intersect on that axis.

8. The poles of the radical axis of two circles taken with respect to each circle, and the two centres of similitude of the circles, are four harmonic points.

9. The radical axis of two circles is equally distant from the two polars of either centre of similitude.

10. If the sides AB, BC, CD, DA, of a quadrilateral circumscribed about a circle whose centre is O touch the circumference at the points E, F, G, H, respectively, and if the chords HE and GF meet in P, the line PO is perpendicular to the diagonal AC.

11. If a quadrilateral is divided into two other quadrilaterals by any secant, the intersections of the diagonals in the three quadrilaterals are in a straight line.

12. The anharmonic ratio of four points on the circumference of a circle is equal to the ratio of the products of the opposite sides of the quadrilateral determined by these points.

13. If a series of circles having their centres in a given straight line cut a given circle orthogonally, they have a common radical axis, which is the perpendicular let fall from the centre of the given circle upon the given straight line.

14. The three circles described upon the diagonals of a complete quadrilateral as diameters have a common radical axis and cut orthogonally the circle described about the triangle formed by the three diagonals.

15. Three circles O_1, O_2, O_3, being given, any fourth circle Q is described and the radical axes of Q and each of the given circles are drawn forming a triangle ABC. Another circle Q' being drawn, a second triangle $A'B'C'$ is formed in the same manner. Prove that the triangles ABC and $A'B'C'$ are homological (24), (70).

16. If two triangles are reciprocal polars with respect to a circle, they are homological (51), (62), (20).

17. If from the vertices of a triangle ABC perpendiculars Aa, Bb, Cc, are let fall upon the opposite sides, the three pairs of sides BC and bc, AC and ac, AB and ab, intersect on the radical axis of the circles circumscribed about the triangles ABC and abc (64, 67).

18. Any common tangent to two circles is divided harmonically by any circle which has a common radical axis with the two given circles (41).

19. If the sides of a quadrilateral $ABCD$ inscribed in a circle are produced to meet in E and F, forming a complete quadrilateral, the square of the third diagonal EF is equal to the sum of the squares of the tangents from E and F; and the tangent from the middle point of EF is equal to one-half of EF.

20. Given the three diagonals of a complete quadrilateral inscribed in a given circle, it is required to construct the quadrilateral (4, 48, 49, 57).

21. Given three circles, it is required to describe a fourth such that the three radical axes of this circle, combined successively with each of the given ones, shall pass through three given points (Exercises 3 and 15).

www.ingramcontent.com/pod-product-compliance
Lightning Source LLC
Chambersburg PA
CBHW020319240426
43673CB00039B/855